Progress in Nonlinear Differential Equations and Their Applications
Volume 58

Editor
Haim Brezis
Université Pierre et Marie Curie
Paris
and
Rutgers University
New Brunswick, N.J.

Piermarco Cannarsa
Carlo Sinestrari

Semiconcave Functions, Hamilton–Jacobi Equations, and Optimal Control

Birkhäuser
Boston • Basel • Berlin

Piermarco Cannarsa
Università di Roma "Tor Vergata"
Dipartimento di Matematica
00133 Roma
Italy

Carlo Sinestrari
Università di Roma "Tor Vergata"
Dipartimento di Matematica
00133 Roma
Italy

Library of Congress Cataloging-in-Publication Data
Cannarsa, Piermarco, 1957-
  Semiconcave functions, Hamilton–Jacobi equations, and optimal control / Piermarco
Cannarsa, Carlo Sinestrari.
    p. cm. – (Progress in nonlinear differential equations and their applications ; v. 58)
  Includes bibliographical references and index.
  ISBN 0-8176-4084-3 (alk. paper)
    1. Concave functions. 2. Hamilton–Jacobi equations. 3. Control theory. 4. Mathematical
optimization. I. Sinestrari, Carlo, 1970- II. Title. III. Series.

  QA353.C64C36 2004
  515'.355–dc22   2004043695
                                                               CIP

AMS Subject Classifications: Primary: 35F20, 49J52, 49-XX, 49-01; Secondary: 35D10, 26B25,
49L25, 35A21, 49J15, 49K15, 49Lxx, 49L20

ISBN 0-8176-4336-2   Printed on acid-free paper.

©2004 Birkhäuser Boston   *Birkhäuser*

Printed in the United States of America.   (TXQ/HP)

9 8 7 6 5 4 3 2 1   SPIN 10982358

www.birkhauser.com

To Francesca

Preface

A gifted British crime novelist[1] once wrote that mathematics is "like one of those languages that is simple, straightforward and logical in the early stages, but which rapidly spirals out of control in a frenzy of idioms, oddities, idiosyncrasies and exceptions to the rule which even native speakers cannot always get right, never mind explain." In fact, providing evidence to contradict such a statement has been one of our guides in writing this monograph. It may then be recommended to describe, right from the beginning, the essential object of our interest, that is, semiconcavity, a property that plays a central role in optimization.

There are various possible ways to introduce semiconcavity. For instance, one can say that a function u is semiconcave if it can be represented, locally, as the sum of a concave function plus a smooth one. Thus, semiconcave functions share many regularity properties with concave functions, but include several other significant examples. Roughly speaking, semiconcave functions can be obtained as envelopes of smooth functions, in the same way as concave functions are envelopes of linear functions. Typical examples of semiconcave functions are the distance function from a closed set $S \subset \mathbb{R}^n$, the least eigenvalue of a symmetric matrix depending smoothly on parameters, and the so-called "inf-convolutions." Another class of examples we are particularly interested in are viscosity solutions of Hamilton–Jacobi–Bellman equations.

At this point, the reader may wonder why we consider semiconcavity rather than the symmetric—yet more usual—notion of semiconvexity. The thing is that as far as optimization is concerned, in this book we focus our attention on minimization rather than maximization. This makes semiconcavity the natural property to look at.

Interest in semiconcave functions was initially motivated by research on nonlinear partial differential equations. In fact, it was exactly in classes of semiconcave functions that the first global existence and uniqueness results were obtained for Hamilton–Jacobi–Bellman equations, see Douglis [69] and Kruzhkov [99, 100, 102]. Afterwards, more powerful uniqueness theories, such as viscosity solutions and minimax solutions, were developed. Nevertheless, semiconcavity maintains its impor-

---

[1] M. Dibdin, *Blood rain*, Faber and Faber, London, 1999.

tance even in modern PDE theory, being the maximal type of regularity that can be expected for certain nonlinear problems. As such, it has been investigated in modern textbooks on Hamilton–Jacobi equations such as Lions [110], Bardi and Capuzzo-Dolcetta [20], Fleming and Soner [81], and Li and Yong [109]. In the context of nonsmooth analysis and optimization, semiconcave functions have also received attention under the name of lower C^k functions, see, e.g., Rockafellar [123].

Compared to the above references, the perspective of this book is different. First, in Chapters 2, 3 and 4, we develop the theory of semiconcave functions without aiming at one specific application, but as a topic in nonsmooth analysis of interest in its own right. The exposition ranges from well-known properties for the experts—analyzed here for the first time in a comprehensive way—to recent results, such as the latest developments in the analysis of singularities. Then, in Chapters 5, 6, 7 and 8, we discuss contexts in which semiconcavity plays an important role, such as Hamilton–Jacobi equations and control theory. Moreover, the book opens with an introductory chapter studying a model problem from the calculus of variations: this allows us to present, in a simple situation, some of the main ideas that will be developed in the rest of the book. A more detailed description of the contents of this work can be found in the introduction at the beginning of each chapter.

In our opinion, an attractive feature of the present exposition is that it requires, on the reader's part, little more than a standard background in real analysis and PDEs. Although we do employ notions and techniques from different fields, we have nevertheless made an effort to keep this book as self-contained as possible. In the appendix we have collected all the definitions we needed, and most proofs of the basic results. For the more advanced ones—not too many indeed—we have given precise references in the literature.

We are confident that this book will be useful for different kinds of readers. Researchers in optimal control theory and Hamilton–Jacobi equations will here find the recent progress of this theory as well as a systematic collection of classical results—for which a precise citation may be hard to recover. On the other hand, for readers at the graduate level, learning the basic properties of semiconcave functions could also be an occasion to become familiar with important fields of modern analysis, such as control theory, nonsmooth analysis, geometric measure theory and viscosity solutions.

We will now sketch some shortcuts for readers with specific interests. As we mentioned before, Chapter 1 is introductory to the whole text; it can also be used on its own to teach a short course on calculus of variations. The first section of Chapter 2 and most of Chapter 3 are essential for the comprehension of anything that follows. On the contrary, Chapter 4, devoted to singularities, could be omitted on a first reading. The PDE-oriented reader could move on to Chapter 5 on Hamilton–Jacobi equations, and then to Chapter 6 on the calculus of variations, where sharp regularity results are obtained for solutions to suitable classes of equations. On the other hand, the reader who wishes to follow a direct path to dynamic optimization, without including the classical calculus of variations, could go directly from Chapter 3 to Chapters 7 and 8 where finite horizon optimal control problems and optimal exit time problems are considered.

We would like to express our gratitude for all the assistance we have received for the realization of this project. The first author is indebted to Sergio Campanato for inspiring his interest in regularity theory, to Giuseppe Da Prato for communicating his taste for functional analysis, to Wendell Fleming and Craig Evans for opening powerful views on optimal control and viscosity solutions, and to his friend Mete Soner for sharing with him the initial enthusiasm for semiconcavity and variational problems. The subsequent collaboration with Halina Frankowska acquainted him with set-valued analysis. Luigi Ambrosio revealed to him enlightening connections with geometric measure theory. The second author is grateful to Alberto Tesei and Roberto Natalini, who first introduced him in the study of nonlinear first order equations. He is also indebted to Constantine Dafermos and Alberto Bressan for their inspiring teachings about hyperbolic conservation laws and control theory.

A significant part of the topics of the book was conceived or refined in the framework of the Graduate School in Mathematics of the University of Rome Tor Vergata, as material developed in graduate courses, doctoral theses, and research papers. We wish to thank all the ones who participated in these activities, in particular Paolo Albano, Cristina Pignotti, and Elena Giorgieri. Special thanks are due to our friends Italo Capuzzo-Dolcetta and Francis Clarke who read parts of the manuscript improving it with their comments. Helpful suggestions were also offered by many other friends and colleagues, such as Giovanni Alberti, Martino Bardi, Nick Barron, Pierre Cardaliaguet, Giovanni Colombo, Alessandra Cutrì, Robert Jensen, Vilmos Komornik, Andrea Mennucci, Roberto Peirone. Finally, we wish to express our warmest thanks to Mariano Giaquinta, whose interest gave us essential encouragement in starting this book, and to Ann Kostant, who followed us with patience during the writing of this work.

Piermarco Cannarsa
Carlo Sinestrari

> The power of doing anything with quickness is always much prized by the possessor, and often without any attention to the imperfection of the performance.
> —JANE AUSTEN, *Pride and Prejudice*

Contents

Semiconcave Functions,
Hamilton–Jacobi Equations,
and Optimal Control

1

A Model Problem

The purpose of this chapter is to outline some of the main topics of the book through the analysis of a simple problem in the calculus of variations. The study of this model problem allows us to introduce the dynamic programming approach and to show how the class of semiconcave functions naturally appears in this context.

In Section 1.1 we introduce semiconcave functions and give some equivalent definitions. Then, in Section 1.2 we state our variational problem, give the *dynamic programming principle* and define the *value function* associated with the problem. In Section 1.3, we restrict our attention to the case where the integrand has no explicit (t, x)-dependence; in this case the value function admits a simple representation formula due to Hopf.

In Section 1.4 we observe that the value function is a solution of a specific partial differential equation, called the *Hamilton–Jacobi* (or sometimes *Hamilton–Jacobi–Bellman*) equation. However, the equation is not satisfied in a classical sense. In fact, the value function in general is not differentiable everywhere, but only Lipschitz continuous, and the equation holds at the points of differentiability. Such a property is not sufficient to characterize the value function, since a Hamilton–Jacobi equation may have infinitely many Lipschitz continuous solutions taking the same initial data.

Before seeing how to handle this difficulty, we give in Section 1.5 an account of the classical method of characteristics for Hamilton–Jacobi equations. This technique gives in an elementary way a local existence result for smooth solutions, and at the same time shows that no global smooth solution exists in general. Although the method is completely independent of the control-theoretic interpretation of the equation, there is an interesting connection between the solutions of the characteristic system and the optimal trajectories of the corresponding problem in control or calculus of variations.

In Section 1.6 we use the semiconcavity property to characterize the value function among the many possible solutions of the Hamilton–Jacobi equation. In fact, we prove that the value function is semiconcave, and that semiconcave Lipschitz continuous solutions of Hamilton–Jacobi equations are unique.

We conclude the chapter by describing, in Section 1.7, the connection between Hamilton–Jacobi equations and another class of partial differential equations, called *hyperbolic conservation laws.* In the one-dimensional case the two classes of equations are strictly related; in particular, we show that semiconcavity corresponds to a well-known estimate for solutions of conservation laws due to Oleinik.

Let us mention that a more general treatment of the problem in the calculus of variations introduced here, including a detailed analysis of the singularities of the value function, will be given in Chapter 6.

1.1 Semiconcave functions

Before starting the analysis of our variational problem, let us introduce semiconcave functions, which are the central topic in this monograph and will play an important role later in this chapter. It is convenient to consider, first, a special class of semiconcave functions, while the general definition will be given in Chapter 2.

Here and in what follows we write $[x, y]$ to denote the segment with endpoints x, y, for any $x, y \in \mathbb{R}^n$. Moreover, we denote by $x \cdot y$, or by $\langle x, y \rangle$, the Euclidean scalar product, and by $|x|$ the usual norm in $\mathbb{R}^n$. Furthermore, $B_r(x)$—and, at times, $B(x, r)$—stands for the open ball centered at x with radius r. We will also use the abbreviated notation B_r for $B_r(0)$.

Definition 1.1.1 *Let $A \subset \mathbb{R}^n$ be an open set. We say that a function $u : A \to \mathbb{R}$ is semiconcave with linear modulus if it is continuous in A and there exists $C \geq 0$ such that*

$$u(x + h) + u(x - h) - 2u(x) \leq C|h|^2, \tag{1.1}$$

for all $x, h \in \mathbb{R}^n$ such that $[x - h, x + h] \subset A$. The constant C above is called a semiconcavity constant for u in S.

Remark 1.1.2 The above definition is often taken in the literature as the definition of a semiconcave function. For us, instead, it is a particular case of Definition 2.1.1, where the right-hand side of (1.1) is replaced by a term of the form $|h|\omega(|h|)$ for some function $\omega(\cdot)$ such that $\omega(\rho) \to 0$ as $\rho \to 0$. The function ω is called *modulus of semiconcavity*, and therefore we say that a function which satisfies (1.1) is semiconcave with a linear modulus.

Semiconcave functions with a linear modulus admit some interesting characterizations, as the next result shows.

Proposition 1.1.3 *Given $u : A \to \mathbb{R}$, with $A \subset \mathbb{R}^n$ open convex, and given $C \geq 0$, the following properties are equivalent:*

(a) u is semiconcave with a linear modulus in A with semiconcavity constant C;
(b) u satisfies

$$\lambda u(x) + (1 - \lambda)u(y) - u(\lambda x + (1 - \lambda)y) \leq C \frac{\lambda(1 - \lambda)}{2} |x - y|^2, \tag{1.2}$$

for all x, y such that $[x, y] \subset A$ and for all $\lambda \in [0, 1]$;

(c) *the function* $x \to u(x) - \frac{C}{2}|x|^2$ *is concave in A;*

(d) *there exist two functions $u_1, u_2 : A \to \mathbb{R}$ such that $u = u_1 + u_2$, u_1 is concave, $u_2 \in C^2(A)$ and satisfies $||D^2 u_2||_\infty \leq C$;*

(e) *for any $v \in \mathbb{R}^n$ such that $|v| = 1$ we have $\frac{\partial^2 u}{\partial v^2} \leq C$ in A in the sense of distributions, that is*

$$\int_A u(x) \frac{\partial^2 \phi}{\partial v^2}(x)\, dx \leq C \int_A \phi(x)\, dx, \qquad \forall \phi \in C_0^\infty(A),\ \phi \geq 0;$$

(f) *u can be represented as $u(x) = \inf_{i \in \mathcal{I}} u_i(x)$, where $\{u_i\}_{i \in \mathcal{I}}$ is a family of functions of $C^2(A)$ such that $||D^2 u_i||_\infty \leq C$ for all $i \in \mathcal{I}$.*

Proof — Let us set $v(x) = u(x) - \frac{C}{2}|x|^2$. Using the identity

$$|x + h|^2 + |x - h|^2 - 2|x|^2 = 2|h|^2,$$

we see that (1.1) is equivalent to

$$v(x + h) + v(x - h) - 2v(x) \leq 0$$

for all x, h such that $[x - h, x + h] \subset A$. It is well known (see Proposition A. 1.2) that such a property, together with continuity, is equivalent to the concavity of v, and so (a) and (c) are equivalent.

The equivalence of (b) and (c) is proved analogously. In fact, using the identity

$$\lambda|x|^2 + (1 - \lambda)|y|^2 - |\lambda x + (1 - \lambda)y|^2 = \lambda(1 - \lambda)|x - y|^2,$$

we see that inequality (1.2) is equivalent to

$$\lambda v(x) + (1 - \lambda)v(y) - v(\lambda x + (1 - \lambda)y) \leq 0$$

for all x, y such that $[x, y] \subset A$ and for all $\lambda \in [0, 1]$.

Now let us show the equivalence of (c) and (d). If (c) holds, then (d) immediately follows taking $u_1(x) = u(x) - \frac{C}{2}|x|^2$ and $u_2(x) = \frac{C}{2}|x|^2$. Conversely, if (d) holds, then for any unit vector v we have

$$\frac{\partial^2}{\partial v^2}\left(u_2 - \frac{C}{2}|x|^2\right) = \frac{\partial^2 u_2}{\partial v^2} - C \leq 0,$$

which implies that $u_2(x) - \frac{C}{2}|x|^2$ is concave. Thus, $u(x) - \frac{C}{2}|x|^2$ is concave since it is the sum of the two concave functions $u_1(x)$ and $u_2(x) - \frac{C}{2}|x|^2$.

The equivalence between (c) and (e) is an easy consequence of the characterization of concave functions as the functions having nonpositive distributional hessian.

Finally, let us prove the equivalence of (c) and (f). We recall that any concave function can be written as the infimum of linear functions (see Corollary A. 1.14). Thus, if (c) holds, we have that $u(x) - \frac{C}{2}|x|^2 = \inf_{i \in \mathcal{I}} v_i(x)$, where the v_i's are linear. Therefore, $u(x) = \inf_{i \in \mathcal{I}} u_i(x)$, where $u_i(x) = v_i(x) + \frac{C}{2}|x|^2$, and this proves (f).

Conversely, assume that (f) is satisfied. Then, setting $v_i(x) = u_i(x) - \frac{C}{2}|x|^2$, we see that $\partial^2_{\nu\nu} v_i \leq 0$ for all $\nu \in \mathbb{R}^n$, and so v_i is concave. Therefore $u(x) - \frac{C}{2}|x|^2$ is concave, being the infimum of concave functions, and this proves (c). ∎

From the previous proposition one can have an intuitive idea of the behavior of semiconcave functions with a linear modulus. Property (e) shows that they are the functions whose second derivatives are bounded above, in contrast with concave functions whose second derivatives are nonpositive. Property (d) shows that a semiconcave function can be regarded as a smooth perturbation of a concave function: thus, its graph can have a nonconcave shape in the smooth parts, but any corner points "upwards," as for concave functions. Property (f) gives a first explanation of why semiconcave functions naturally occur in minimization problems.

Examples of semiconcave functions will be given throughout the book and in particular in Chapter 2. We conclude the section with a typical example of a function which is not semiconcave.

Example 1.1.4 The function $u(x) = |x|$ is not semiconcave in any open set containing 0. In fact, inequality (1.1) is violated for any $C > 0$ if one takes $x = 0$ and h small enough. More generally, we find that $u(x) = |x|^\alpha$, is not semiconcave with a linear modulus if $\alpha < 2$; we will see, however, that, if $\alpha > 1$, it is semiconcave according to the general definition which will be given in Chapter 2.

1.2 A problem in the calculus of variations

We now start the analysis of our model problem. Given $0 < T \leq +\infty$, we set $Q_T =]0, T[\times \mathbb{R}^n$. We suppose that two continuous functions

$$L : \overline{Q}_T \times \mathbb{R}^n \to \mathbb{R}, \qquad u_0 : \mathbb{R}^n \to \mathbb{R}$$

are given. The function L will be called the *running cost*, or *lagrangian*, while u_0 is called the *initial cost*. We assume that both functions are bounded from below.

For fixed $(t, x) \in \overline{Q}_T$, we introduce the set of *admissible arcs*

$$\mathcal{A}(t, x) = \{ y \in W^{1,1}([0, t]; \mathbb{R}^n) \ : \ y(t) = x \}$$

and the *cost functional*

$$J_t[y] = \int_0^t L(s, y(s), \dot{y}(s)) \, ds + u_0(y(0)).$$

Then we consider the following problem:

$$\text{minimize } J_t[y] \text{ over all arcs } y \in \mathcal{A}(t, x). \tag{1.3}$$

Problems of this form are classical in the calculus of variations. In the case we are considering the initial endpoint of the admissible trajectories is free, and the terminal one is fixed. Cases where the endpoints are both fixed or both free are also interesting and could be studied by similar techniques, but will not be considered here.

The first step in the *dynamic programming approach* to the above problem is the introduction of the *value function*.

Definition 1.2.1 *The function* $u : \overline{Q}_T \to \mathbb{R}$ *defined as*

$$u(t, x) = \inf_{y \in \mathcal{A}(t,x)} J_t[y] \tag{1.4}$$

is called the value function of the minimization problem (1.3).

By our assumptions u is finite everywhere. In addition we have

$$u(0, x) = u_0(x). \tag{1.5}$$

The basic idea of the approach is to show that u admits an alternative characterization as the solution of a suitable partial differential equation, and thus it can be obtained without referring directly to the definition. Once u is known, the minimization problem is substantially simplified.

The following result is called *Bellman's optimality principle* or *dynamic programming principle* and is the starting point for the study of the properties of u.

Theorem 1.2.2 *Let* $(t, x) \in \overline{Q}_T$ *and* $y \in \mathcal{A}(t, x)$. *Then, for all* $t' \in [0, t]$,

$$u(t, x) \leq \int_{t'}^{t} L(s, y(s), \dot{y}(s)) \, ds + u(t', y(t')). \tag{1.6}$$

In addition, the arc y *is a minimizer for problem* (1.3) *if and only if equality holds in* (1.6) *for all* $t' \in [0, t]$.

Proof — For fixed $t' \in [0, t]$, let z be any arc in $W^{1,1}([0, t']; \mathbb{R}^n)$ such that $z(t') = y(t')$. If we set

$$\xi(s) = \begin{cases} z(s), & s \in [0, t'], \\ y(s), & s \in [t', t], \end{cases}$$

we have that $\xi \in \mathcal{A}(t, x)$ and therefore

$$u(t, x) \leq J_t[\xi] = \int_{t'}^{t} L(s, y, \dot{y}) \, ds + \int_{0}^{t'} L(s, z, \dot{z}) \, ds + u_0(z(0)).$$

Taking the infimum over all $z \in \mathcal{A}(t', y(t'))$ we obtain (1.6).

If (1.6) holds as an equality for all $t' \in [0, t]$, then choosing $t' = 0$ yields that y is a minimizer for J_t. Conversely, if y is a minimizer we find, by the definition of u and by inequality (1.6),

$$\int_{0}^{t} L(s, y, \dot{y}) \, ds + u_0(y(0)) = u(t, x) \leq \int_{t'}^{t} L(s, y, \dot{y}) \, ds + u(t', y(t')) \tag{1.7}$$

for any given $t' \in [0, t]$. This implies that $J_{t'}[y] \leq u(t', y(t'))$. Since by definition $J_{t'}[y] \geq u(t', y(t'))$, we must have equality in (1.7), and therefore also in (1.6). ∎

We can give a sharper formulation of the dynamic programming principle, as in the following result.

Theorem 1.2.3 *Let* $(t, x) \in \overline{Q}_T$. *Then, for all* $t' \in [0, t]$,

$$u(t, x) = \inf_{y \in \mathcal{A}(t, x)} \left\{ \int_{t'}^{t} L(s, y(s), \dot{y}(s)) \, ds + u(t', y(t')) \right\}. \tag{1.8}$$

Proof — Given $\varepsilon > 0$, let $y \in \mathcal{A}(t, x)$ be such that

$$u(t, x) + \varepsilon \geq \int_{0}^{t} L(s, y, \dot{y}) \, ds + u_0(y(0)).$$

Then

$$u(t, x) \geq \int_{t'}^{t} L(s, y, \dot{y}) \, ds + \int_{0}^{t'} L(s, y, \dot{y}) \, ds + u_0(y(0)) - \varepsilon$$

$$\geq \int_{t'}^{t} L(s, y, \dot{y}) \, ds + u(t', y(t')) - \varepsilon.$$

By the arbitrariness of ε we deduce that $u(t, x)$ is greater than or equal to the right-hand side of (1.8). The converse inequality follows from Theorem 1.2.2. ∎

1.3 The Hopf formula

From now on we consider the special case of $L(t, x, q) = L(q)$ and $T = +\infty$. We assume that

$$\begin{cases} \text{(i) } L \text{ is convex and } \lim_{|q| \to \infty} \dfrac{L(q)}{|q|} = +\infty \\ \\ \text{(ii) } u_0 \in \text{Lip}\,(\mathbb{R}^n). \end{cases} \tag{1.9}$$

Then we can show that the value function of our problem admits a simple representation formula called *Hopf's formula*.

Theorem 1.3.1 *Under hypotheses* (1.9) *the value function u satisfies*

$$u(t, x) = \min_{z \in \mathbb{R}^n} \left[t L \left(\frac{x - z}{t} \right) + u_0(z) \right] \tag{1.10}$$

for all $(t, x) \in Q_T$.

Proof — Observe that the minimum in (1.10) exists thanks to hypotheses (1.9). Let us denote by $v(t, x)$ the left-hand side of (1.10).

For fixed $(t, x) \in Q_T$ and $z \in \mathbb{R}^n$, let us set

$$y(s) = z + \frac{s}{t}(x - z), \qquad 0 \le s \le t.$$

Then $y \in \mathcal{A}(t, x)$ and therefore

$$u(t, x) \le J_t[y] = tL\left(\frac{x - z}{t}\right) + u_0(z).$$

Taking the infimum over z we obtain that $u(t, x) \le v(t, x)$.

To prove the opposite inequality, let us take $\zeta \in \mathcal{A}(t, x)$. From Jensen's inequality it follows that

$$L\left(\frac{x - \zeta(0)}{t}\right) = L\left(\frac{1}{t}\int_0^t \dot{\zeta}(s)\, ds\right) \le \frac{1}{t}\int_0^t L(\dot{\zeta}(s))\, ds$$

and therefore

$$v(t, x) \le u_0(\zeta(0)) + tL\left(\frac{x - \zeta(0)}{t}\right) \le J_t[\zeta].$$

Taking the infimum over $\zeta \in \mathcal{A}(t, x)$ we conclude that $v(t, x) \le u(t, x)$. ∎

Using Hopf's formula we can prove a first regularity property of u.

Theorem 1.3.2 *Under the assumptions* (1.9) *the value function u is Lipschitz continuous in* $\overline{Q}_T$. *More precisely, we have*

$$|u(t', x') - u(t, x)| \le L_0|x - x'| + L_1|t - t'|, \tag{1.11}$$

where $L_0 = \mathrm{Lip}\,(u_0)$ and $L_1 \ge 0$ is a suitable constant.

Proof — Let us first observe that (1.10) implies

$$\frac{u(t, x) - u_0(x)}{t} \le L(0) \tag{1.12}$$

for all (t, x) with $t > 0$. Let us now take $(t, x), (t', x') \in Q_T$ and let $y \in \mathbb{R}^n$ be such that

$$u(t, x) = tL\left(\frac{x - y}{t}\right) + u_0(y).$$

Then we find, using (1.12),

$$L\left(\frac{x - y}{t}\right) = \frac{u(t, x) - u_0(x)}{t} + \frac{u_0(x) - u_0(y)}{t}$$

$$\le L(0) + L_0\frac{|x - y|}{t}.$$

Since L is superlinear, we can find a constant $C_1 > 0$ depending on L_0 and $L(0)$ such that $L(q) \geq L(0) + L_0|q|$ if $|q| > C_1$. Then the previous inequality implies

$$\left|\frac{x-y}{t}\right| \leq C_1. \tag{1.13}$$

Let us now set

$$y' = x' - t'\frac{x-y}{t}.$$

Then we have

$$\frac{x-y}{t} = \frac{x'-y'}{t'}, \qquad y' - y = x' - x + (t - t')\frac{x-y}{t},$$

and so Hopf's formula (1.10), together with (1.13), yields

$$u(t', x') - u(t, x) \leq (t' - t)L\left(\frac{x-y}{t}\right) + u_0(y') - u_0(y)$$

$$\leq |t - t'| \max_{|q| \leq C_1} L(q) + L_0|y' - y|$$

$$\leq \left(L_0 C_1 + \max_{|q| \leq C_1} L(q)\right)|t - t'| + L_0|x' - x|.$$

By interchanging the role of (x, t) and (x', t'), we obtain the reverse inequality. This proves the conclusion in the case of $t, t' > 0$. If we have, for instance, $t' = 0$, we can estimate

$$|u(t, x) - u(0, x')| = \left|tL\left(\frac{x-y}{t}\right) + u_0(y) - u_0(x')\right|$$

$$\leq t \max_{|q| \leq C_1} L(q) + L_0(|y - x| + |x - x'|)$$

$$\leq t\left(\max_{|q| \leq C_1} L(q) + L_0 C_1\right) + L_0|x - x'|. \qquad \blacksquare$$

A well-known theorem due to Rademacher (see e.g., [71, 72, 14]) asserts that a Lipschitz continuous function is differentiable almost everywhere, and so the previous result immediately implies

Corollary 1.3.3 *Under hypotheses (1.9), the value function u is differentiable a.e. in Q_T.*

On the other hand, it is easy to see that, in general, u fails to be everywhere differentiable, even if u_0 and L are differentiable.

Example 1.3.4 Let us consider problem (1.4) with $n = 1$, $L(q) = q^2/2$ and the initial cost given by

$$u_0(z) = \begin{cases} -z^2 & \text{if } |z| < 1 \\ 1 - 2|z| & \text{if } |z| \geq 1. \end{cases}$$

Observe that u_0 is of class C^1 but not C^2. This is not essential for the behavior we are going to describe; one can build examples with C^∞ data exhibiting similar properties (see Example 6.3.5).

Let us compute the value function using Hopf's formula. We find that the minimum in (1.10) is attained at

$$\begin{cases} z = \dfrac{x}{1 - 2t} & \text{if } t < 1/2 \text{ and } |x| < 1 - 2t \\ z = x + \operatorname{sgn}(x)2t & \text{if } |x| \geq 1 - 2t \geq 0 \text{ or if } t \geq 1/2, \end{cases}$$

whence

$$u(t, x) = \begin{cases} -\dfrac{x^2}{1 - 2t} & \text{if } t < 1/2 \text{ and } |x| < 1 - 2t \\ 1 - 2(|x| + t) & \text{if } |x| \geq 1 - 2t \geq 0 \text{ or if } t \geq 1/2. \end{cases}$$

Therefore $u(t, x)$ is not differentiable at the points of the form $(t, 0)$ with $t \geq 1/2$. ∎

1.4 Hamilton–Jacobi equations

In this section we introduce a partial differential equation which is solved by the value function of our variational problem. We assume throughout that hypotheses (1.9) are satisfied. We use the notation

$$u_t = \frac{\partial u}{\partial t}, \qquad \nabla u = \left(\frac{\partial u}{\partial x_1}, \dots, \frac{\partial u}{\partial x_n} \right).$$

Theorem 1.4.1 *Let u be differentiable at a point* $(t, x) \in Q_T$. *Then*

$$u_t(t, x) + H(\nabla u(t, x)) = 0, \tag{1.14}$$

where

$$H(p) = \sup_{q \in \mathbb{R}^n} [p \cdot q - L(q)]. \tag{1.15}$$

Equation (1.14) is called the *Hamilton–Jacobi equation* of our problem in the calculus of variations. In the terminology of control theory, such an equation is also called *Bellman's equation* or *dynamic programming equation*. The function H is called the *hamiltonian*. In general, a function defined as in (1.15) is called the *Legendre transform* of L (see Appendix A.1).

Proof — Let (t, x) be a point at which u is differentiable. Given $q \in \mathbb{R}^n$, $s > 0$, let us set $y(\tau) = x + (\tau - t)q$. Then we obtain from (1.6)

$$u(t + s, x + sq) \leq \int_t^{t+s} L(\dot{y}(\tau))d\tau + u(t, x),$$

which implies

$$\frac{u(t + s, x + sq) - u(t, x)}{s} \leq L(q).$$

Letting $s \downarrow 0$ we obtain

$$u_t(t, x) + q \cdot \nabla u(t, x) - L(q) \leq 0,$$

and we deduce, by the arbitrariness of q, that

$$u_t(t, x) + H(\nabla u(t, x)) \leq 0.$$

To prove the converse inequality, let us take $\hat{x} \in \mathbb{R}^n$ such that

$$u(t, x) = tL\left(\frac{x - \hat{x}}{t}\right) + u_0(\hat{x}).$$

Such an $\hat{x}$ exists by Hopf's formula. We then set, for $s \in [0, t[$,

$$x_s = \frac{s}{t}x + \left(1 - \frac{s}{t}\right)\hat{x}.$$

Since $(x - \hat{x})/t = (x_s - \hat{x})/s$, we deduce from Hopf's formula

$$u(t, x) - u(s, x_s) \geq tL\left(\frac{x - \hat{x}}{t}\right) + u_0(\hat{x}) - sL\left(\frac{x_s - \hat{x}}{s}\right) + u_0(\hat{x})$$

$$= (t - s)L\left(\frac{x - \hat{x}}{t}\right).$$

Dividing by $t - s$ and letting $s \uparrow t$ we obtain

$$u_t(t, x) + \frac{x - \hat{x}}{t} \cdot \nabla u(t, x) - L\left(\frac{x - \hat{x}}{t}\right) \geq 0,$$

and we conclude that $u_t(t, x) + H(\nabla u(t, x)) \geq 0$. ∎

From Corollary 1.3.3 and Theorem 1.4.1 it follows that the value function u given in (1.10) satisfies

$$\begin{cases} u_t + H(\nabla u) = 0 & (t, x) \in \mathbb{R}_+ \times \mathbb{R}^n \text{ a.e.} \\ u(0, x) = u_0(x) & x \in \mathbb{R}^n. \end{cases} \tag{1.16}$$

The involutive character of the Legendre transform implies that the correspondence between calculus of variations and Hamilton–Jacobi equations is valid in both directions. More precisely, let H, u_0 be given such that

$$\begin{cases} \text{(i) } H \text{ is convex and } \lim_{|p| \to \infty} \dfrac{H(p)}{|p|} = +\infty \\[2mm] \text{(ii) } u_0 \in \text{Lip}(\mathbb{R}^n) \end{cases} \tag{1.17}$$

and suppose that we want to solve problem (1.16). To this purpose, we can define L to be the Legendre transform of H:

$$L(q) = \max_{p \in \mathbb{R}^n} [p \cdot q - H(p)].$$

Then (see Theorem A. 2.3) L satisfies property (1.9)–(i) and H is the Legendre transform of L, i.e. (1.15) holds. Therefore, Hopf's formula (1.10) yields a Lipschitz function u that solves (1.16).

The property of solving problem (1.16) almost everywhere, however, is not enough to characterize the value function u. Indeed, such a problem can have more than one solution in the class $\text{Lip}(\mathbb{R}_+ \times \mathbb{R}^n)$, as the next example shows.

Example 1.4.2 The problem

$$\begin{cases} u_t + u_x^2 = 0 & (t, x) \in \mathbb{R}_+ \times \mathbb{R} \text{ a.e.} \\ u(0, x) = 0 & x \in \mathbb{R} \end{cases} \tag{1.18}$$

admits the solution $u \equiv 0$. However, for any $a > 0$, the function u_a defined as

$$u_a(t, x) = \begin{cases} 0 & \text{if } |x| \geq at \\ a|x| - a^2 t & \text{if } |x| < at \end{cases}$$

is a Lipschitz function satisfying the equation almost everywhere together with its initial condition.

Simple examples like the one above show that the property of solving the equation almost everywhere is too weak and does not suffice to provide a satisfactory notion of generalized solution. It is therefore desirable to find additional conditions to ensure uniqueness and characterize the value function among the Lipschitz continuous solutions of the equation. A possible way of doing this relies on the semi-concavity property and will be pursued in Section 1.6.

1.5 Method of characteristics

We describe in this section the method of characteristics, which is a classical approach to the study of first order partial differential equations like the Hamilton–Jacobi equation (1.16). This method explains why such equations do not possess in general smooth solutions for all times, and has some interesting connections with the variational problem associated to the equation. A more general treatment of these topics will be given in Section 5.1.

Suppose that H, u_0 are in $C^2(\mathbb{R}^n)$, and suppose that we already know that problem (1.16) has a solution u of class C^2 in some strip Q_T. For fixed $z \in \mathbb{R}^n$, let us denote by $X(t; z)$ the solution of the ordinary differential equation (here the dot denotes differentiation with respect to t)

$$\dot{X} = DH(\nabla u(t, X)), \qquad X(0) = z. \tag{1.19}$$

Such a solution is defined in some maximal interval $[0, T_z[$ (although it will later turn out that $T_z = T$ for all z). The curve $t \to (t, X(t; z))$ is called the *characteristic curve* associated with u and starting from the point $(0, z)$. Let us now set

$$U(t; z) = u(t, X(t; z)), \qquad P(t; z) = \nabla u(t, X(t; z)). \tag{1.20}$$

Then, using the fact that u solves problem (1.16) we find that

$$\dot{U} = u_t(t, X) + \nabla u(t, X) \cdot \dot{X} = -H(P) + DH(P) \cdot P,$$

$$\dot{P} = \nabla u_t(t, X) + \nabla^2 u(t, X)\dot{X} = \nabla(u_t + H(\nabla u))(t, X) = 0.$$

Therefore P is constant, and so the right-hand side of (1.19) is also constant. Thus, X is defined in $[0, T[$ and we can compute explicitly X, U, P obtaining

$$\begin{cases} P(t; z) = Du_0(z) \\ X(t; z) = z + t DH(Du_0(z)) \\ U(t; z) = u_0(z) + t[DH(Du_0(z)) \cdot Du_0(z) - H(Du_0(z))]. \end{cases} \tag{1.21}$$

Observe that the right-hand side of (1.21) is no longer defined in terms of the solution u, but only depends on the initial value u_0. This suggests that, even without assuming in advance the existence of a solution, one can use these formulas to define one. As we are now going to show, such a construction can be in general carried out only locally in time.

We need the following classical result about the global invertibility of maps (see e.g., [11, Th. 3.1.8]).

Theorem 1.5.1 *Let $F : \mathbb{R}^n \to \mathbb{R}^n$ be of class C^1 and proper (that is, $F^{-1}(K)$ is compact whenever K is compact). If $\det DF(x) \neq 0$ for all $x \in \mathbb{R}^n$, then F is a global C^1-diffeomorphism from $\mathbb{R}^n$ onto itself.*

As a consequence, we obtain the following invertibility result for maps depending on a parameter.

Theorem 1.5.2 *Let $\Phi : [a, b] \times \mathbb{R}^n \to \mathbb{R}^n$ be of class C^1. Suppose that there exists $M > 0$ such that $|\Phi(t, z) - z| \leq M$ for all $(t, z) \in [a, b] \times \mathbb{R}^n$. Suppose also that the jacobian with respect to the z-variable $D_z F(t, z)$ has nonzero determinant for all (t, z). Then there exists a unique map $\Psi : [a, b] \times \mathbb{R}^n \to \mathbb{R}^n$ of class C^1 such that $\Phi(t, \Psi(t, x)) = x$ for all $(t, x) \in [a, b] \times \mathbb{R}^n$.*

Proof — For fixed $t \in [a, b]$, define $F(z) = \Phi(t, z)$. Then $F : \mathbb{R}^n \to \mathbb{R}^n$ is of class C^1 and $\det F(z) \neq 0$ for all z by our assumptions. To show that F is proper, let us take any compact set $K \subset \mathbb{R}^n$. Then $F^{-1}(K)$ is closed by the continuity of F. To prove that it is also bounded, let us denote by d the diameter of K and take any z_1, z_2 in $F^{-1}(K)$. Then we have, by our assumptions on Φ,

$$|z_2 - z_1| \leq |z_2 - F(z_2)| + |z_1 - F(z_1)| + |F(z_2) - F(z_1)| \leq 2M + d,$$

showing that $F^{-1}(K)$ is bounded. Hence F is proper.

Now we can apply Theorem 1.5.1 to obtain that F has a global inverse. Since the same holds for all t, we obtain that there exists $\Psi(t, x)$ such that $\Phi(t, \Psi(t, x)) = x$ for all $(t, x) \in [a, b] \times \mathbb{R}^n$.

It remains to prove that Ψ is regular with respect to both arguments. For this purpose, we set $\tilde{\Phi}(t, z) = (t, \Phi(t, z))$ and $\tilde{\Psi}(t, x) = (t, \Phi(t, x))$. Then $\tilde{\Phi}$ and $\tilde{\Psi}$ are reciprocal inverse. Since $\tilde{\Phi}$ is of class C^1, we have that $\tilde{\Psi}$ (and therefore Ψ) is also of class C^1 provided the jacobian of $\tilde{\Phi}$ has nonzero determinant. But this follows from our hypotheses on Φ, since we have

$$D\tilde{\Phi}(t, z) = \begin{pmatrix} 1 & 0 \\ \partial_t \Phi(t, z) & D_z \Phi(t, z) \end{pmatrix}. \qquad \blacksquare$$

We can now give the local existence result for classical solutions based on the method of characteristics.

Theorem 1.5.3 *Let $u_0, H \in C^2(\mathbb{R}^n)$ be given, and suppose that Du_0 and $D^2 u_0$ are bounded. Let us set*

$$T^* = \sup\{t > 0 : I + t D^2 H(Du_0(z))D^2 u_0(z) \text{ is invertible for all } z \in \mathbb{R}^n\}.$$

Then problem (1.16) has a unique solution $u \in C^2([0, T^*[\times \mathbb{R}^n)$.

Proof — Let X, U, P be defined as in (1.21). Then, for any $T < T^*$, we can apply Theorem 1.5.2 to the map X in $[0, T] \times \mathbb{R}^n$. In fact, the jacobian $X_z(t; z) = I + t D^2 H(Du_0(z))D^2 u_0(z)$ is invertible by hypothesis, and the quantity $X(t; z) - z = t DH(Du_0(z))$ is uniformly bounded. Therefore we can find a map $Z(t; x)$ of class C^1 such that $X(t; Z(t; x)) = x$ for all $(t, x) \in [0, T] \times \mathbb{R}^n$. After setting

$$u(t, x) = U(t; Z(t; x)), \qquad (t, x) \in [0, T] \times \mathbb{R}^n,$$

let us prove that u is a C^2 solution of problem (1.16).

To check that the initial condition is satisfied, it suffices to observe that $X(0; z) = z$ and $U(0; z) = u_0(z)$ for all z; therefore $Z(0; x) = x$ and $u(x, 0) = u_0(x)$ for all x. The verification that u is of class C^2 and satisfies the equation is technical but straightforward, as we now show. From the definition it is clear that u is of class at least C^1. Let us compute its derivatives. By (1.21) we have

$$U_z(t; z) = Du_0(z) + t\, Du_0(z)\, D^2 H(Du_0(z))\, D^2 u_0(z)$$
$$= P(t; z)\, X_z(t; z),$$

where the subscript denotes partial differentiation. This implies, by the definition of Z, that

$$\nabla u(t, x) = U_z(t; Z(t; x)) Z_x(t; x) = P(t; Z(t, x)). \tag{1.22}$$

We also find that

$$u_t(t, x) = U_t(t; Z(t, x)) + U_z(t; Z(t, x)) Z_t(t; x)$$
$$= DH(P) \cdot P - H(P) + P X_z Z_t,$$

where we have written for simplicity P, Z, etc. instead of $P(t; Z(t, x))$, $Z = Z(t; x)$, respectively. Taking into account that $X(t; Z(t; x)) \equiv x$ we obtain

$$X_t(t; Z) + X_z(t; Z) Z_t = 0.$$

Therefore, since $X_t = DH(P)$ by (1.21),

$$u_t(t, x) = X_t \cdot P - H(P) + P X_z Z_t = -H(P).$$

This equality, together with (1.22), implies that $u \in C^2$ and satisfies problem (1.16).

Uniqueness follows from the remarks at the beginning of this section. Indeed, if we have another solution v, we can define the characteristic curves associated to v, which are also given by (1.21), since they only depend on H and u_0. In particular, we have $v(t, X(t; z)) = U(t, z)$ for all t, z. Therefore,

$$v(t, x) = v(t, X(t; Z(t; x))) = U(t; Z(t, x)) = u(t, x)$$

for all $(t, x) \in [0, T] \times \mathbb{R}^n$, showing that u and v coincide.   ∎

Let us recall an elementary property from linear algebra.

Lemma 1.5.4 *Let B be a symmetric positive semidefinite $n \times n$-matrix. Then, for any $v \in \mathbb{R}^n$ we have $\langle Bv, v \rangle = 0$ if and only if $Bv = 0$.*

Proof — Let $e_1, \ldots, e_n$ be an orthonormal basis of eigenvectors of B, and let $\lambda_1, \ldots, \lambda_n$ be the corresponding eigenvalues. If we set $v_i = v \cdot e_i$, we have

$$\langle Bv, v \rangle = \sum_{i=1}^n \lambda_i v_i^2.$$

Since B is positive semidefinite, we have $\lambda_i \geq 0$ for all i and so all terms in the above sum are nonnegative. If $\langle Bv, v \rangle = 0$, then $\lambda_i v_i^2 = 0$ for all $i = 1, \ldots, n$. But then also $\lambda_i^2 v_i^2 = 0$ for all i, and we deduce

$$|Bv|^2 = \langle Bv, Bv \rangle = \sum_{i=1}^n \lambda_i^2 v_i^2 = 0.$$ ∎

Corollary 1.5.5 *Let u_0 and H be as in Theorem 1.5.3. Set*

$$M_0 = \sup_{z \in \mathbb{R}^n} |Du_0(z)|, \quad M_1 = \sup_{z \in \mathbb{R}^n} ||D^2 u_0(z)||, \quad M_2 = \sup_{|p| \le M_0} ||D^2 H(p)||.$$

Then problem (1.16) has a C^2 solution at least for time $t \in [0, (M_1 M_2)^{-1}[$. If H and u_0 are both convex (or both concave) then the problem has a C^2 solution for all positive times.

Proof — If M_1, M_2 are defined as above, the matrix $D^2 H(Du_0(z)) D^2 u_0(z)$ has norm less than $M_1 M_2$. Thus, if $t < (M_1 M_2)^{-1}$ we have

$$|t D^2 H(Du_0(z)) D^2 u_0(z) v| < |v|$$

for all $v \in \mathbb{R}^n$, $v \ne 0$, showing that the matrix $I + t D^2 H(Du_0(z)) D^2 u_0(z)$ is invertible.

Let us now prove the last part of the statement in the case when H, u_0 are convex (the concave case is completely analogous). It suffices to show that the matrix $I + t D^2 H(Du_0(z)) D^2 u_0(z)$ is invertible for all t, z. Let us argue by contradiction and suppose that there exist t, z and a nonzero vector v such that

$$v + t D^2 H(Du_0(z)) D^2 u_0(z) v = 0.$$

Let us set for simplicity $A = D^2 H(Du_0(z))$, $B = D^2 u_0(z)$. Then A, B are both positive semidefinite, and we have $v = -t A B v$. But then

$$0 \le t \langle A B v, B v \rangle = -\langle v, B v \rangle \le 0,$$

which implies that $\langle v, B v \rangle = 0$. By Lemma 1.5.4, we deduce that $B v = 0$, in contradiction with the property that $v = -t A B v$. $\blacksquare$

The time T^* given by Theorem 1.5.3 is optimal, as shown by the next result.

Theorem 1.5.6 *Let H, u_0 and T^* be as in Theorem 1.5.3. If $T' > T^*$, then no C^2 solution of problem (1.16) exists in $[0, T'[\times \mathbb{R}^n$.*

We give two different proofs of this statement; the argument of the second one is more intuitive, but can be applied only in the case $n = 1$.

First proof — We argue by contradiction and suppose that there exists a solution $u \in C^2([0, T'[\times \mathbb{R}^n)$, with $T' > T^*$. By definition of T^*, there exist $t^* \in]T^*, T'[$, $z^* \in \mathbb{R}^n$ and $\theta \in \mathbb{R}^n \setminus \{0\}$ such that

$$X_z(t^*; z^*) \theta = \theta + t^* D^2 H(Du_0(z^*)) D^2 u_0(z^*) \theta = 0.$$

This implies in particular that

$$P_z(t^*; z^*) \theta = D^2 u_0(z^*) \theta \ne 0.$$

Differentiating (1.20) with respect to z we obtain

$$\nabla^2 u(t^*, X(t^*, z^*)) X_z(t^*, z^*) = P_z(t^*, z^*).$$

Taking the scalar product with θ of both sides of the equality yields a contradiction.

∎

Second proof — Here we restrict ourselves to the case $n = 1$. As in the first proof, we argue by contradiction and suppose that there exists a solution u up to some time $T' > T^*$. By definition of T^* we can find $t^* < T'$ and $z^* \in \mathbb{R}$ such that $1 + t^* H''(u_0'(z^*))u_0''(z^*) = 0$. Then $H''(u_0'(z^*))u_0''(z^*) < 0$ and so, if we fix any $\hat{t} \in]t^*, T'[$, we have

$$1 + \hat{t} H''(u_0'(z^*))u_0''(z^*) < 1 + t^* H''(u_0'(z^*))u_0''(z^*) = 0.$$

This shows that the function $z \to X(\hat{t}; z)$ has a negative derivative at $z = z^*$ and so is decreasing in some neighborhood of z^*. Thus, there exists $z_1 < z_2$ such that $X(\hat{t}; z_1) > X(\hat{t}; z_2)$. On the other hand, we have $X(0; z_1) = z_1 < z_2 = X(0; z_2)$. By continuity, there exists $s \in]0, \hat{t}[$ such that $X(s; z_1) = X(s; z_2)$. It follows that

$$u_0'(z_i) = P(s; z_i) = u_x(s, X(s; z_i)), \qquad i = 1, 2,$$

which implies $u_0'(z_1) = u_0'(z_2)$. But then we deduce from (1.21) that $X(s; z_2) - X(s; z_1) = z_2 - z_1 \neq 0$, in contradiction with our choice of s.   ∎

Remark 1.5.7 As it is clear from the second proof, the critical time T^* is the time at which the characteristic lines start to cross. Intuitively speaking, since u_x is constant along any of these lines, the crossing of characteristics corresponds to the formation of discontinuities in the gradient of u. Let us do some observations about this behavior. From the computations of the second proof we see that the map $z \to X(t; z)$ is strictly increasing if $t < T^*$ and not monotone for $t > T^*$. Thus, the map $z \to X(T^*; z)$ is increasing; it is either strictly increasing or it is constant on some interval. In the first case we can intuitively say that the characteristics are not yet crossing at the critical time $t = T^*$, but start crossing immediately afterwards. In the latter case there is a family of characteristics which are all converging to the same point at $t = T^*$, as in Example 1.3.4. However, it is easily seen that this second behavior is nongeneric, in the sense that it is not stable under small perturbations of the initial value. In fact, it only occurs if the derivative of $z \to X(T^*; z)$ vanishes identically on an interval.   ∎

Let us now focus on the case of a Hamilton–Jacobi equation coming from a variational problem of the form considered in the previous sections. We have seen that the value function is Lipschitz continuous and solves the equation almost everywhere; we now study the relationship between the value function and the smooth solution obtained by the method of characteristics. We will show that the two functions agree as long as the latter exists, and that there is a connection between the value function and the characteristics which is valid for all times.

We make some more regularity assumptions on L, u_0 with respect to the previous sections; namely we assume, in addition to (1.9), that L is strictly convex and that u_0 is differentiable. Then, if we denote by H the Legendre transform of L, we have that H is differentiable (see Theorem A. 2.4 in the Appendix). Observe that such properties of H and u_0 are weaker than the ones needed in Theorem 1.5.3, but are enough to ensure that the functions P, X, U in (1.21) are well defined.

Our first remark is that minimizers for the problem in the calculus of variations are also characteristic curves for the Hamilton–Jacobi equation.

Proposition 1.5.8 *Let $L, u_0 : \mathbb{R}^n \to \mathbb{R}$ satisfy assumptions (1.9); suppose in addition that L is strictly convex and that u_0 is of class C^1. Given $(t, x) \in [0, \infty[\times \mathbb{R}^n$, let $y(\cdot)$ be a minimizer for problem (1.3). Then $y(\cdot) = X(\cdot, z)$ for some $z \in \mathbb{R}^n$.*

Proof — Let $y(\cdot)$ be a minimizer for problem (1.3) and let $z = y(0)$. In the proof of Hopf's formula (Theorem 1.3.1) we have seen that y has the form $y(s) = z + s(x - z)/t$ for some $z \in \mathbb{R}^n$. In addition, z is a minimizer for the right-hand side of (1.10), which implies that $DL((x - z)/t) = Du_0(z)$. By well-known properties of the Legendre transform (see (A. 19)), this implies that $DH(Du_0(z)) = (x - z)/t$, and so $y(s) = X(s; z)$. ∎

Let us set for $(t, x) \in [0, +\infty[\times \mathbb{R}^n$,

$$\mathcal{Z}(t; x) = \{z \in \mathbb{R}^n \ : \ X(t; z) = x\}. \tag{1.23}$$

If the hypotheses of Theorem 1.5.3 are satisfied, then such a set is a singleton for small times; in general it is a compact set, as a consequence of the definition of X and of the properties of u_0, H.

Proposition 1.5.9 *Given L, u_0 as in the previous proposition, let $u(t, x)$ be the value function of problem (1.4). Then*

$$u(t, x) = \min_{z \in \mathcal{Z}(t;x)} U(t; z). \tag{1.24}$$

Proof — As we have seen in the proof of the previous proposition, when one takes the minimum in Hopf's formula (1.10) it is enough to consider the values of z such that $x = X(t; z)$. Therefore

$$u(t, x) = \min_{z \in \mathcal{Z}(t;x)} \left[u_0(z) + tL \left(\frac{x - z}{t} \right) \right].$$

The condition $z \in \mathcal{Z}(t; x)$ is equivalent to $DH(Du_0(z)) = (x - z)/t$. Recalling identity (A. 21), we have, for $z \in \mathcal{Z}(t; x)$,

$$U(t; z) = u_0(z) + tL \left(\frac{x - z}{t} \right)$$

and so the two sides of (1.24) coincide.   ■

In other words, the solution given by Hopf's formula is a suitable selection of the multivalued function obtained by the method of characteristics. Let us point out that such a behavior is peculiar to Hamilton–Jacobi equations with convex hamiltonian: if H is neither convex nor concave there is in general no way of finding a weak solution by taking such a selection.

Corollary 1.5.10 *Let $L, u_0 \in C^2(\mathbb{R}^n)$ satisfy assumptions* (1.9). *Suppose in addition that D^2L is positive definite everywhere and that D^2u_0 is uniformly bounded. Then there exists $T^* > 0$ such that the value function of problem* (1.4) *is of class C^2 in $[0, T^*[\times \mathbb{R}^n$. If u_0 is convex, then $T^* = +\infty$ and so the value function is C^2 everywhere. In addition, given (t, x) with $t < T^*$, there exists a unique minimizer for problem* (1.3), *which is given by*

$$y(s) = x + (s - t)DH(\nabla u(t, x)). \tag{1.25}$$

Proof — By Theorem A. 2.5 the Legendre transform of L is of class C^2 and strictly convex. Hence, Theorem 1.5.3 ensures the existence of $T^* > 0$ such that $\mathcal{Z}(t, x)$ is a singleton for $t < T^*$ and such that the method of characteristics yields a smooth solution of the Hamilton–Jacobi equation in $[0, T^*[\times \mathbb{R}^n$. By the previous theorem, this smooth solution coincides with the value function. As observed in Corollary 1.5.5, $T^* = +\infty$ if u_0 is convex. Finally, (1.25) follows from Proposition 1.5.8, from (1.22) and (1.21).   ■

The previous theorem shows how the dynamic programming approach can provide a solution to the problem in the calculus of variations we are considering. Such a result, however, relies on the smoothness of the value function and we have seen that such a property does not hold in general for all times. When one tries to extend this approach to the cases when the value function is only Lipschitz, several difficulties arise. We have already mentioned the nonuniqueness of Lipschitz continuous solutions of the Hamilton–Jacobi equation. It is also not obvious how to restate equation (1.25) at the points where u is not differentiable. These issues will be discussed in detail in the remainder of the book.

1.6 Semiconcavity of Hopf's solution

In this section we show that the semiconcavity property characterizes the value function among all possible Lipschitz continuous solutions of the Hamilton–Jacobi equation (1.16).

Theorem 1.6.1 *Let L, u_0 satisfy assumptions* (1.9). *Suppose in addition that*

$$\text{(i) } L \in C^2(\mathbb{R}^n), \quad D^2L(q) \le \frac{2}{\alpha}I \quad \forall q \in \mathbb{R}^n \tag{1.26}$$

$$\text{(ii) } u_0(x + h) + u_0(x - h) - 2u_0(x) \le C_0|h|^2, \quad \forall x, h \in \mathbb{R}^n$$

for suitable constants $\alpha > 0$, $C_0 \geq 0$. Then there exists a constant $C_1 \geq 0$ such that

$$u(t+s, x+h) + u(t-s, x-h) - 2u(t, x)$$
$$\leq \frac{2tC_0}{2t + \alpha(t^2 - s^2)C_0}(|h| + C_1|s|)^2 \qquad (1.27)$$

for all $t > 0$, $s \in]-t, t[$, $x, h \in \mathbb{R}^n$.

Proof — For fixed t, s, x, h as in the statement of the theorem, let us choose $\hat{x} \in \mathbb{R}^n$ such that

$$u(t, x) = tL\left(\frac{x - \hat{x}}{t}\right) + u_0(\hat{x}). \qquad (1.28)$$

Such a $\hat{x}$ exists by Hopf's formula; in addition, by (1.13), there exists C_1, depending only on L, such that

$$\frac{|x - \hat{x}|}{t} \leq C_1. \qquad (1.29)$$

We set, for $\lambda \geq 0$,

$$x_\lambda^+ = \hat{x} + \lambda\left(h - s\frac{x - \hat{x}}{t}\right), \qquad x_\lambda^- = \hat{x} - \lambda\left(h - s\frac{x - \hat{x}}{t}\right).$$

Then we have

$$\frac{x_\lambda^+ + x_\lambda^-}{2} = \hat{x}, \qquad \frac{x_\lambda^+ - x_\lambda^-}{2} = \lambda\left(h - s\frac{x - \hat{x}}{t}\right). \qquad (1.30)$$

By (1.29) we have

$$\frac{|x_\lambda^+ - x_\lambda^-|}{2} \leq \lambda(|h| + C_1|s|). \qquad (1.31)$$

By Hopf's formula (1.10) we have

$$u(t \pm s, x \pm h) \leq (t \pm s)L\left(\frac{x \pm h - x_\lambda^\pm}{t \pm s}\right) + u_0(x_\lambda^\pm).$$

Thus, keeping into account (1.28), we can estimate

$$u(t+s, x+h) + u(t-s, x-h) - 2u(t, x)$$
$$\leq 2t\left[\frac{t+s}{2t}L\left(\frac{x+h-x_\lambda^+}{t+s}\right) + \frac{t-s}{2t}L\left(\frac{x-h-x_\lambda^-}{t-s}\right) - L\left(\frac{x-\hat{x}}{t}\right)\right]$$
$$+ u_0(x_\lambda^+) + u_0(x_\lambda^-) - 2u_0(\hat{x}). \qquad (1.32)$$

From (1.31) and (1.26)(ii) it follows that

$$u_0(x_\lambda^+) + u_0(x_\lambda^-) - 2u_0(\hat{x}) \leq C_0\lambda^2(|h| + C_1|s|)^2. \tag{1.33}$$

We now take $q_0, q_1 \in \mathbb{R}^n$, $\theta \in [0, 1]$ and set $q_\theta = \theta q_1 + (1-\theta)q_0$. Using assumption (1.26)(i) and the equivalence between (b) and (e) in Proposition 1.1.3, we have

$$\theta L(q_1) + (1 - \theta)L(q_0) - L(q_\theta) \leq \frac{\theta(1 - \theta)}{\alpha}|q_1 - q_0|^2. \tag{1.34}$$

Therefore, observing that

$$\frac{t + s}{2t} \frac{x + h - x_\lambda^+}{t + s} + \frac{t - s}{2t} \frac{x - h - x_\lambda^-}{t - s} = \frac{x - \hat{x}}{t},$$

we obtain from (1.34)

$$\frac{t + s}{2t} L\left(\frac{x + h - x_\lambda^+}{t + s}\right) + \frac{t - s}{2t} L\left(\frac{x - h - x_\lambda^-}{t - s}\right) - L\left(\frac{x - \hat{x}}{t}\right)$$
$$\leq \frac{t^2 - s^2}{\alpha(2t)^2}\left|\frac{x + h - x_\lambda^+}{t + s} - \frac{x - h - x_\lambda^-}{t - s}\right|^2. \tag{1.35}$$

In addition

$$\frac{x + h - x_\lambda^+}{t + s} - \frac{x - h - x_\lambda^-}{t - s}$$
$$= \frac{(t - s)(x + h) - (t + s)(x - h)}{t^2 - s^2} - \frac{(t - s)x_\lambda^+ - (t + s)x_\lambda^-}{t^2 - s^2}$$
$$= 2\frac{th - sx + s\hat{x}}{t^2 - s^2} - \frac{2\lambda t}{t^2 - s^2}\left(h - s\frac{x - \hat{x}}{t}\right)$$
$$= \frac{2(1 - \lambda)t}{t^2 - s^2}\left(h - s\frac{x - \hat{x}}{t}\right),$$

which implies, by (1.29)

$$\left|\frac{x + h - x_\lambda^+}{t + s} - \frac{x - h - x_\lambda^-}{t - s}\right| \leq \frac{2(1 - \lambda)t}{t^2 - s^2}(|h| + C_1|s|). \tag{1.36}$$

From (1.32), (1.33), (1.35) and (1.36) we obtain, for all $\lambda \geq 0$,

$$u(t + s, x + h) + u(t - s, x - h) - 2u(t, x)$$
$$\leq C(\lambda)[|h| + C_1|s|]^2,$$

where

$$C(\lambda) = \frac{2t}{\alpha(t^2 - s^2)}(1 - \lambda)^2 + C_0\lambda^2.$$

It is easily checked that

$$\min_{\lambda \geq 0} C(\lambda) = \frac{2tC_0}{2t + \alpha C_0(t^2 - s^2)}, \tag{1.37}$$

and this proves (1.27). ∎

Estimate (1.30) of the previous theorem easily implies that u is semiconcave with a linear modulus. Actually, to have a semiconcavity estimate for u it is not necessary that both assumptions in (1.26) are satisfied, but it is enough to assume one of the two, as shown in the next corollary.

Corollary 1.6.2 *Let assumptions* (1.9) *be satisfied and let u be the function defined by Hopf's formula* (1.10).

(i) If L satisfies property (1.26)(i) *for some $\alpha > 0$, then there exists C_1 such that*

$$u(t + s, x + h) + u(t - s, x - h) - 2u(t, x) \leq \frac{2t}{\alpha(t^2 - s^2)}(|h| + C_1|s|)^2.$$

(ii) If u_0 satisfies (1.26)(ii) *for some $C_0 \geq 0$, then there exists C_1 such that*

$$u(t + s, x + h) + u(t - s, x - h) - 2u(t, x) \leq C_0(|h| + C_1|s|)^2.$$

(iii) If u_0 is concave, then u is concave (jointly in t, x).

Proof — Formally, statements (i) and (ii) are obtained by letting respectively $C_0 \to \infty$ and $\alpha \downarrow 0$ in estimate (1.27). A precise motivation can be obtained by a suitable adaptation of the proof of the previous theorem. To prove (i), we apply inequality (1.32) with $\lambda = 0$. With this choice of λ we have $x_\lambda^+ = x_\lambda^- = \hat{x}$; thus, the terms with u_0 drop out while the terms involving L can be estimated as before. To prove (ii) we choose instead $\lambda = 1$; then it is easily checked that the total contribution of the terms with L in (1.32) is zero, while the remaining terms can be estimated as before. Finally, statement (iii) follows from (ii) taking $C_0 = 0$. ∎

If both hypotheses in (1.26) are violated, i.e., L is not twice differentiable and u_0 is not semiconcave, then u may fail to be semiconcave, as shown by the next example.

Example 1.6.3 Consider a one-dimensional problem with lagrangian and initial cost given respectively by

$$L(q) = \frac{q^2}{2} + |q|, \qquad u_0(x) = \frac{|x|}{2}.$$

Clearly, hypotheses (1.9) are satisfied. On the other hand, assumptions (1.26) are both violated, since L is not twice differentiable at $q = 0$ and u_0 is not semiconcave (see Example 1.1.4).

By Hopf's formula the value function is given by

$$u(t, x) = \min_{z \in \mathbb{R}} \left\{ \frac{|z|}{2} + \frac{(x - z)^2}{2t} + |x - z| \right\}.$$

It is easily found that the quantity to minimize is a decreasing function of z for $z < x$ and increasing for $z > x$. Hence the minimum is attained for $z = x$, i.e.,

$$u(t, x) = \frac{|x|}{2}, \qquad \forall t \geq 0, \ x \in \mathbb{R}.$$

Thus, the value function is not semiconcave.   ∎

We now show that the semiconcavity property singles out Hopf's solution among all Lipschitz continuous solutions of the Hamilton–Jacobi equation.

Theorem 1.6.4 *Let $H \in C^2(\mathbb{R}^n)$ be convex and let $u_1, u_2 \in \text{Lip}\,(\overline{Q}_T)$ be solutions of* (1.16) *such that, for any $t > 0$ and $i = 1, 2$,*

$$u_i(t, x + h) + u_i(t, x - h) - 2u_i(t, x) \leq C\left(1 + \frac{1}{t}\right)|h|^2, \qquad x, h \in \mathbb{R}^n \quad (1.38)$$

for some $C > 0$. Then $u_1 = u_2$ everywhere in $\overline{Q}_T$.

We first recall an elementary algebraic property.

Lemma 1.6.5 *Let A, B be two symmetric $n \times n$ matrices. Suppose that $0 \leq A \leq \Lambda I$ and $B \leq kI$ for some $\Lambda, k > 0$. Then*

$$\text{trace}(AB) \leq nk\Lambda.$$

Proof — Let $\{e_1, \ldots, e_n\}$ be on orthonormal basis of eigenvectors of B and let $\{\mu_1, \ldots, \mu_n\}$ be the corresponding eigenvalues. Then our assumptions imply

$$\mu_i \leq k, \qquad 0 \leq \langle e_i, Ae_i \rangle \leq \Lambda, \qquad \forall i = 1, \ldots, n.$$

It follows that

$$\text{trace}(AB) = \sum_{i=1}^{n} \langle e_i, ABe_i \rangle = \sum_{i=1}^{n} \mu_i \langle e_i, Ae_i \rangle$$

$$\leq k \sum_{i=1}^{n} \langle e_i, Ae_i \rangle \leq nk\Lambda. \qquad\qquad ∎$$

Proof of Theorem 1.6.4 — Setting $\bar{u}(t, x) = u_1(t, x) - u_2(t, x)$ we have, for $(t, x) \in Q_T$ a.e.,

$$\bar{u}_t(t, x) = H(\nabla u_2(t, x)) - H(\nabla u_1(t, x)) = -b(t, x)\nabla\bar{u}(t, x),$$

where

$$b(t, x) = \int_0^1 DH(r\nabla u_2(t, x) + (1 - r)\nabla u_1(t, x))dr.$$

Let $\phi : \mathbb{R} \to [0, \infty[$ be a C^1 function to be fixed later. Setting $v = \phi(\bar{u})$ we obtain, for $(t, x) \in Q_T$ a.e.,

$$v_t(t, x) + b(t, x) \cdot \nabla v(t, x) = 0. \tag{1.39}$$

Let $k \in C^\infty(\mathbb{R}^n)$ be a nonnegative function whose support is contained in the unit sphere and whose integral is 1. We define, for $i = 1, 2, \varepsilon > 0$ and $(t, x) \in Q_T$,

$$u_i^\varepsilon(t, x) = \frac{1}{\varepsilon^n} \int_{\mathbb{R}^n} u_i(t, y) k \left(\frac{x - y}{\varepsilon} \right) dy.$$

By well-known properties of convolutions the functions u_i^ε are of class C^∞ with respect to x and satisfy

(i) $|\nabla u_i^\varepsilon| \leq \mathrm{Lip}\,(u_i)$

(ii) for all $t > 0$, $\nabla u_i^\varepsilon(t, \cdot) \to \nabla u_i(t, \cdot)$ a.e. as $\varepsilon \to 0$.

$$\tag{1.40}$$

In addition, u_i^ε satisfies the semiconcavity estimate (1.38) for all $\varepsilon > 0$. Therefore, using property (e) in Proposition 1.1.3,

$$\nabla^2 u_i^\varepsilon(t, x) \leq C \left(1 + \frac{1}{t} \right) I, \qquad (t, x) \in Q_T. \tag{1.41}$$

Setting

$$b_\varepsilon(t, x) = \int_0^1 DH(r \nabla u_2^\varepsilon(t, x) + (1 - r) \nabla u_1^\varepsilon(t, x)) dr,$$

we can rewrite equation (1.39) in the form

$$v_t + \mathrm{div}\,(v b_\varepsilon) = (\mathrm{div}\, b_\varepsilon) v + (b_\varepsilon - b) \cdot \nabla v.$$

Let us set

$$R = \max\{|DH(p)| \ : \ |p| \leq \max(\mathrm{Lip}\,(u_1), \mathrm{Lip}\,(u_2))\},$$

$$\Lambda = \max\{\|D^2 H(p)\| \ : \ |p| \leq \max(\mathrm{Lip}\,(u_1), \mathrm{Lip}\,(u_2))\}.$$

By (1.41) and Lemma 1.6.5 it follows that

$$\mathrm{div}\, b_\varepsilon =$$

$$= \int_0^1 \sum_{k,l=1}^n \frac{\partial^2 H}{\partial p_k \partial p_l} (r \nabla u_2^\varepsilon + (1 - r) \nabla u_1^\varepsilon) \left(r \frac{\partial^2 u_2^\varepsilon}{\partial x_k \partial x_l} + (1 - r) \frac{\partial^2 u_1^\varepsilon}{\partial x_k \partial x_l} \right) dr$$

$$\leq n \Lambda C (1 + 1/t). \tag{1.42}$$

This is the estimate where the semiconcavity assumption on the u_i's is used.

Given $(t_0, x_0) \in Q_T$, let us introduce the function

$$E(t) = \int_{B(x_0, R(t_0 - t))} v(t, x) dx, \qquad 0 \leq t \leq t_0$$

and the cone

$$\mathcal{C} = \{(t, x) \; : \; 0 \le t \le t_0, \; x \in B(x_0, R(t_0 - t))\},$$

where we use the notation

$$B(x_0, R(t_0 - t)) = \{x \in \mathbb{R}^n \; : \; |x - x_0| < R(t_0 - t)\}.$$

Then E is Lipschitz continuous and satisfies, for a.e. $t > 0$,

$$
\begin{aligned}
E'(t) &= \int_{B(x_0, R(t_0-t))} v_t dx - R \int_{\partial B(x_0, R(t_0-t))} v \, dS \\
&= \int_{B(x_0, R(t_0-t))} \{-\operatorname{div}(v b_\varepsilon) + (\operatorname{div} b_\varepsilon) v + (b_\varepsilon - b) \cdot \nabla v\} dx \\
&\quad - R \int_{\partial B(x_0, R(t_0-t))} v \, dS \\
&= -\int_{\partial B(x_0, R(t_0-t))} v(b_\varepsilon \cdot \nu + R) dS \\
&\quad + \int_{B(x_0, R(t_0-t))} \{(\operatorname{div} b_\varepsilon) v + (b_\varepsilon - b) \cdot \nabla v\} dx \\
&\le \int_{B(x_0, R(t_0-t))} \{(\operatorname{div} b_\varepsilon) v + (b_\varepsilon - b) \cdot \nabla v\} dx \\
&\le n \Lambda C \left(1 + \frac{1}{t}\right) E(t) + \int_{B(x_0, R(t_0-t))} (b_\varepsilon - b) \cdot \nabla v \, dx.
\end{aligned}
$$

Letting $\varepsilon \to 0$ we obtain, by (1.40)(ii),

$$E'(t) \le n \Lambda C \left(1 + \frac{1}{t}\right) E(t), \qquad t \in \,]0, t_0[\text{ a.e.} \tag{1.43}$$

We now choose the function ϕ in the definition of v. We fix $\eta > 0$ and take ϕ such that $\phi(z) = 0$ if $|z| \le \eta[\operatorname{Lip}(u_1) + \operatorname{Lip}(u_2)]$ and $\phi(z) > 0$ otherwise. Then the assumption that $u_1 = u_2$ for $t = 0$ implies that $v(t, x) = 0$ if $t \le \eta$. Thus we obtain, from (1.43) and Gronwall's inequality,

$$0 \le E(t) \le E(\eta) \exp \left(\int_\eta^t n \Lambda C \left(1 + \frac{1}{s}\right) ds\right) = 0$$

for any $t \in [0, t_0]$. Hence

$$|u_2 - u_1| \le \eta \, [\operatorname{Lip}(u_1) + \operatorname{Lip}(u_2)] \text{ on } \mathcal{C}.$$

By the arbitrariness of $\eta > 0$, we deduce that u_1 and u_2 coincide in $\mathcal{C}$, and, in particular, $u_1(t_0, x_0) = u_2(t_0, x_0)$. $\blacksquare$

We can summarize the results we have obtained about problem (1.16) as follows.

Corollary 1.6.6 *Let H, u_0 satisfy hypotheses (1.17). Suppose in addition that $H \in C^2(\mathbb{R}^n)$ and that either H is uniformly convex or u_0 is semiconcave with a linear modulus. Then there exists a unique $u \in \text{Lip}([0, \infty[\times \mathbb{R}^n)$ which solves problem (1.16) almost everywhere and which satisfies*

$$u(t, x + h) + u(t, x - h) - 2u(t, x) \leq C\left(1 + \frac{1}{t}\right)|h|^2, \quad x, h \in \mathbb{R}^n, t > 0,$$

$$(1.44)$$

for a suitable $C > 0$. In addition, u is given by Hopf's formula (1.10), taking as L the Legendre transform of H.

Proof — We recall that, if H is uniformly convex and L is the Legendre transform of H, then L satisfies property (1.26)(i) (see (A. 20) in the appendix). The result then follows from Theorem 1.4.1, Corollary 1.6.2 and Theorem 1.6.4. ∎

Let us point out that, nowadays, using semiconcavity to obtain uniqueness results is a procedure that has mainly an historical interest, since it has been later incorporated in the more general theory of viscosity solutions. A stronger motivation for the study of the semiconcavity lies in the consequences for the regularity of the value function. As we will see, the generalized differential of a semiconcave function enjoys special properties, which play in important role in applications to the calculus of variations and optimal control theory.

1.7 Semiconcavity and entropy solutions

In this section we discuss briefly the connection between Hamilton–Jacobi equations and another class of partial differential equations called the *hyperbolic conservation laws*.

Given $H \in C^2(\mathbb{R})$ and $T \in]0, \infty]$, let $u \in C^2([0, T[\times \mathbb{R})$ be a solution of

$$u_t + H(u_x) = 0. \tag{1.45}$$

Then, if we set $v(t, x) := u_x(t, x)$ and differentiate (1.45) with respect to x, we see that v solves

$$v_t + H(v)_x = 0. \tag{1.46}$$

Conversely, if v is a C^1 solution of (1.46) and we set, for a given $x_0 \in \mathbb{R}$,

$$u(t, x) := \int_{x_0}^x v(t, y) \, dy - \int_0^t H(v(s, x_0)) \, ds \tag{1.47}$$

then u is a solution of (1.45).

Equation (1.46) is called a hyperbolic conservation law. We observe that the above transformation can no longer be done when one considers equations in several space dimensions or systems of equations.

Just like Hamilton–Jacobi equations, conservation laws do not possess in general global smooth solutions. It is interesting therefore to investigate the relation between the two equations when dealing with generalized solutions. Conservation laws are in divergence form and it is natural to consider solutions in the sense of distributions. More precisely, let us consider equation (1.46) with initial data

$$v(0, x) = v_0(x), \qquad x \in \mathbb{R} \text{ a.e.,} \tag{1.48}$$

where $v_0 \in L^\infty(\mathbb{R})$. Then we say that $v \in L^\infty([0, T[\times \mathbb{R})$ is a *weak solution* of problem (1.46)–(1.48) if it satisfies

$$\int_0^T \int_{\mathbb{R}} [v(t, x)\phi_t(t, x) + H(v(t, x))\phi_x(t, x)] \, dx dt$$
$$= -\int_{\mathbb{R}} v_0(x)\phi(0, x) \, dx, \qquad \forall \phi \in C_0^\infty([0, T[\times \mathbb{R}). \tag{1.49}$$

Here C_0^∞ denotes the class of C^∞ functions with compact support.

It can be proved that the weak solutions to hyperbolic conservation laws defined above correspond to Lipschitz continuous almost everywhere solutions to Hamilton–Jacobi equations, as stated in the next result.

Theorem 1.7.1 *Let $u \in \text{Lip}([0, T[\times \mathbb{R})$ be an almost everywhere solution of (1.45) and let $v(t, x) = u_x(t, x)$. Then v is a weak solution of (1.46)–(1.48) with $v_0(x) = u_x(0, x)$.*

Conversely, let $v \in L^\infty([0, T[\times \mathbb{R})$ be a weak solution of (1.46)–(1.48) for some $v_0 \in L^\infty(\mathbb{R})$. Then, there exists $u \in \text{Lip}([0, T[\times \mathbb{R})$ which satisfies (1.45) almost everywhere and is such that $u_x(t, x) = v(t, x)$ for a.e. $(t, x) \in [0, T[\times \mathbb{R}$ and $u_x(0, x) = v_0(x)$ for a.e. $x \in \mathbb{R}$.

Proof — See [63, Theorem 2, p. 955]. ∎

We have seen that, for Hamilton–Jacobi equations, the notion of almost everywhere solution is too weak and a Cauchy problem may possess in general infinitely many such solutions. By the equivalence result stated above, the same is true for the distributional solutions of scalar conservation laws: in order to obtain a uniqueness result one needs to restrict the class of solutions by adding some suitable admissibility condition. In the case where H is convex, we can extend to conservation laws the existence and uniqueness result for Hamilton–Jacobi equations of the previous chapter, provided we require that the solutions satisfy a "differentiated" form of the semiconcavity inequality.

Corollary 1.7.2 *Given $H \in C^2(\mathbb{R})$ such that $H''(p) \geq c > 0$ and given $v_0 \in L^\infty(\mathbb{R})$ there exists a unique $v \in L^\infty([0, \infty[\times \mathbb{R})$ which is a weak solution of problem (1.46)–(1.48) and satisfies a.e.*

$$v(y, t) - v(x, t) \leq \frac{y - x}{ct}, \qquad x < y, \, t > 0. \tag{1.50}$$

Proof — We observe that inequality (1.50) is equivalent to the property that for all $t > 0$, $x \to v(x, t) - x/(ct)$ is nonincreasing. Therefore, if $u \in \mathrm{Lip}([0, \infty[\times \mathbb{R})$ is such that $u_x = v$ a.e., then v satisfies (1.50) if and only if, for all $t > 0$, $x \to u(x, t) - x^2/(2ct)$ is concave. The result then follows from Corollary 1.6.2, Theorem 1.6.4 and Proposition 1.7.1, taking into account also the characterizations of semiconcavity given in Proposition 1.1.3.   ∎

Condition (1.50) is called the *Oleinik one-sided Lipschitz inequality*. A weak solution of (1.45)–(1.48) satisfying the Oleinik inequality is called an *entropy solution*. There are other equivalent definitions of entropy solutions of scalar conservation laws. The most powerful one was given by Kruzhkov [101] replacing equality (1.49) by a family of integral inequalities. Kruzhkov's definition also covers the cases where H is not convex and where the space dimension is greater than one. In the case of Hamilton–Jacobi equations with a general H, the appropriate class of solutions is given by the *viscosity solutions* of Crandall and Lions (see Chapter 5), which coincide with semiconcave solutions for the equations with convex Hamiltonian considered in this chapter. It can be shown (see e.g., [52]) that the two theories are equivalent in the one-dimensional case; that is, for a general nonconvex H, a function u is a viscosity solution of (1.45) if and only if $v = u_x$ is an entropy solution of (1.46).

Bibliographical notes

The dynamic programming approach for the analysis of control problems was started by Bellman [25] in the 1950s and it gradually became a standard tool in control theory (see e.g., the books [88, 111, 80]). However, in the special case of calculus of variations considered in this chapter, many of the ideas we have described were present in much older works. For instance, the connection between variational problems and Hamilton–Jacobi equations is well known in classical mechanics. In this context, an approach based on the introduction of the value function was already used by Carathéodory [49].

The representation formula (1.10) was given by Conway and Hopf in [63, 90]. An analogous formula had been previously found for the solution of conservation laws by Lax [106] and by Oleinik [114].

The first uniqueness results for semiconcave solutions of Hamilton–Jacobi equations were obtained by Kruzhkov and Douglis (see e.g., [99, 69, 100]). Our presentation of Theorem 1.6.4 follows [71].

For Hamilton–Jacobi equations of evolutionary type the semiconcavity of solutions was initially studied only with respect to the space variables since such a property is strong enough for obtaining a uniqueness result. More recently, the work of Cannarsa and Soner [45, 46] underlined the interest of semiconcavity with respect to the whole set of variables for the analysis of the singular set.

The connection between hyperbolic conservation laws and Hamilton–Jacobi equations has been known since the beginning of the study of the two theories. The uniqueness statement of Corollary 1.7.2 was proved by Oleinik in [114]. In our exposition we have reversed the historical order since Oleinik's uniqueness theorem was proved before the corresponding statement for Hamilton–Jacobi equations. Oleinik's paper inspired the aforementioned studies by Kruzhkov and Douglis and in particular suggested that the semiconcavity property could provide a uniqueness criterion for Hamilton–Jacobi equations.

2

Semiconcave Functions

This chapter and the following two are devoted to the general properties of semiconcave functions. We begin here by studying the direct consequences of the definition and some basic examples, while the next chapters deal with generalized differentials and singularities. At this stage we study semiconcave functions without referring to specific applications; later in the book we show how the results obtained here can be applied to Hamilton–Jacobi equations and optimization problems.

The chapter is structured as follows. In Section 2.1 we define semiconcave functions in full generality, and study some direct consequences of the definition, like the Lipschitz continuity and the relationship with the differentiability. Then we consider some examples in Section 2.2, like the distance function from a set, or the solutions to certain partial differential equations. We give an account of the vanishing viscosity method for Hamilton–Jacobi equations, where semiconcavity estimates play an important role. In Section 2.3 we recall some properties which are peculiar to semiconcave functions with a linear modulus, like Alexandroff's theorem or Jensen's lemma. In Section 2.4 we investigate the relation between viscous Hamilton–Jacobi equations and the heat equation induced by the Cole–Hopf transformation, showing that semiconcavity corresponds to the Li–Yau differential Harnack inequality for the heat equation. Finally, in Section 2.5 we analyze the relation between semiconcavity and a generalized one-sided estimate, a property which will be applied later in the book to prove semiconcavity of viscosity solutions.

2.1 Definition and basic properties

Throughout the section S will be a subset of $\mathbb{R}^n$.

Definition 2.1.1 *We say that a function* $u : S \to \mathbb{R}$ *is* semiconcave *if there exists a nondecreasing upper semicontinuous function* $\omega : \mathbb{R}_+ \to \mathbb{R}_+$ *such that* $\lim_{\rho \to 0^+} \omega(\rho) = 0$ *and*

$$\lambda u(x) + (1 - \lambda)u(y) - u(\lambda x + (1 - \lambda)y) \leq \lambda(1 - \lambda)|x - y|\omega(|x - y|) \quad (2.1)$$

for any pair $x, y \in S$, such that the segment $[x, y]$ is contained in S and for any $\lambda \in [0, 1]$. We call ω a modulus of semiconcavity for u in S. A function v is called semiconvex in S if $-v$ is semiconcave.

In the case of ω linear, we recover the class of semiconcave functions introduced in the previous chapter (see Definition 1.1.1 and Proposition 1.1.3). We recall that, if $\omega(\rho) = \frac{C}{2}\rho$, for some $C \geq 0$, then C is called a *semiconcavity constant* for u in S.

We denote by SC (S) the space of all semiconcave functions in S and by SCL (S) the functions which are semiconcave in S with a linear modulus. A usual, we use the notation SC $_{loc}(S)$ or SCL $_{loc}(S)$ for the functions which are semiconcave (with a linear modulus) locally in S, i.e., on every compact subset of S.

As we have remarked in Chapter 1, semiconcave functions with a linear modulus are the most common in the literature. Although they are a smaller class, they are sufficient for many applications; in addition, they enjoy stronger properties than general semiconcave functions and are easier to analyze, since they are more closely related to concave functions. Nevertheless, it is interesting to consider semiconcave functions with a general modulus, since they are a larger class, sharing many of the properties of the case of a linear modulus.

An interesting consequence of the general definition of semiconcavity given above is that any C^1 function is semiconcave, without any assumption on its second derivatives, as the next result shows.

Proposition 2.1.2 Let $u \in C^1(A)$, with A open. Then both u and $-u$ are locally semiconcave in A with modulus equal to the modulus of continuity of Du.

Proof — Let ω_K denote the modulus of continuity of Du in a compact set $K \subset A$ and let $x, y \in K$ be such that the segment $[x, y] \subset K$. We have

$$
\begin{aligned}
&\lambda u(x) + (1 - \lambda)u(y) - u(\lambda x + (1 - \lambda)y) \\
&= \lambda[u(x) - u(x + (1 - \lambda)(y - x)] + (1 - \lambda)[u(y) - u(y + \lambda(x - y))] \\
&= -\lambda\langle Du(\xi_1), (1 - \lambda)(y - x)\rangle - (1 - \lambda)\langle Du(\xi_2), \lambda(x - y)\rangle,
\end{aligned}
$$

for suitable points ξ_1 and ξ_2 belonging to $[x, y]$. Since $|\xi_1 - \xi_2| \leq |x - y|$ we obtain

$$
\begin{aligned}
&|\lambda u(x) + (1 - \lambda)u(y) - u(\lambda x + (1 - \lambda)y)| \\
&= \lambda(1 - \lambda)|\langle Du(\xi_1) - Du(\xi_2), x - y\rangle| \\
&\leq \lambda(1 - \lambda)|x - y|\omega_K(|x - y|). \qquad \blacksquare
\end{aligned}
$$

In the following (see Theorem 3.3.7) we will see that the converse result is also true, that is, a function which is both semiconcave and semiconvex is continuously differentiable. Another immediate consequence of Proposition 2.1.2 is the following.

Corollary 2.1.3 If $u : A \to \mathbb{R}$, with A open convex, is such that $u = u_1 + u_2$, where $u_1 \in C^1(A)$ and u_2 is concave on A, then $u \in$ SC $_{loc}(A)$.

Proof — The function u is semiconcave on any compact subset of A since it is the sum of semiconcave functions. ∎

Such a result can be regarded as a generalization of the implication $(d) \Rightarrow (a)$ in Proposition 1.1.3. We have seen there that semiconcave functions with a linear modulus also satisfy the converse implication, that is, they can be written as the sum of a concave function and a smooth one. This implies that many regularity properties of concave functions (e.g., local Lipschitz continuity) extend immediately to semi-concave functions with linear modulus. The next result shows that this is no longer true for semiconcave functions with arbitrary modulus.

Proposition 2.1.4 *Given $\alpha \in]0, 1[$, there exists $u : I \to \mathbb{R}$, with $I \subset \mathbb{R}$, which is semiconcave with modulus $\omega(r) = Cr^\alpha$ for some $C > 0$ and that cannot be written in the form $u = u_1 + u_2$ with u_1 concave and $u_2 \in C^1(I)$.*

Proof — Let us choose $\beta > 1$ such that $\alpha\beta < 1$. Let us set

$$s_0 = 0, \qquad s_n = \sum_{k=1}^{n} k^{-\beta} \quad (n \geq 1), \qquad L = \lim_{n \to \infty} s_n.$$

We then define a function $v : [0, L] \to \mathbb{R}$ by setting

$$v(x) = (x - s_n)^\alpha, \qquad \text{if } x \in [s_n, s_{n+1}[.$$

Let us also set $v(x) = 0$ for $x \notin [0, L]$. The function v is discontinuous at s_n for all $n \geq 1$ and continuous elsewhere. Let us observe that

$$\sum_{n=1}^{\infty} [v(s_n-) - v(s_n+)] = \sum_{n=1}^{\infty} n^{-\alpha\beta} = +\infty. \tag{2.2}$$

It is also easy to see that v satisfies the one-sided Hölder estimate

$$v(y) - v(x) \leq (y - x)^\alpha, \qquad \forall x, y \quad x \leq y. \tag{2.3}$$

Let us now set

$$u(x) = \int_0^x v(y)\, dy, \qquad x \in]0, L+1[.$$

Using (2.3) we obtain, for any x, y with $0 < x \leq y < L+1$,

$$\lambda u(x) + (1 - \lambda)u(y) - u(\lambda x + (1 - \lambda)y)$$

$$= \lambda(1 - \lambda)(y - x) \int_0^1 [v(y - \theta\lambda(y - x)) - v(x + \theta(1 - \lambda)(y - x))]\, d\theta$$

$$\leq \lambda(1 - \lambda)(y - x) \int_0^1 (1 - \theta)(y - x)^\alpha d\theta = \lambda(1 - \lambda)(y - x)^{1+\alpha}/2.$$

Thus, u is semiconcave with modulus $\omega(r) = r^\alpha/2$. Let us now assume that there exist two functions $u_1, u_2 :]0, L+1[\to \mathbb{R}$, u_1 concave and $u_2 \in C^1$ such that $u =$

$u_1 + u_2$. Then setting $v_i = u_i'$ for $i = 1, 2$, we have that $v_2 \in C(\,]0, L + 1[\,)$, while v_1 is defined for every $x \neq s_n$, $x \neq L$ and is decreasing. In addition, $v_1 + v_2 = v$. Since v_2 is continuous, we have for every n

$$v(s_n-) - v(s_n+) = v_1(s_n-) - v_1(s_n+).$$

Since v_1 is decreasing, we deduce from (2.2) that $v_1(x) \to -\infty$ as $x \to L^-$. This is a contradiction since v_1 is defined and decreasing in $]0, L + 1[$. Therefore there exist no functions u_1, u_2 as above. This proves the proposition. ∎

Let us derive another important consequence of the definition of semiconcavity.

Proposition 2.1.5 *Let $\{u_\alpha\}_{\{\alpha \in \mathcal{A}\}}$ be a family of functions defined in S and semiconcave with the same modulus ω. Then the function $u := \inf_{\alpha \in \mathcal{A}} u_\alpha$ is also semiconcave in S with modulus ω provided $u > -\infty$.*

Proof — Let us take x, y such that $[x, y] \subset S$ and $\lambda \in [0, 1]$. Given any $\varepsilon > 0$, we can find α such that

$$u_\alpha(\lambda x + (1 - \lambda)y) \leq u(\lambda x + (1 - \lambda)y) + \varepsilon.$$

Then we have

$$\lambda u(x) + (1 - \lambda)u(y) - u(\lambda x + (1 - \lambda)y)$$
$$\leq \lambda u_\alpha(x) + (1 - \lambda)u_\alpha(y) - u_\alpha(\lambda x + (1 - \lambda)y) + \varepsilon$$
$$\leq \lambda(1 - \lambda)|x - y|\omega(|x - y|) + \varepsilon.$$

Since $\varepsilon > 0$ is arbitrary, we obtain the assertion. ∎

Combining Propositions 2.1.2 and 2.1.5 we obtain the following result.

Corollary 2.1.6 *The infimum of a family of C^1 functions which are uniformly bounded from below and have equicontinuous gradients is a semiconcave function. If the functions of the family are of class C^2 and have second derivatives uniformly bounded, then the infimum is semiconcave with linear modulus.*

In the case of a linear modulus the above result was already obtained in Proposition 1.1.3, where we also proved the converse implication. In the following (see Theorem 3.4.2), we will see that the converse implication is true for an arbitrary modulus; that is, any semiconcave function can be represented as the infimum of a family of C^1 functions. Functions defined as infimum of smooth functions occur in the study of optimization problems (see the "lower–C^1" and "lower–C^2" functions in [123] or the "marginal functions" in [20]). We will analyze in more detail the relation between semiconcave functions and infima of smooth functions in Section 3.4.

Like concavity, the semiconcavity property implies local Lipschitz continuity, as shown in the next theorem.

Theorem 2.1.7 *A semiconcave function $u : S \to \mathbb{R}$ is locally Lipschitz continuous in the interior of S.*

Proof — We divide the proof in different steps. It is not restrictive to assume that S is open.

STEP 1: u is locally bounded from below.

Given $x_0 \in S$, we take a closed cube Q centered at x_0 and contained in S. We denote by L the diameter of Q and by $x_1, x_2, \ldots, x_{2^n}$ the vertices of Q. Let us set

$$m_0 = \min\{u(x_i) \ : \ i = 1, 2, \ldots, 2^n\}.$$

Let x_i and x_j be two consecutive vertices of Q. Using the semiconcavity inequality (2.1) we obtain, for any $\lambda \in [0, 1]$,

$$
\begin{aligned}
&u(\lambda x_i + (1 - \lambda)x_j) \\
&\geq \lambda u(x_i) + (1 - \lambda)u(x_j) - \lambda(1 - \lambda)|x_i - x_j|\omega(|x_i - x_j|) \\
&\geq m_0 - \frac{L\omega(L)}{4}.
\end{aligned}
$$

This shows that u is bounded from below on the 1-dimensional faces of Q. We can repeat the procedure by taking any convex combination of two points lying on different 1-dimensional faces, and we obtain that $u \geq m_0 - L\omega(L)/2$ on the 2-dimensional faces. Iterating the procedure n times, we obtain that u is bounded from below on the whole cube Q.

STEP 2: u is locally bounded from above.

Given $x_0 \in S$, we consider now $R > 0$ such that $\overline{B(x_0, R)} \subset S$. By the first step, u is greater than some constant m on $\overline{B(x_0, R)}$. Using (2.1) we find, for any $x \in \overline{B(x_0, R)}$,

$$
\begin{aligned}
&\frac{|x - x_0|}{R + |x - x_0|} u\left(x_0 - R\frac{x - x_0}{|x - x_0|}\right) + \frac{R}{R + |x - x_0|} u(x) - u(x_0) \\
&\leq \frac{R|x - x_0|}{R + |x - x_0|} \omega(R + |x - x_0|).
\end{aligned}
$$

Therefore

$$
\begin{aligned}
u(x) &\leq -\frac{|x - x_0|}{R} m + \frac{R + |x - x_0|}{R} u(x_0) + |R|\omega(R + |x - x_0|) \\
&\leq |m| + 2|u(x_0)| + R\omega(2R),
\end{aligned}
$$

which shows that u is bounded from above in $\overline{B(x_0, R)}$.

STEP 3: u is locally Lipschitz continuous.

We first observe that the semiconcavity inequality (2.1) implies the following: if z belongs to the segment $[x, y] \subset S$, then

$$\frac{u(x) - u(z)}{|x - z|} - \frac{u(z) - u(y)}{|z - y|} \le \omega(|x - y|).$$

Given $x_0 \in S$, let us fix $R > 0$ such that $\overline{B(x_0, R)} \subset S$. By the first two steps, there exist constants m, M such that $m \le u \le M$ in $\overline{B(x_0, R)}$. Given any $x, y \in B(x_0, R/2)$, consider the straight line through x, y and call x', y' the points of this line at distance R from x_0, ordered in such a way that $x \in [x', y]$ and $y \in [x, y']$. Then we have

$$\frac{u(x') - u(x)}{|x' - x|} - \omega(|x' - y|) \le \frac{u(x) - u(y)}{|x - y|} \le \frac{u(y) - u(y')}{|y - y'|} + \omega(|y' - x|),$$

which implies

$$\frac{|u(x) - u(y)|}{|x - y|} \le \frac{2(M - m)}{R} + \omega(2R). \qquad \blacksquare$$

Remark 2.1.8 It follows from the last step of the above proof that the Lipschitz constant of u in a given set $B \subset\subset B' \subset S$ depends only on B, on $\sup_{B'} |u|$ and on the modulus of semiconcavity of u.

As we have remarked before, the previous theorem is immediate for semiconcave function with linear modulus. In fact, it is well known that concave functions are locally Lipschitz continuous and, using Proposition 1.1.3(c), this property can be directly extended to semiconcave function with linear modulus. However, Proposition 2.1.4 shows that such an argument cannot be applied to general semiconcave functions. Therefore we have given an independent proof of the Lipschitz continuity, which is actually quite similar to the one for the concave case. Similar remarks apply to many of the properties considered in the following.

The next result shows that the linear growth near 0 is a threshold for a modulus of semiconcavity, because the only semiconcave functions whose modulus tends to zero faster than linear are the concave functions.

Proposition 2.1.9 *Let* $u : A \to \mathbb{R}$ *be a semiconcave function, with A open, and with a modulus ω such that*

$$\lim_{\rho \to 0^+} \frac{\omega(\rho)}{\rho} = 0.$$

Then u is concave on all convex subsets of A.

Proof — We first observe that it suffices to prove the result in the case when u is smooth. In fact, the general case can be obtained by regularizing u since the convolution of a semiconcave function with a mollifier is semiconcave with the same modulus.

We may therefore assume that $u \in C^2(A)$. Then, for any $x \in A$ and $v \in S^{n-1}$,

$$\partial^2_{vv} u(x) = \lim_{h \to 0^+} \frac{u(x + hv) + u(x - hv) - 2u(x)}{h^2}$$

$$\le \lim_{h \to 0^+} \frac{h\omega(2h)}{h^2} = 0.$$

This shows that u is concave in A. ∎

Checking inequality (2.1) for any $\lambda \in [0, 1]$ may sometimes increase the technical difficulty of proving that a given function is semiconcave. The next result shows that, for a large class of moduli, it suffices to prove the inequality for the midpoint of any segment.

Theorem 2.1.10 *Let* $u : S \rightarrow \mathbb{R}$ *be a continuous function. Suppose that there exists a nondecreasing upper semicontinuous function* $\tilde{\omega} : \mathbb{R}_+ \rightarrow \mathbb{R}_+$ *such that* $\lim_{\rho \rightarrow 0+} \tilde{\omega}(\rho) = 0$ *and*

$$u(x) + u(y) - 2u\left(\frac{x+y}{2}\right) \le \frac{|x-y|}{2}\tilde{\omega}(|x-y|) \tag{2.4}$$

for any x, y *such that the segment* $[x, y]$ *is contained in* S. *Then* u *is semiconcave in* S *with modulus*

$$\omega(\rho) = \sum_{h=0}^{\infty} \tilde{\omega}\left(\frac{\rho}{2^h}\right), \tag{2.5}$$

provided the right-hand side is finite. If $\tilde{\omega}$ *is linear, then we can take* $\omega = \tilde{\omega}$.

Proof — Let x_0, x_1 be such that $[x_0, x_1] \subset S$. We set, for $\lambda \in]0, 1[$,

$$x_\lambda = \lambda x_1 + (1 - \lambda)x_0.$$

We also define, for $k \ge 1$,

$$D_k := \{x_\lambda \ : \ \lambda = \frac{j}{2^k} \text{ for some } j = 0, 1, \ldots, 2^k\}.$$

We claim that

$$\lambda u(x_1) + (1 - \lambda)u(x_0) - u(x_\lambda) \le \lambda(1 - \lambda)L\sum_{h=0}^{k-1}\tilde{\omega}\left(\frac{L}{2^h}\right), \ \forall x_\lambda \in D_k, \tag{2.6}$$

where we have set $L := |x_1 - x_0|$. We proceed by induction on k. In the case $k = 1$ inequality (2.6) is easily checked; in fact it holds by hypothesis if $\lambda = 1/2$, while it is trivial if $\lambda = 0, 1$. We now assume that (2.6) is satisfied for a certain $k \ge 1$ and take $x_\lambda \in D_{k+1} \setminus D_k$. If we set $\mu = \lambda - 2^{-k-1}$, $\nu = \lambda + 2^{-k-1}$, then $x_\mu, x_\nu \in D_k$. Therefore, we deduce from (2.6)

$$(\mu + \nu)u(x_1) + (2 - \mu - \nu)u(x_0) - u(x_\mu) - u(x_\nu)$$
$$\le (\mu(1 - \mu) + \nu(1 - \nu))L\sum_{h=0}^{k-1}\tilde{\omega}\left(\frac{L}{2^h}\right),$$

which implies, by the definition of μ, ν, that

$$2\lambda u(x_1) + 2(1-\lambda)u(x_0) - u(x_\mu) - u(x_\nu)$$

$$\leq 2\left(\lambda(1-\lambda) - \frac{1}{2^{2k+2}}\right) L \sum_{h=0}^{k-1} \tilde{\omega}\left(\frac{L}{2^h}\right). \tag{2.7}$$

On the other hand, we have by (2.4)

$$u(x_\mu) + u(x_\nu) - 2u(x_\lambda) \leq \frac{L}{2^{k+1}}\tilde{\omega}\left(\frac{L}{2^k}\right). \tag{2.8}$$

It follows that

$$\lambda u(x_1) + (1-\lambda)u(x_0) - u(x_\lambda) \leq \frac{L}{2^{k+2}}\tilde{\omega}\left(\frac{L}{2^k}\right) + \lambda(1-\lambda)L \sum_{h=0}^{k-1} \tilde{\omega}\left(\frac{L}{2^h}\right). \tag{2.9}$$

Our hypotheses on λ imply that

$$\lambda(1-\lambda) \geq \frac{1}{2^{k+1}}\left(1 - \frac{1}{2^{k+1}}\right) > \frac{1}{2^{k+2}}.$$

Thus, we conclude from (2.9)

$$\lambda u(x_1) + (1-\lambda)u(x_0) - u(x_\lambda) \leq \lambda(1-\lambda)L \sum_{h=0}^{k} \tilde{\omega}\left(\frac{L}{2^h}\right)$$

which is the induction step required to prove (2.6). Since the union of the sets D_j is dense in $[x_0, x_1]$ and u is supposed continuous, this proves that u is semiconcave with the desired modulus.

If $\tilde{\omega}$ is linear we can use the same procedure to prove, instead of (2.6), the sharper estimate

$$\lambda u(x_1) + (1-\lambda)u(x_0) - u(x_\lambda) \leq \lambda(1-\lambda)L\tilde{\omega}(L), \qquad \forall x_\lambda \in D_k. \tag{2.10}$$

As before, the case $k = 1$ it holds by hypothesis. When we proceed from k to $k+1$, inequality (2.7) is replaced by

$$2\lambda u(x_1) + 2(1-\lambda)u(x_0) - u(x_\mu) - u(x_\nu)$$

$$\leq 2\left(\lambda(1-\lambda) - \frac{1}{2^{2k+2}}\right) L\tilde{\omega}(L),$$

while estimate (2.8) becomes, thanks to the linearity of $\tilde{\omega}$,

$$u(x_\mu) + u(x_\nu) - 2u(x_\lambda) \leq \frac{L}{2^{k+1}}\tilde{\omega}\left(\frac{L}{2^k}\right) = \frac{L}{2^{2k+1}}\omega(L).$$

Summing the two inequalities we obtain (2.10). We conclude by a density argument as in the previous case. ∎

Remark 2.1.11 The previous result does not yield a full equivalence between inequalities (2.1) and (2.4), except in the case of a linear modulus; in fact, the modulus ω defined in (2.5) is greater than $\tilde{\omega}$ and it is even infinite for some choices of $\tilde{\omega}$. We do not know whether such a result is sharp or it could be improved by a better technique of proof. However, it is satisfactory in most applications, where we have a power-like modulus of the form $\tilde{\omega}(\rho) = C\rho^\alpha$ for some $0 < \alpha \leq 1$. In this case, in fact, the right-hand side of (2.5) is finite and has again the form $\omega(\rho) = C'\rho^\alpha$ for some $C' > C$.

We conclude this section analyzing under which conditions the composition of a semiconcave function with another function is also semiconcave.

Proposition 2.1.12 *Let* $u : A \to \mathbb{R}$ *be a locally semiconcave function, with* A *open.*

(i) If $f : \mathbb{R} \to \mathbb{R}$ *is increasing and locally semiconcave, then* $f \circ u$ *is locally semiconcave on* A. *If the moduli of semiconcavity of* u *and* f *are both linear, then so is the modulus of* $f \circ u$.

(ii) If $\phi : V \to A$ *is a* C^1 *function, with* $V \subset \mathbb{R}^m$ *open, then* $u \circ \phi$ *is locally semiconcave in* V. *If* $D\phi$ *is locally Lipschitz in* V *and the semiconcavity modulus of* u *is linear, then the modulus of* $u \circ \phi$ *is also linear.*

Proof — (i) Let S be any compact subset of A, and let us denote by ω_1, ω_2 the semiconcavity moduli and by L_1 and L_2 the Lipschitz constants of u in S and of f in $u(S)$, respectively. We have, for any x, y such that $[x, y] \subset S$ and for any $\lambda \in [0, 1]$,

$$f(u(\lambda x + (1 - \lambda)y))$$
$$\geq f(\lambda u(x) + (1 - \lambda)u(y) - \lambda(1 - \lambda)|x - y|\omega_1(|x - y|))$$
$$\geq f(\lambda u(x) + (1 - \lambda)u(y)) - L_2\lambda(1 - \lambda)|x - y|\omega_1(|x - y|)$$

and therefore

$$\lambda f(u(x)) + (1 - \lambda)f(u(y)) - f(u(\lambda x + (1 - \lambda)y))$$
$$\leq \lambda f(u(x)) + (1 - \lambda)f(u(y)) - f(\lambda u(x) + (1 - \lambda)u(y))$$
$$+ L_2\lambda(1 - \lambda)|x - y|\omega_1(|x - y|)$$
$$\leq \lambda(1 - \lambda)|x - y|(L_1\omega_2(L_1|x - y|) + L_2\omega_1(|x - y|)).$$

Thus, $f \circ u$ is semiconcave with modulus $\omega(\rho) := L_1\omega_2(L_1\rho) + L_2\omega_1(\rho)$.
(ii) Given $x_0 \in V$, let S be a convex neighborhood of x_0 such that $\bar{S} \subset V$ and $\mathrm{co}\,\phi(\bar{S}) \subset A$. Let us denote by ω_1 the modulus of continuity of $D\phi$ in S and by ω_2 the semiconcavity modulus of u in $\mathrm{co}\,\phi(S)$. Moreover, let L_1 and L_2 be the Lipschitz constants of ϕ in S and of u in $\mathrm{co}\,\phi(S)$. We have, for any x, y such that $[x, y] \subset S$ and for any $\lambda \in [0, 1]$,

$$u(\lambda\phi(x) + (1 - \lambda)\phi(y)) - u(\phi(\lambda x + (1 - \lambda)y))$$
$$\leq L_2|\lambda\phi(x) + (1 - \lambda)\phi(y) - \phi(\lambda x + (1 - \lambda)y))|$$
$$\leq L_2\lambda(1 - \lambda)|x - y|\omega_1(|x - y|)$$

and therefore

$$\lambda u(\phi(x)) + (1 - \lambda)u(\phi(y)) - u(\phi(\lambda x + (1 - \lambda)y))$$
$$\leq \lambda u(\phi(x)) + (1 - \lambda)u(\phi(y)) - u(\lambda\phi(x) + (1 - \lambda)\phi(y))$$
$$+u(\lambda\phi(x) + (1 - \lambda)\phi(y)) - u(\phi(\lambda x + (1 - \lambda)y))$$
$$\leq \lambda(1 - \lambda)|x - y|(L_1\omega_2(L_1|x - y|) + L_2\omega_1(|x - y|)). \qquad \blacksquare$$

2.2 Examples

A first interesting example of a semiconcave function is provided by the distance function. We recall that the distance function from a given nonempty closed set $C \subset \mathbb{R}^n$ is defined by

$$d_C(x) = \min_{y \in C} |y - x|, \qquad (x \in \mathbb{R}^n).$$

As we show below, d_C is not semiconcave in the whole space $\mathbb{R}^n$, but is semiconcave on the complement of C, at least locally. On the other hand, the square of the distance function is semiconcave in $\mathbb{R}^n$. Before proving this result, let us introduce a property of sets which is useful for the analysis of the semiconcavity of d_C.

Definition 2.2.1 *We say that a set $C \subset \mathbb{R}^n$ satisfies an* interior sphere condition *for some $r > 0$ if C is the union of closed spheres of radius r, i.e., for any $x \in C$ there exists y such that $x \in \overline{B_r(y)} \subset C$.*

Proposition 2.2.2 *Let $C \subset \mathbb{R}^n$ be a closed set, $C \neq \emptyset, \mathbb{R}^n$. Then the distance function d_C satisfies the following properties:*

(i) $d_C^2 \in SCL(\mathbb{R}^n)$ with semiconcavity constant 2.

(ii) $d_C \in SCL_{loc}(\mathbb{R}^n \setminus C)$. More precisely, given a set S (not necessarily compact) such that $\mathrm{dist}(S, C) > 0$, d_C is semiconcave in S with semiconcavity constant equal to $\mathrm{dist}(S, C)^{-1}$.

(iii) If C satisfies an interior sphere condition for some $r > 0$, then $d_C \in SCL(\overline{\mathbb{R}^n \setminus C})$ with semiconcavity constant equal to r^{-1}.

(iv) d_C is not locally semiconcave in the whole space $\mathbb{R}^n$.

Proof — **(i)** For any $x \in \mathbb{R}^n$ we have

$$d_C^2(x) - |x|^2 = \inf_{y \in C} |x - y|^2 - |x|^2 = \inf_{y \in C} |y|^2 - 2\langle x, y \rangle.$$

Since the infimum of linear functions is concave we deduce, by Proposition 1.1.3, that property (i) holds.

(ii) Let us first observe that, given $z, h \in \mathbb{R}^n$, $z \neq 0$, we have

$$(|z + h| + |z - h|)^2$$

$$\leq 2(|z + h|^2 + |z - h|^2) = 4(|z|^2 + |h|^2) \leq \left(2|z| + \frac{|h|^2}{|z|}\right)^2.$$

Thus

$$|z + h| + |z - h| - 2|z| \leq \frac{|h|^2}{|z|}. \tag{2.11}$$

Let S be a set with positive distance from C. For any x, h such that $[x-h, x+h] \subset S$, let $\bar{x} \in C$ be such that $d_C(x) = |x - \bar{x}|$. Then

$$d_C(x + h) + d_C(x - h) - 2d_C(x)$$
$$\leq |x + h - \bar{x}| + |x - h - \bar{x}| - 2|x - \bar{x}|$$
$$\leq \frac{|h|^2}{|x - \bar{x}|}.$$

Moreover $|x - \bar{x}| = d_C(x) \geq \text{dist}(S, C)$. By Theorem 2.1.10 we conclude that d_C satisfies the desired property.

(iii) By hypothesis C is the union of closed spheres with radius r. Thus, the distance from C is the infimum of the distance from such spheres. By Proposition 2.1.5 it suffices to prove the result in the case $C = \overline{B}_r$. We have, for any $x \in \mathbb{R}^n \setminus C$,

$$d_C(x) = |x| - r = d_{\{0\}}(x) - r.$$

Since the distance between $\{0\}$ and the complement of C is r, we obtain from part (ii) that d_C is semiconcave with the desired modulus.

(iv) Let y be any point in the complement of C and let x be a projection of y onto C. Then, if we set

$$v = \frac{y - x}{|y - x|}$$

we have that $d_C(x + hv) = h$ for all $h \in [0, d_C(y)]$. Therefore, since d_C is nonnegative, we obtain

$$d_C(x + hv) + d_C(x - hv) - 2d_C(x) \geq h,$$

showing that d_C is not semiconcave in any neighborhood of x. ∎

Remark 2.2.3 Any compact set C with a smooth C^2 boundary satisfies an interior sphere condition. However, even sets with corners may satisfy such a condition, provided the corners are pointing "inwards." For instance, a set of the form $C = B_r(x_1) \cup B_r(x_2)$, with $x_1, x_2 \in \mathbb{R}^n$ and $r > |x_2 - x_1|/2$ is nonsmooth and satisfies the interior sphere condition with radius r. The condition does not hold instead if C is a convex polyhedron (the corners of such a polyhedron C are not contained in any sphere contained in C) or a set of dimension lower than n, e.g., a point. If a set C does not satisfy an interior sphere condition, the distance function d_C is locally semiconcave in the complement of C, as stated in part (ii) of the above proposition, but the semiconcavity constant can in general blow up as we approach the boundary of C. A typical example is the case $C = \{0\}$; then $d_C(x) = |x|$ is not semiconcave in any neighborhood of $\{0\}$ (see Example 3.3.9).

Another function which is not differentiable in general, but which is semiconcave, is the smallest eigenvalue of a symmetric matrix.

Proposition 2.2.4 *Let $M(x)$ be a $k \times k$ symmetric matrix whose entries are functions in $C^1(A)$, with $A \subset \mathbb{R}^n$ open. Then the smallest eigenvalue of $M(x)$ is a locally semiconcave function in A.*

Proof — Let $\lambda_1(x)$ denote the smallest eigenvalue of $M(x)$. From linear algebra it is known that

$$\lambda_1(x) = \min_{\xi \in S^{k-1}} \langle M(x)\xi, \xi \rangle,$$

where S^{k-1} is the unit sphere in $\mathbb{R}^k$. Since, for any $\xi \in S^{k-1}$, the function $x \to \langle M(x)\xi, \xi \rangle$ is of class C^1, our assertion follows from Propositions 2.1.5 and 2.1.2. ∎

In a similar way, one can prove that the sum of the smallest h eigenvalues $\lambda_1, \ldots, \lambda_h$ of $A(x)$ is semiconcave for all $h = 1, \ldots, k$, recalling that

$$\sum_{i=1}^{h} \lambda_i(x) = \min_{\xi_1, \ldots, \xi_h \in S^{k-1}} \left\{ \sum_{i=1}^{h} \langle M(x)\xi_i, \xi_i \rangle \ : \ \langle \xi_i, \xi_j \rangle = \delta_{ij} \ \forall i, j \right\}.$$

Example 2.2.5 For $x \in \mathbb{R}$, define

$$A(x) = \begin{pmatrix} 0 & x \\ x & 0 \end{pmatrix}.$$

It is easily checked that the eigenvalues of $A(x)$ are $\pm x$. Thus the smallest eigenvalue is $\lambda_1(x) = \min\{-x, x\} = -|x|$, which is a concave function of x.

We now give examples of partial differential equations whose solutions satisfy a semiconcavity estimate.

Proposition 2.2.6 *Let $u \in C^4([0, T] \times \mathbb{R}^n)$ be a solution of the equation*

$$\partial_t u(t, x) + H(\nabla u(t, x)) = a \Delta u(t, x), \tag{2.12}$$

where $a > 0$ and $H \in C^2(\mathbb{R}^n)$ is uniformly convex. Suppose that the first and second derivatives of u are bounded. Then, for any $t \in]0, T]$, the function $u(t, \cdot)$ is semiconcave in $\mathbb{R}^n$ with semiconcavity constant $(\lambda t)^{-1}$, where $\lambda > 0$ is a lower bound on the eigenvalues of $D^2 H$.

Proof — Let us set $w(t, x) = t\partial^2_{x_1 x_1} u(t, x)$. We have

$$\partial_t w = \frac{w}{t} + t \partial^2_{x_1 x_1} (a \Delta u - H)$$

$$= \frac{w}{t} + a \Delta w - \langle DH, \nabla w \rangle - t \langle D^2 H(\nabla \partial_{x_1} u), \nabla \partial_{x_1} u \rangle.$$

Now by assumption

$$t\langle D^2 H(\nabla \partial_{x_1} u), \nabla \partial_{x_1} u\rangle \geq t\lambda |\nabla \partial_{x_1} u|^2 \geq \lambda w^2/t.$$

Therefore

$$\partial_t w \leq \frac{w}{t}(1 - \lambda w) + a\Delta w - \langle DH, \nabla w\rangle.$$

Since $w(t, \cdot) \to 0$ uniformly as $t \to 0$, we deduce by the maximum principle that $\lambda w(t, x) \leq 1$ everywhere. This implies that $\partial^2_{x_1 x_1} u(t, x) \leq (\lambda t)^{-1}$. The same reasoning can be applied to $\partial^2_{\nu\nu} u$ for any unit vector $\nu \in \mathbb{R}^n$. We conclude by Proposition 1.1.3. ∎

Remark 2.2.7 The interesting part of the previous statement is not the semiconcavity of u in itself (which would be a consequence of the smoothness) but the fact that the semiconcavity constant is independent of the diffusion coefficient a. This property is of great interest in the *vanishing viscosity method* for Hamilton–Jacobi equations. In this approach one wants to obtain a solution to equation

$$\partial_t u(t, x) + H(\nabla u(t, x)) = 0 \qquad (2.13)$$

by taking the limit as $a \to 0^+$ of the solution u_a to (2.12). Equation (2.12) is called the *parabolic regularization* of (2.13) and is in some respect easier to study since its leading order term is linear and the classical theory for semilinear parabolic equations can be applied. The uniform semiconcavity estimate of Proposition 2.2.6 yields compactness of the u_a's (as we will see in Theorem 3.3.3). In addition, it shows that $u = \lim u_a$ is also semiconcave, and this allows us to apply the uniqueness result of Theorem 1.6.4. This method is an alternative to the one in the previous chapter based on the representation of u as the value function of the associated variational problem. Both methods yield the same generalized solution, i.e., the unique semiconcave solution.

The name "vanishing viscosity method" has been given in analogy with the theory of conservation laws in fluid dynamics, where the second order terms usually represent the viscosity of the fluid. Let us observe that in the modern theory of viscosity solutions this approach can be carried out under much weaker estimates on the approximate solutions u_a's than the semiconcavity estimate considered here (see Theorem 5.2.3). Let us also mention that another second order equation for which a similar semiconcavity estimate holds is the equation for the pressure in the theory of porous media (see e.g., [16])

$$\partial_t v(t, x) = av(t, x)\Delta v(t, x) + |\nabla v(t, x)|^2.$$

2.3 Special properties of SCL (A)

While many properties of semiconcave functions are valid in the case of an arbitrary modulus of semiconcavity, there are some results which are peculiar to the case of a linear modulus; we collect in this section some important ones, in addition to those already given in Proposition 1.1.3.

We have seen in Proposition 1.1.3 that semiconcave functions with a linear modulus can be regarded as C^2 perturbations of concave functions. This allows to extend immediately some well-known properties of concave functions, such as the following.

Theorem 2.3.1 *Let $u \in$ SCL (A), with $A \subset \mathbb{R}^n$ open. Then the following properties hold.*

(i) (**Alexandroff's Theorem**) *u is twice differentiable a.e.; that is, for a.e. every $x_0 \in A$, there exist a vector $p \in \mathbb{R}^n$ and a symmetric matrix B such that*

$$\lim_{x \to x_0} \frac{u(x) - u(x_0) - \langle p, x - x_0 \rangle + \langle B(x - x_0), x - x_0 \rangle}{|x - x_0|^2} = 0.$$

(ii) The gradient of u, defined almost everywhere in A, belongs to the class $BV_{loc}(A, \mathbb{R}^n)$.

Proof — Properties (i) and (ii) hold for a convex function (see e.g., [72, Ch. 6.3]). Since u is the difference of a smooth function and a convex one, u also satisfies these properties. ∎

The following result shows that semiconcave functions with linear modulus exhibit a behavior similar to C^2 functions near a minimum point.

Theorem 2.3.2 *Let $u \in$ SCL (A), with $A \subset \mathbb{R}^n$ open, and let $x_0 \in A$ be a point of local minimum for u. Then there exist a sequence $\{x_h\} \subset A$ and an infinitesimal sequence $\{\varepsilon_h\} \subset \mathbb{R}_+$ such that u is twice differentiable in x_h and that*

$$\lim_{h \to \infty} x_h = x_0, \quad \lim_{h \to \infty} Du(x_h) = 0, \quad D^2 u(x_h) \geq -\varepsilon_h I \quad \forall h.$$

The proof of this theorem is based on the following result.

Theorem 2.3.3 (Jensen's Lemma) *Let $u \in$ SCL (A), with $A \subset \mathbb{R}^n$ open, and let $x_0 \in A$ be a strict local minimum point for u. Given $p \in \mathbb{R}^n$, let A_p denote the set of points where the function $x \to u(x) + \langle p, x \rangle$ has a local minimum. Then, for any $r > 0$ and $\delta > 0$, the set*

$$E = \{x \in B_r(x_0) : x \in A_p \ \text{for some} \ p \in B_\delta\}$$

has positive measure.

Proof — Let C be a semiconcavity constant for u. Let us fix $r_0 \leq r$ such that $B(x_0, r_0) \subset A$ and that $\min_{\partial B(x_0, r_0)} u > u(x_0)$. Then we take $\delta_0 \leq \delta$ such that

$$r_0 \delta_0 < \min_{\partial B(x_0, r_0)} u - u(x_0). \tag{2.14}$$

Let us first consider the case when u is of class C^∞. Given $p \in B_{\delta_0}$, we obtain from (2.14) that any point $x \in A_p \cap \overline{B(x_0, r_0)}$ belongs in fact to the open ball $B(x_0, r_0)$. Recalling also Proposition 1.1.3 we find, at such an x,

$$Du(x) = -p, \qquad 0 \le D^2 u(x) \le CI.$$

Let us define

$$E_0 = \{x \in B_{r_0}(x_0) \ : \ x \in A_p \text{ for some } p \in B_{\delta_0}\}.$$

Then we obtain that Du is a surjective mapping from E_0 to B_{δ_0}, and in addition $\|\det D^2 u\| \le C^n$ on E_0. We conclude that

$$\text{meas}(B_{\delta_0}) = \text{meas}(Du(E_0)) \le \int_{E_0} \|\det D^2 u(x)\| \, dx \le \text{meas}(E_0) C^n,$$

and so we obtain that E_0 has measure greater than some constant M_0 depending only on C and δ_0.

In the case where u is not smooth, we proceed by approximation. Using convolutions we obtain a family u_h of C^∞ functions, semiconcave with the same modulus of u, and converging uniformly to u. Then, for h large enough, the functions u_h also satisfy inequality (2.14). If we call E_h the set defined as E_0 with u replaced by u_h, we obtain that the measure of E_h is also greater than M_0. Next we observe that if some x belongs to E_h for infinitely many h, then it also belongs to E_0. This implies that E_0 has measure at least M_0. Since E contains E_0, this proves the result.   ∎

Proof of Theorem 2.3.2 — Let $B \subset\subset \Omega$ be a bounded open set containing x_0 such that $u(x) \le u(x_0)$ for $x \in \bar{B}$, and let $w(x) = u(x) + |x - x_0|^4$. Then $w : B \to \mathbb{R}$ satisfies the assumptions of Jensen's Lemma; we deduce that there exist a sequence $\{p_h\} \subset \mathbb{R}^n$ converging to 0 and a sequence $\{x_h\} \subset B$ such that $w(x) + \langle p_h, x \rangle$ has a minimum at x_h. In view of Alexandroff's Theorem it is not restrictive to assume that w is twice differentiable at x_h. Then we have $Dw(x_h) = p_h$ and $D^2 w(x_h) \ge 0$ for any h. Any limit point of $\{x_h\}$ is a minimum for w in $\bar{B}$; since x_0 is the only minimum point for w in $\bar{B}$, this implies that $x_h \to x_0$. Since we have $Du(x_h) = Dw(x_h) - 4|x_h - x_0|^2 (x_h - x_0)$ and $D^2 u(x_h) = D^2 w(x_h) - 4|x_h - x_0|^2 I - 8(x_h - x_0) \otimes (x_h - x_0)$ the assertion follows with $\varepsilon_h = 12|x_h - x_0|^2$.   ∎

Theorem 2.3.2 plays an important role in the theory of viscosity solutions to second order degenerate elliptic equations (see [66]).

2.4 A differential Harnack inequality

Let us consider the parabolic Hamilton–Jacobi equation

$$\partial_t u(t, x) + |\nabla u(t, x)|^2 = \Delta u(t, x), \qquad t \ge 0, x \in \mathbb{R}^n. \tag{2.15}$$

We have seen in Proposition 2.2.6 that the solutions to this equation are semiconcave. We now show how such a semiconcavity result is related to the classical Harnack inequality for the heat equation.

A remarkable feature of equation (2.15) is that it can be reduced to the heat equation by a change of unknown called the *Cole–Hopf transformation*, or logarithmic transformation. In fact, if we set $w(t, x) = \exp(-u(t, x))$, a direct computation shows that u satisfies (2.15) if and only if $\partial_t w = \Delta w$. Let us investigate the properties of w which follow from the semiconcavity of u.

Proposition 2.4.1 *Let w be a positive solution of the heat equation in $[0, T] \times \mathbb{R}^n$ whose first and second derivatives are bounded. Then w satisfies*

$$\nabla^2 w + \frac{w}{2t} I - \frac{\nabla w \otimes \nabla w}{w} \geq 0; \tag{2.16}$$

$$\partial_t w + \frac{nw}{2t} - \frac{|\nabla w|^2}{w} \geq 0. \tag{2.17}$$

Here $\nabla^2 w$ denotes the hessian matrix of w with respect to the space variables; inequality (2.16) means that the matrix on the left-hand side is positive semidefinite.

Proof — It is not restrictive to assume that w is greater than some positive constant; if this is not the case, we can replace w by $w + \varepsilon$ and then let $\varepsilon \to 0^+$. Let us set $u(t, x) = -\ln(w(t, x))$. Then u is a solution of equation (2.15). In addition, u is bounded together with its first and second derivatives. Therefore, by Proposition 2.2.6, $u(t, \cdot)$ is semiconcave with modulus $\omega(\rho) = \rho/(4t)$. Using the equivalent formulations of Proposition 1.1.3, we can restate this property as

$$\nabla^2 u \leq \frac{1}{2t} I.$$

On the other hand, an easy computation shows that

$$\nabla^2 u = -\frac{\nabla^2 w}{w} + \frac{\nabla w \otimes \nabla w}{w^2}$$

and this proves (2.16). Taking the trace of the left-hand side of (2.16), we obtain

$$\Delta w + \frac{nw}{2t} - \frac{|\nabla w|^2}{w} \geq 0,$$

which implies (2.17), since w solves the heat equation. ∎

Inequality (2.17) is called a *differential Harnack estimate*. The connection with the classical Harnack inequality is explained by the following result.

Corollary 2.4.2 *Let w be as in the previous proposition. Then w satisfies*

$$w(t_2, y) \geq w(t_1, x) \left(\frac{t_1}{t_2}\right)^{n/2} \exp\left(-\frac{|y - x|^2}{4(t_2 - t_1)}\right)$$

for all $x, y \in \mathbb{R}^n$, $t_2 > t_1 > 0$. *Therefore, given any bounded set* $\Omega \subset \mathbb{R}^n$ *and* $t_2 > t_1 > 0$, *we have*

$$\max_{\Omega} w(t_1, \cdot) \leq C \min_{\Omega} w(t_2, \cdot) \qquad (2.18)$$

for some constant $C > 0$ *depending on* t_1, t_2 *and the diameter of* Ω *but not on* w.

Proof — Let us set

$$\gamma(s) = x + \frac{s - t_1}{t_2 - t_1}(y - x), \qquad s \in [t_1, t_2].$$

Then we have, using (2.17),

$$\frac{d}{ds} w(s, \gamma(s)) = \partial_t w + \nabla w \cdot \dot{\gamma}$$

$$\geq \partial_t w - \frac{|\nabla w|^2}{w} - \frac{w|\dot{\gamma}|^2}{4}$$

$$\geq -\frac{nw}{2t} - \frac{w|\dot{\gamma}|^2}{4}.$$

It follows that

$$\ln\left(\frac{w(t_2, y)}{w(t_1, x)}\right) = \int_{t_1}^{t_2} \frac{d}{ds} \ln w(s, \gamma(s))\, ds$$

$$\geq \int_{t_1}^{t_2} \left(-\frac{n}{2t} - \frac{|y - x|^2}{4(t_2 - t_1)^2}\right) ds$$

$$= -\frac{n}{2} \ln\left(\frac{t_2}{t_1}\right) - \frac{|y - x|^2}{4(t_2 - t_1)},$$

which proves our statement.   ∎

We recall that the Harnack estimate (2.18) holds in a much more general context than the one considered here; our aim was to point out the relationship with semiconcavity induced by the Hopf–Cole transformation.

2.5 A generalized semiconcavity estimate

In this section we compare the semiconcavity estimate with another one-sided estimate, a priori weaker, which was introduced in [46]. We prove here that the two estimates are in some sense equivalent, and this has applications for the study of certain Hamilton–Jacobi equations, as we will see in the following (see Theorem 5.3.7).

Let us consider a function $u : A \to \mathbb{R}$, with $A \subset \mathbb{R}^n$ open. Given $x_0 \in A$, we set, for $0 < \delta < \text{dist}(x_0, \partial A)$, $x \in B_1$,

$$u_{x_0, \delta}(x) = \frac{u(x_0 + \delta x) - u(x_0)}{\delta}. \qquad (2.19)$$

Definition 2.5.1 *Let $C \subset A$ be a compact set. We say that u satisfies a* generalized one-sided estimate *in C if there exist $K \geq 0$, $\delta_0 \in \,]0$, dist$(C, \partial A)[$ and a nondecreasing upper semicontinuous function $\widetilde{\omega} : [0, 1] \to \mathbb{R}_+$, such that $\lim_{h \to 0} \widetilde{\omega}(h) = 0$ and*

$$\lambda u_{x_0, \delta}(x) + (1 - \lambda) u_{x_0, \delta}(y) - u_{x_0, \delta}(\lambda x + (1 - \lambda)y)$$
$$\leq \lambda(1 - \lambda)|x - y|\{K\delta + \widetilde{\omega}(|x - y|)\} \tag{2.20}$$

for all $x_0 \in C$, $\delta \in \,]0, \delta_0[$, $x, y \in B_1$, $\lambda \in [0, 1]$.

It is easily seen that, if u is semiconcave in A, then the above property is satisfied taking $\widetilde{\omega}$ equal to a modulus of semiconcavity of u in A and $K = 0$. Conversely, semiconcavity can be deduced from the one-sided estimate above, as the next result shows.

Theorem 2.5.2 *Let $u : A \to \mathbb{R}$, with A open and let C be a compact subset of A. If u satisfies a generalized one-sided estimate in C, then u is semiconcave in C.*

Proof — By hypothesis inequality (2.20) holds for some $K, \delta_0, \widetilde{\omega}$. Let us take $x, y \in C$ such that $[x, y] \subset C$ and $\lambda \in [0, 1]$. It is not restrictive to assume $|x - y| < \delta_0/2$. For any δ with $|x - y| < \delta < \delta_0$, we set

$$x_0 = \lambda x + (1 - \lambda)y, \ x' = \delta^{-1}(1 - \lambda)(x - y), \ y' = \delta^{-1}\lambda(y - x).$$

From (2.19) and (2.20) we obtain

$$\lambda u(x) + (1 - \lambda)u(y) - u(\lambda x + (1 - \lambda)y)$$
$$= \delta\{\lambda u_{x_0, \delta}(x') + (1 - \lambda)u_{x_0, \delta}(y') - u_{x_0, \delta}(\lambda x' + (1 - \lambda)y')\}$$
$$\leq \delta\lambda(1 - \lambda)|x' - y'|\{K\delta + \widetilde{\omega}(|x' - y'|)\}$$
$$= \lambda(1 - \lambda)|x - y|\{K\delta + \widetilde{\omega}(\delta^{-1}|x - y|)\}.$$

Therefore

$$\lambda u(x) + (1 - \lambda)u(y) - u(\lambda x + (1 - \lambda)y) \leq \lambda(1 - \lambda)|x - y| \, \omega(|x - y|)$$

where $\omega(\rho) := \inf_{\delta \in \,]\rho, \delta_0[}\{K\delta + \widetilde{\omega}(\delta^{-1}\rho)\}$. The function ω is upper semicontinuous and nondecreasing. The conclusion will follow if we show that $\lim_{h \to 0} \omega(h) = 0$. Given $\varepsilon \in \,]0, 2K\delta_0[$, we choose $\eta \in \,]0, 1[$ such that $\widetilde{\omega}(s) < \varepsilon/2$ for $0 < s < \eta$. For any $\rho < \epsilon\eta(2K)^{-1}$ we have $\varepsilon(2K)^{-1} \in \,]\rho, \delta_0[$; therefore

$$\omega(\rho) \leq \left\{K\frac{\varepsilon}{2K} + \widetilde{\omega}\left(\frac{2K}{\varepsilon}\rho\right)\right\} < \varepsilon.$$

This shows that $\lim_{\rho \to 0} \omega(\rho) = 0$ and concludes the proof. ∎

Remark 2.5.3 From the above proof it is clear that the modulus of semiconcavity ω found for u is different from the function $\widetilde{\omega}$ that appears in the one-sided estimate (2.20). In some cases it is easy to compute explicitly ω. Let us suppose, for instance, that $\widetilde{\omega}(\rho) = C\rho^{\alpha}$, with $C, \alpha > 0$. Differentiating with respect to δ the quantity $K\delta + Ch^{\alpha}\delta^{-\alpha}$ we find that, for small ρ,

$$\omega(\rho) = C'\rho^{\frac{\alpha}{\alpha+1}}.$$

Thus, ω is also power-like, but with a lower exponent. Let us also observe that ω cannot be linear except in the case where $\widetilde{\omega} \equiv 0$.

Bibliographical notes

The notion of semiconcavity has been used in several different contexts in the last decades. We have already mentioned in the previous chapter applications to the uniqueness of solutions to Hamilton–Jacobi equations. In the framework of control theory, the role of semiconcavity for the characterization of the value function was underlined by Hrustalev [92].

Most authors have considered the notion of semiconcavity corresponding to a linear modulus in Definition 2.1.1. Moduli of a power form have been considered by Capuzzo Dolcetta and Ishii [48] and by Cannarsa and Soner [45]. The general definition used here appears in [32]. The Lipschitz continuity for a general modulus was proved in [10]. Theorem 2.1.10 generalizes a result in [45]. The counterexample of Proposition 2.1.4 is new.

The examples of Section 2.2 are well known in the literature. More references on the distance function will be given in next chapter. A detailed treatment of the vanishing viscosity method for Hamilton–Jacobi equations can be found in [100, 102, 110].

Theorems 2.3.2 and 2.3.3 are due to R. Jensen [98]. Our exposition here follows [66].

The Cole–Hopf transformation was introduced in [89] and [59]. The idea of stating the Harnack estimate in the form of a differential inequality of the type (2.17) is due to Li and Yau [108] and was motivated by the study of parabolic equations on manifolds. The matrix inequality (2.16) is due to R. Hamilton [85].

The generalized semiconcavity estimate of Section 2.5 was introduced in [46]. Theorem 2.5.2 is taken from [127].

3

Generalized Gradients and Semiconcavity

In the last decades a branch of mathematics has developed called *nonsmooth analysis*, whose object is to generalize the basic tools of calculus to functions that are not differentiable in the classical sense. For this purpose, one introduces suitable notions of generalized differentials, which are extensions of the usual gradient; the best known example is the subdifferential of convex analysis. The motivation for this study is that in more and more fields of analysis, like the optimization problems considered in this book, the functions that come into play are often nondifferentiable.

For semiconcave functions, the analysis of generalized gradients is important in view of applications to control theory. As we have already seen in a special case (Corollary 1.5.10), if the value function of a control problem is smooth, then one can design the optimal trajectories knowing the differential of the value function. In the general case, where the value function is not smooth but only semiconcave, one can try to follow a similar procedure starting from its generalized gradient.

In Section 3.1 we define the generalized differentials which are relevant for our purposes and recall basic properties and equivalent characterizations of these objects. Then, we restrict ourselves to semiconcave functions. In Section 3.2 we show that semiconcave functions possess one-sided directional derivatives everywhere, while in Section 3.3 we describe the special properties of the superdifferential of a semiconcave function; in particular, we show that it is nonempty at every point and that it is a singleton exactly at the points of differentiability. These properties are classical in the case of concave functions; here we prove that they hold for semiconcave functions with arbitrary modulus.

Section 3.4 is devoted to the so-called marginal functions, which are obtained as the infimum of smooth functions. We show that semiconcave functions can be characterized as suitable classes of marginal functions. In addition, we describe the semi-differentials of a marginal function using the general results of the previous sections. In Section 3.5 we study the so-called inf-convolutions. They are marginal functions defined by a process which is a generalization of Hopf's formula, and provide approximations to a given function which enjoy useful properties. Finally, in Section 3.6 we introduce proximal gradients and proximally smooth sets, and we analyze how these notions are related to semiconcavity.

3.1 Generalized differentials

We begin with the definitions of some generalized differentials and derivatives from nonsmooth analysis. In this section u is a real-valued function defined on an open set $A \subset \mathbb{R}^n$.

Definition 3.1.1 *For any $x \in A$, the sets*

$$D^-u(x) = \left\{ p \in \mathbb{R}^n : \liminf_{y \to x} \frac{u(y) - u(x) - \langle p, y - x \rangle}{|y - x|} \geq 0 \right\} \tag{3.1}$$

$$D^+u(x) = \left\{ p \in \mathbb{R}^n : \limsup_{y \to x} \frac{u(y) - u(x) - \langle p, y - x \rangle}{|y - x|} \leq 0 \right\} \tag{3.2}$$

are called, respectively, the (Fréchet) superdifferential *and* subdifferential *of u at x.*

From the definition it follows that, for any $x \in A$,

$$D^-(-u)(x) = -D^+u(x). \tag{3.3}$$

Example 3.1.2

Let $A = \mathbb{R}$ and let $u(x) = |x|$. Then it is easily seen that $D^+u(0) = \emptyset$ whereas $D^-u(0) = [-1, 1]$.

Let $A = \mathbb{R}$ and let $u(x) = \sqrt{|x|}$. Then, $D^+u(0) = \emptyset$ whereas $D^-u(0) = \mathbb{R}$.

Let $A = \mathbb{R}^2$ and $u(x, y) = |x| - |y|$. Then, $D^+u(0, 0) = D^-u(0, 0) = \emptyset$.

Definition 3.1.3 *Let $x \in A$ and $\theta \in \mathbb{R}^n$. The upper and lower Dini derivatives of u at x in the direction θ are defined as*

$$\partial^+u(x, \theta) = \limsup_{h \to 0^+, \theta' \to \theta} \frac{u(x + h\theta') - u(x)}{h}$$

and

$$\partial^-u(x, \theta) = \liminf_{h \to 0^+, \theta' \to \theta} \frac{u(x + h\theta') - u(x)}{h},$$

respectively.

It is readily seen that, for any $x \in A$ and $\theta \in \mathbb{R}^n$

$$\partial^-(-u)(x, \theta) = -\partial^+u(x, \theta). \tag{3.4}$$

Remark 3.1.4 Whenever u is Lipschitz continuous in a neighborhood of x, the lower Dini derivative reduces to

$$\partial^-u(x, \theta) = \liminf_{h \to 0^+} \frac{u(x + h\theta) - u(x)}{h} \tag{3.5}$$

for any $\theta \in \mathbb{R}^n$. Indeed, if $L > 0$ is the Lipschitz constant of u we have

$$\left| \frac{u(x + h\theta') - u(x)}{h} - \frac{u(x + h\theta) - u(x)}{h} \right| \leq L|\theta' - \theta|,$$

and (3.5) easily follows. A similar property holds for the upper Dini derivative.

Proposition 3.1.5 *Let $u : A \to \mathbb{R}$ and $x \in A$. Then the following properties hold true.*

(a) We have

$$D^+u(x) = \{ p \in \mathbb{R}^n : \partial^+u(x, \theta) \le \langle p, \theta \rangle \quad \forall \theta \in \mathbb{R}^n \},$$

$$D^-u(x) = \{ p \in \mathbb{R}^n : \partial^-u(x, \theta) \ge \langle p, \theta \rangle \quad \forall \theta \in \mathbb{R}^n \}.$$

(b) $D^+u(x)$ and $D^-u(x)$ are closed convex sets (possibly empty).
(c) $D^+u(x)$ and $D^-u(x)$ are both nonempty if and only if u is differentiable at x; in this case we have that

$$D^+u(x) = D^-u(x) = \{Du(x)\}.$$

Proof — **(a)** The fact that, for every vector $p \in D^+u(x)$ and any $\theta \in \mathbb{R}^n$,

$$\partial^+u(x, \theta) \le \langle p, \theta \rangle \tag{3.6}$$

is a direct consequence of (3.1). The converse can be proved by contradiction. Indeed, let $p \in \mathbb{R}^n$ be a vector satisfying (3.6) for all θ, but suppose that $p \notin D^+u(x)$. Then, a number $\varepsilon > 0$ and a sequence $\{x_k\} \subset A$ exist such that $x_k \to x$ as $k \to \infty$, and

$$u(x_k) - u(x) \ge \langle p, x_k - x \rangle + \epsilon|x_k - x| \tag{3.7}$$

for all $k \in \mathbb{N}$. Moreover, possibly taking a subsequence, we may assume that the sequence of unit vectors

$$\theta_k := \frac{x_k - x}{|x_k - x|}$$

converges to some unit vector θ as $k \to \infty$. Then, (3.7) yields

$$\varepsilon + \langle p, \theta \rangle \le \limsup_{k \to \infty} \frac{u(x_k) - u(x)}{|x_k - x|}$$
$$= \limsup_{k \to \infty} \frac{u(x + |x_k - x|\theta_k) - u(x)}{|x_k - x|} \le \partial^+u(x, \theta),$$

which contradicts (3.6). The analogous property for $D^-u(x)$ immediately follows recalling (3.3) and (3.4).
(b) This property is a consequence of (a).
(c) If u is differentiable at x, then it is easily seen that

$$Du(x) \in D^+u(x) \cap D^-u(x).$$

Conversely, let $p_1 \in D^+u(x)$ and $p_2 \in D^-u(x)$. Then, property (i) yields

$$\langle p_2, \theta \rangle \le \partial^-u(x, \theta) \le \partial^+u(x, \theta) \le \langle p_1, \theta \rangle,$$

and so $\langle p_1 - p_2, \theta \rangle \ge 0$ for any $\theta \in \mathbb{R}^n$. This implies that $p_1 = p_2$. So, if $D^+u(x)$ and $D^-u(x)$ are both nonempty, they coincide and reduce to a singleton. Using the definition of $D^{\pm}$ we obtain that u is differentiable at x with gradient $Du(x) = p_1$. Hence the conclusion follows. ∎

If u is continuous, then it turns out that the super– and subdifferential of u can be described in terms of "test" functions. For this purpose, it is useful to introduce the following notion of contact between (the graphs of) two real–valued functions, u and ϕ, defined on A.

Definition 3.1.6 *We say that ϕ touches u from above at a point $x_0 \in A$ if*

$$u(x_0) = \phi(x_0) \tag{3.8}$$

and, for some open ball $B_r(x_0) \subset A$,

$$u(x) \leq \phi(x), \qquad \forall x \in B_r(x_0). \tag{3.9}$$

Similarly, ϕ touches u from below at $x_0 \in A$ if the above properties hold with $\leq$ replaced by $\geq$.

Clearly, if (3.8) holds, then (3.9) is equivalent to requiring that $u - \phi$ has a *local maximum* at x.

Proposition 3.1.7 *Let $u \in C(A)$, $p \in \mathbb{R}^n$, and $x \in A$. Then the following properties are equivalent:*

(a) *$p \in D^+u(x)$ (resp. $p \in D^-u(x)$);*
(b) *$p = D\phi(x)$ for some function $\phi \in C^1(A)$ touching u from above (resp. below) at x;*
(c) *$p = D\phi(x)$ for some function $\phi \in C^1(A)$ such that $u - \phi$ attains a local maximum (resp. minimum) at x.*

Before giving the proof, let us prove a technical lemma.

Lemma 3.1.8 *Let $\omega :]0, +\infty[\to [0, +\infty[$ be an upper semicontinuous function such that $\lim_{r \to 0} \omega(r) = 0$. Then there exists a continuous nondecreasing function $\omega_1 : [0, +\infty[\to [0, +\infty[$ such that*

(i) *$\omega_1(r) \to 0$ as $r \to 0$,*
(ii) *$\omega(r) \leq \omega_1(r)$ for any $r \geq 0$,*
(iii) *the function $\gamma(r) := r\omega_1(r)$ is in $C^1([0, +\infty[)$ and satisfies $\gamma'(0) = 0$.*

Proof — Let us first set

$$\bar{\omega}(r) = \max_{\rho \in]0,r]} \omega(\rho).$$

Then $\bar{\omega}$ is nondecreasing, not smaller than ω and tends to 0 as $r \to 0$. Next we define for $r > 0$

$$\omega_0(r) = \frac{1}{r} \int_r^{2r} \bar{\omega}(\rho)d\rho, \qquad \omega_1(r) = \frac{1}{r} \int_r^{2r} \omega_0(\rho)d\rho,$$

and we set $\omega_1(0) = 0$. We first observe that, since $\bar{\omega}$ is nondecreasing, the same holds for ω_0 and ω_1. Then we have that $\omega(r) \leq \omega_0(r) \leq \bar{\omega}(2r)$, and so $\omega_0(r) \to 0$ as $r \to 0$. Arguing in the same way with ω_1 we deduce that properties (i) and (ii) hold. To prove (iii), let us set $\gamma(r) = r\omega_1(r)$. Then $\gamma \in C^1(]0, +\infty[)$ with derivative $\gamma'(r) = 2\omega_0(2r) - \omega_0(r)$. Thus $\gamma'(r) \to 0$ as $r \to 0$ and so γ is C^1 in the closed half-line $[0, +\infty[$. $\blacksquare$

Proof of Proposition 3.1.7 — The implications (b) $\Rightarrow$ (c) and (c) $\Rightarrow$ (a) are obvious; so, it is enough to prove that (a) implies (b). Given $p \in D^+u(x)$, let us define, for $r > 0$,

$$\omega(r) = \max_{y:|y-x|=r} \left[\frac{u(y) - u(x) - \langle p, y - x \rangle}{r} \right]_+,$$

where $[\cdot]_+$ denotes the positive part. The function ω is continuous and tends to 0 as $r \to 0$, by definition of D^+u. Let ω_1 be the function given by the previous lemma. Then, setting

$$\phi(y) = u(x) + \langle p, y - x \rangle + |y - x|\omega_1(|y - x|),$$

we have that ϕ is in C^1 and touches u from above at x. $\blacksquare$

The above characterization can be used to study the super- and subdifferentiability sets of a nonsmooth function u. A first result in this direction is the following.

Proposition 3.1.9 *Let* $u \in C(A)$. *Then the sets*

$$A^+(u) = \{x \in A : D^+u(x) \neq \emptyset\}$$

and

$$A^-(u) = \{x \in A : D^-u(x) \neq \emptyset\}$$

are both dense in A.

Proof — Given $x_0 \in A$, let us fix $R > 0$ such that $\overline{B}_R(x_0) \subset A$. Given $\varepsilon > 0$, let x_ε be a point of $\overline{B}_R(x_0)$ where the function $u(x) - |x - x_0|^2/\varepsilon$ attains its maximum. Then we have

$$u(x_\varepsilon) - \frac{|x_\varepsilon - x_0|^2}{\varepsilon} \geq u(x_0),$$

which implies that

$$|x_\varepsilon - x_0|^2 \leq \varepsilon \left[\max_{\overline{B}_R(x_0)} u - u(x_0) \right].$$

Thus, $x_\varepsilon \to x_0$ as $\varepsilon \to 0$. It also follows that x_ε lies in the interior of $\overline{B}_R(x_0)$ for ε small enough. For such an ε let us set $\phi_\varepsilon(x) = |x - x_0|^2/\varepsilon$. Then $u - \phi_\varepsilon$ attains a local maximum at x_ε and so, by Proposition 3.1.7(c), $D\phi_\varepsilon(x_\varepsilon) \in D^+u(x_\varepsilon)$. This shows that $A^+(u)$ is dense in A. $\blacksquare$

Let us now introduce another kind of generalized gradient which will be important in our later analysis.

Definition 3.1.10 *Let* $u : A \to \mathbb{R}$ *be locally Lipschitz. A vector* $p \in \mathbb{R}^n$ *is called a* reachable gradient *of* u *at* $x \in A$ *if a sequence* $\{x_k\} \subset A \setminus \{x\}$ *exists such that* u *is differentiable at* x_k *for each* $k \in \mathbb{N}$, *and*

$$\lim_{k \to \infty} x_k = x, \qquad \lim_{k \to \infty} Du(x_k) = p. \tag{3.10}$$

The set of all reachable gradients of u *at* x *is denoted by* $D^*u(x)$.

It is easily seen that $D^*u(x)$ is a compact set: it is closed by definition and it is bounded since we are taking u Lipschitz. From Rademacher's Theorem it follows that $D^*u(x) \neq \emptyset$ for every $x \in A$. Let us now recall some more definitions from nonsmooth analysis.

Definition 3.1.11 *For any fixed* $x \in A$ *and* $\theta \in \mathbb{R}^n$, *the* generalized lower derivative *is defined as*

$$u_-^0(x, \theta) = \liminf_{h \to 0^+, \, y \to x} \frac{u(y + h\theta) - u(y)}{h}. \tag{3.11}$$

Moreover the set

$$\partial_C u(x) = \{p \in \mathbb{R}^n \ : \ u_-^0(x, \theta) \leq \langle p, \theta \rangle \quad \forall \theta \in \mathbb{R}^n\} \tag{3.12}$$

is called Clarke's gradient *of* u *at* x.

The generalized derivative $u_-^0(x, \theta)$ defined above is quite different from the Dini derivatives $\partial^\pm u(x, \theta)$ defined in Definition 3.1.3, as the next example shows.

Example 3.1.12 Let $u(x) = |x|$, for $x \in \mathbb{R}$, and let $\theta = 1$. Then $\partial^+ u(0, \theta) = \partial^- u(0, \theta) = 1$. On the other hand, since we can choose $y = -h$ in (3.11), we easily see that $u_-^0(0, \theta) = -1$. We obtain similarly that $u_-^0(0, -1) = -1$; in general $u_-^0(0, \theta) = -|\theta|$. Therefore

$$\partial_C u(0) = \{p \in \mathbb{R} \ : \ -|\theta| \leq p\theta \quad \forall \theta \in \mathbb{R}\} = [-1, 1].$$

Observe that for this function $\partial_C u(0) = D^- u(0)$. We will see that this is true for all convex and semiconvex functions. ∎

One can also define the *generalized upper derivative* as

$$u_+^0(x, \theta) = \limsup_{h \to 0^+, \, y \to x} \frac{u(y + h\theta) - u(y)}{h}.$$

Since $u_-^0(x, \theta) = -u_+^0(x, -\theta)$, an equivalent definition of Clarke's gradient is therefore

$$\partial_C u(x) = \{p \in \mathbb{R}^n \ : \ u_+^0(x, \theta) \geq \langle p, \theta \rangle \quad \forall \theta \in \mathbb{R}^n\}.$$

It is proved in [54, Th. 2.5.1] that, if u is locally Lipschitz, then the reachable gradients of u at x generate Clarke's gradient in the sense of convex analysis, i.e.,

$$\partial_C u(x) = \text{co } D^* u(x). \tag{3.13}$$

Consequently, $\partial_C u(x)$ is nonempty, convex and compact (see Corollary A. 1.7). Moreover, by Remark 3.1.4,

$$u^0_-(x, \theta) \le \partial^- u(x, \theta) \le \partial^+ u(x, \theta) \le u^0_+(x, \theta)$$

for any $x \in A$ and $\theta \in \mathbb{R}^n$. Then, recalling point (a) of Proposition 3.1.5, we conclude that

$$D^+ u(x) \subset \partial_C u(x) \quad \text{and} \quad D^- u(x) \subset \partial_C u(x)$$

for any $x \in A$. We conclude the section by recalling the following extension of the classical mean value theorem (see [54, Th. 2.3.7]).

Theorem 3.1.13 *Let $u : A \to \mathbb{R}$ be locally Lipschitz. Given $x, y \in A$ such that $[x, y] \subset A$, there exists $\xi \in]x, y[$ and $p \in \partial_C u(\xi)$ such that $u(y) - u(x) = p \cdot (y - x)$.*

3.2 Directional derivatives

We begin our exposition of the differential properties of semiconcave functions showing that they possess (one-sided) directional derivatives

$$\partial u(x, \theta) := \lim_{h \to 0^+} \frac{u(x + h\theta) - u(x)}{h} \tag{3.14}$$

at any point x and in any direction θ.

Theorem 3.2.1 *Let $u : A \to \mathbb{R}$ be semiconcave. Then, for any $x \in A$ and $\theta \in \mathbb{R}^n$,*

$$\partial u(x, \theta) = \partial^- u(x, \theta) = \partial^+ u(x, \theta) = u^0_-(x, \theta). \tag{3.15}$$

Proof — Let $\delta > 0$ be fixed so that $B_{\delta|\theta|}(x) \subset A$. Then, for any pair of numbers h_1, h_2 satisfying $0 < h_1 \le h_2 < \delta$, estimate (2.1) yields

$$\left(1 - \frac{h_1}{h_2}\right) u(x) + \frac{h_1}{h_2} u(x + h_2\theta) - u(x + h_1\theta) \le h_1 \left(1 - \frac{h_1}{h_2}\right) |\theta| \omega(h_2|\theta|).$$

Hence,

$$\frac{u(x + h_1\theta) - u(x)}{h_1} \tag{3.16}$$
$$\ge \frac{u(x + h_2\theta) - u(x)}{h_2} - \left(1 - \frac{h_1}{h_2}\right) |\theta| \omega(h_2|\theta|).$$

Taking the lim inf as $h_1 \to 0^+$ in both sides of the above inequality, we obtain

$$\partial^- u(x, \theta) \geq \frac{u(x + h_2\theta) - u(x)}{h_2} - |\theta|\omega(h_2|\theta|).$$

Now, taking the lim sup as $h_2 \to 0^+$, we conclude that

$$\partial^- u(x, \theta) \geq \partial^+ u(x, \theta).$$

So, $\partial u(x, \theta)$ exists and coincides with the lower and upper Dini derivatives.

To complete the proof of (3.15) it suffices to show that

$$\partial^+ u(x, \theta) \leq u^0_-(x, \theta), \tag{3.17}$$

since the reverse inequality holds by definition and by Remark 3.1.4. For this purpose, let $\varepsilon > 0, \lambda \in]0, \delta[$ be fixed. Since u is continuous, we can find $\alpha \in]0, (\delta - \lambda)\theta[$ such that

$$\frac{u(x + \lambda\theta) - u(x)}{\lambda} \leq \frac{u(y + \lambda\theta) - u(y)}{\lambda} + \varepsilon, \quad \forall y \in B_\alpha(x).$$

Using inequality (3.16) with x replaced by y, we obtain

$$\frac{u(y + \lambda\theta) - u(y)}{\lambda} \leq \frac{u(y + h\theta) - u(y)}{h} + |\theta|\omega(\lambda|\theta|), \quad \forall h \in]0, \lambda[.$$

Therefore,

$$\frac{u(x + \lambda\theta) - u(x)}{\lambda} \leq \inf_{y \in B_\alpha(x), h \in]0, \lambda[} \frac{u(y + h\theta) - u(y)}{h} + |\theta|\omega(\lambda|\theta|) + \varepsilon.$$

This implies, by definition of $u^0_-(x, \theta)$, that

$$\frac{u(x + \lambda\theta) - u(x)}{\lambda} \leq u^0_-(x, \theta) + |\theta|\omega(\lambda|\theta|) + \varepsilon.$$

Hence, taking the limit as $\varepsilon, \lambda \to 0$, we obtain inequality (3.17). ∎

3.3 Superdifferential of a semiconcave function

The superdifferential of a semiconcave function enjoys many properties that are not valid for a general Lipschitz continuous function, and that can be regarded as extensions of analogous properties of concave functions. We start with the following basic estimate. Throughout the section $A \subset \mathbb{R}^n$ is an open set.

Proposition 3.3.1 *Let $u : A \to \mathbb{R}$ be a semiconcave function with modulus ω and let $x \in A$. Then, a vector $p \in \mathbb{R}^n$ belongs to $D^+u(x)$ if and only if*

$$u(y) - u(x) - \langle p, y - x \rangle \leq |y - x|\omega(|y - x|) \tag{3.18}$$

for any point $y \in A$ such that $[y, x] \subset A$.

Proof — If $p \in \mathbb{R}^n$ satisfies (3.18), then, by the very definition of superdifferential, $p \in D^+u(x)$. In order to prove the converse, let $p \in D^+u(x)$. Then, dividing the semiconcavity inequality (2.1) by $(1 - \lambda)|x - y|$, we have

$$\frac{u(y) - u(x)}{|y - x|} \leq \frac{u(x + (1 - \lambda)(y - x)) - u(x)}{(1 - \lambda)|y - x|} + \lambda\omega(|x - y|), \quad \forall\lambda \in \,]0, 1].$$

Hence, taking the limit as $\lambda \to 1^-$, we obtain

$$\frac{u(y) - u(x)}{|y - x|} \leq \frac{\langle p, y - x \rangle}{|y - x|} + \omega(|x - y|),$$

since $p \in D^+u(x)$. Estimate (3.18) follows. ∎

Remark 3.3.2 In particular, if u is concave on a convex set A, we find that $p \in D^+u(x)$ if and only if

$$u(y) \geq u(x) + \langle p, y - x \rangle, \quad \forall y \in A.$$

In convex analysis (see Appendix A. 1) this property is usually taken as the definition of the superdifferential. Therefore, the Fréchet super- and subdifferential coincide with the classical semidifferentials of convex analysis in the case of a concave (resp. convex) function.

Before investigating further properties of the superdifferential, let us show how Proposition 3.3.1 easily yields a compactness property for semiconcave functions.

Theorem 3.3.3 Let $u_n : A \to \mathbb{R}$ be a family of semiconcave functions with the same modulus ω. Given an open set $B \subset\subset A$, suppose that the u_n's are uniformly bounded in B. Then there exists a subsequence u_{n_k} converging uniformly to a function $u : B \to \mathbb{R}$ semiconcave with modulus ω. In addition, $Du_{n_k} \to Du$ a.e. in B.

Proof — From Remark 2.1.8 we deduce that the functions u_n are uniformly Lipschitz in B; thus, by the Ascoli–Arzelà theorem, there exists a subsequence, which we denote again by u_n for simplicity, converging uniformly to some $u : B \to \mathbb{R}$. Since the semiconcavity inequality is preserved by pointwise convergence, u is semiconcave with the same modulus of the u_n's. Let us consider a point $x_0 \in B$ such that u and all u_n are differentiable at x_0, for all n; we observe that all points of B have this property except for a set of measure zero. We claim that $Du_n(x_0) \to Du(x_0)$. To prove this, we argue by contradiction. The sequence $\{Du_n(x_0)\}$ is bounded; if it does not converge to $Du(x_0)$ there exists a subsequence converging to some $p_0 \neq Du(x_0)$. But then, passing to the limit in (3.18), we obtain that $p_0 \in D^+u(x_0)$ and we find a contradiction. This proves that $Du_n \to Du$ a.e. in B. ∎

Let us now derive other important consequences of Proposition 3.3.1.

Proposition 3.3.4 Let $u : A \to \mathbb{R}$ be a semiconcave function with modulus ω, and let $x \in A$. Then the following properties hold true.

(a) *If $\{x_k\}$ is a sequence in A converging to x, and if $p_k \in D^+u(x_k)$ converges to a vector $p \in \mathbb{R}^n$, then $p \in D^+u(x)$.*

(b) *$D^*u(x) \subset \partial D^+u(x)$.*

(c) *$D^+u(x) \neq \emptyset$.*

(d) *If $D^+u(x)$ is a singleton, then u is differentiable at x.*

(e) *If $D^+u(y)$ is a singleton for every $y \in A$, then $u \in C^1(A)$.*

Remark 3.3.5 By $\partial D^+u(x)$ in part (b) above we mean the boundary taken with respect to the standard topology of $\mathbb{R}^n$, and not the relative boundary of convex analysis (see Definition A. 1.8). Observe therefore that every time the dimension of $D^+u(x)$ is less than n, we have $\partial D^+u(x) = D^+u(x)$.

Proof — **(a)** This property follows from Proposition 3.3.1, using also the upper semicontinuity of ω.

(b) First we note that, in light of (a), $D^*u(x) \subset D^+u(x)$. Therefore, the nontrivial part of (b) is that all reachable gradients are boundary points of $D^+u(x)$. Indeed, let $p \in D^*u(x)$ and let $x_k \in A$ be as in (3.10). Without loss of generality we may assume that

$$\lim_{k \to \infty} \frac{x_k - x}{|x_k - x|} = \theta$$

for some unit vector $\theta \in \mathbb{R}^n$. We claim that

$$p - t\theta \notin D^+u(x), \quad \forall t > 0, \tag{3.19}$$

which implies that p is in the boundary of $D^+u(x)$. Indeed, by (3.18),

$$\begin{aligned}
&u(x_k) - u(x) - \langle p - t\theta, x_k - x \rangle \\
&= u(x_k) - u(x) - \langle Du(x_k), x_k - x \rangle \\
&\quad + \langle Du(x_k) - p, x - x_k \rangle + t\langle \theta, x - x_k \rangle \\
&\geq -|x_k - x|\omega(|x_k - x|) - |Du(x_k) - p||x_k - x| + t\langle \theta, x - x_k \rangle.
\end{aligned}$$

Therefore,

$$\liminf_{k \to \infty} \frac{u(x_k) - u(x) - \langle p - t\theta, x_k - x \rangle}{|x_k - x|} \geq t,$$

and our claim (3.19) follows.

(c) This property is an immediate consequence of (b). Indeed, u is locally Lipschitz by Theorem 2.1, and so $D^*u(x) \neq \emptyset$.

(d) Suppose that $D^+u(x) = \{p\}$ for some $p \in \mathbb{R}^n$, and let $\{x_k\} \subset A$ be any sequence such that $x_k \to x$. By (c), we can pick a point $p_k \in D^+u(x_k)$ for any k. On the other hand, by (a), sequence $\{p_k\}$ can only admit p as a cluster point. Therefore, $p_k \to p$. Moreover, by Proposition 3.3.1,

$$\begin{aligned}
&u(x_k) - u(x) - \langle p, x_k - x \rangle \\
&= u(x_k) - u(x) + \langle p_k, x - x_k \rangle + \langle p_k - p, x_k - x \rangle \\
&\geq -|x_k - x|\omega(|x_k - x|) - |p_k - p||x_k - x|.
\end{aligned}$$

Hence,

$$\liminf_{k\to\infty} \frac{u(x_k) - u(x) - \langle p, x_k - x \rangle}{|x_k - x|} \geq 0 \,.$$

Since $\{x_k\}$ is arbitrary, we conclude that $p \in D^- u(x)$. Then, Proposition 3.1.5(c) implies that u is differentiable at x.

(e) This assertion is an easy consequence of (a) and (d). ∎

In the terminology of set-valued analysis, property (a) above is usually expressed by saying that the set valued map $D^+ u$ is *upper semicontinuous* (see Definition A. 5.1).

Theorem 3.3.6 *Let $u : A \to \mathbb{R}$ be a semiconcave function and let $x \in A$. Then*

$$D^+ u(x) = \text{co} D^* u(x) = \partial_C u(x) \,. \tag{3.20}$$

Moreover the directional derivatives of u defined in (3.14) satisfy

$$\partial u(x, \theta) = \min_{p \in D^+ u(x)} \langle p, \theta \rangle = \min_{p \in D^* u(x)} \langle p, \theta \rangle \tag{3.21}$$

for any $\theta \in \mathbb{R}^n$.

Proof — We prove (3.21) first. From Theorem 3.2.1, Proposition 3.1.5(a) and Proposition 3.3.4(b) we deduce that

$$\partial u(x, \theta) \leq \min_{p \in D^+ u(x)} \langle p, \theta \rangle \leq \min_{p \in D^* u(x)} \langle p, \theta \rangle \,, \quad \forall \theta \in \mathbb{R}^n \,.$$

Therefore, to obtain (3.21) it suffices to show that

$$\min_{p \in D^* u(x)} \langle p, \theta \rangle \leq \partial u(x, \theta) \,, \quad \forall \theta \in \mathbb{R}^n \,. \tag{3.22}$$

Indeed, let $\theta \in \mathbb{R}^n$ be a fixed unit vector. Since u is a.e. differentiable, one can find a sequence $\{x_k\}$ of differentiability points of u, such that

$$\theta_k := \frac{x_k - x}{|x_k - x|} \to \theta \,, \quad \text{as } k \to \infty \,,$$

and $Du(x_k)$ converges to some vector $p_0 \in D^* u(x)$. Let ω be a modulus of semi-concavity for u. Then, by (3.18),

$$\langle Du(x_k), \theta_k \rangle \leq \frac{u(x + |x_k - x|\theta_k) - u(x)}{|x_k - x|} + \omega(|x_k - x|) \,.$$

Taking the limit as $k \to \infty$, we obtain

$$\langle p_0, \theta \rangle \leq \partial u(x, \theta) \,,$$

whence (3.22) follows. Now, since by Proposition A. 1.17,

$$\min_{p \in D^* u(x)} \langle p, \theta \rangle = \min_{p \in \text{co} D^* u(x)} \langle p, \theta \rangle \,,$$

equality (3.21) implies that the two convex sets $D^+u(x)$ and $\text{co}D^*u(x)$ have the same support function. By well-known properties of convex sets (see Corollary A. 1.7 and Proposition A. 1.18) they coincide. Taking into account also (3.13), we obtain (3.20).

∎

Theorem 3.3.7 *If $u : A \to \mathbb{R}$ is both semiconvex and semiconcave in A, then $u \in C^1(A)$. In addition, on each compact subset of A the modulus of continuity of Du is of the form $\omega_1(r) = c_1\omega(c_2r)$, where ω is a modulus of semiconvexity and of semiconcavity for u and $c_1, c_2 > 0$ are constants.*

Proof — We recall that, in the case of a modulus of the form $\omega(r) = r^\alpha$, with $0 < \alpha < 1$, the above result can be deduced from [133, Prop. V.8–V.9]. We give here an independent proof, which is valid for any modulus.

At any point of A both u and $-u$ have nonempty superdifferential. Since $D^+(-u) = -D^-(u)$, this implies that the subdifferential of u is also nonempty everywhere. Therefore, by Proposition 3.1.5(c) and 3.3.4(e), u belongs to $C^1(A)$.

To prove the second part of the statement, observe first that if u is semiconcave and semiconvex with modulus ω, then by Proposition 3.3.1

$$|u(y) - u(x) - \langle Du(x), y - x \rangle| \leq |y - x|\omega(|y - x|) \qquad (3.23)$$

for all $x, y \in A$ such that $[x, y] \subset A$

Now take any $x \in A$ and let $r > 0$ such that $B_{2r}(x) \subset A$. Given any $v, w \in \mathbb{R}^n$ with $|v|, |w| \leq r$ we have, using (3.23) and the semiconcavity and semiconvexity of u,

$$\begin{aligned}
\langle Du&(x + w) - Du(x), v \rangle \\
&\leq u(x + w + v) - u(x + w) - u(x + v) + u(x) + 2|v|\omega(|v|) \\
&= u(x + w + v) - \frac{1}{2}u(x + 2v) - \frac{1}{2}u(x + 2w) \\
&\quad -u(x + w) + \frac{1}{2}u(x + 2w) + \frac{1}{2}u(x) \\
&\quad -u(x + v) + \frac{1}{2}u(x + 2v) + \frac{1}{2}u(x) + 2|v|\omega(|v|) \\
&\leq \frac{1}{2}|v + w|\omega(2|v + w|) + \frac{1}{2}|w|\omega(2|w|) + \frac{1}{2}|v|\omega(2|v|) + 2|v|\omega(|v|).
\end{aligned}$$

It follows that, for all $w \in B_r$,

$$\begin{aligned}
|Du(x + w) - Du(x)| &= \frac{1}{|w|} \max_{|v|=|w|} \langle Du(x + w) - Du(x), v \rangle \\
&\leq \omega(4|w|) + \omega(2|w|) + 2\omega(|w|) \\
&\leq 4\omega(4|w|).
\end{aligned}$$

∎

In the case of a linear modulus we can give a sharper estimate of the modulus of continuity of Du.

Corollary 3.3.8 *Let $A \subset \mathbb{R}^n$ be open convex and let $u : A \to \mathbb{R}$ be both semiconvex and semiconcave with a linear modulus and constant C. Then $u \in C^{1,1}(A)$ and the Lipschitz constant of Du is equal to C.*

Proof — The previous theorem already implies that Du is locally Lipschitz in A, so we only have to estimate its Lipschitz constant. By Proposition 1.1.3, all directional second derivatives of u of the form $\partial_{vv} u$ with $|v| = 1$ satisfy $\|\partial_{vv} u\| \leq C$. Now any mixed distributional second derivative is the linear combination of directional ones, as follows:

$$\partial^2_{\xi\eta} u = \frac{1}{4} \left(\partial^2_{(\xi+\eta)(\xi+\eta)} u - \partial^2_{(\xi-\eta)(\xi-\eta)} u \right), \qquad \forall \xi, \eta \in \mathbb{R}^n.$$

We deduce that

$$\|D^2 u\| = \sup_{|\xi|=|\eta|=1} |\partial^2_{\xi\eta} u| \leq \sup_{|\xi|=|\eta|=1} \frac{C}{4} \left(|\xi + \eta|^2 + |\xi - \eta|^2 \right) = C$$

in the sense of distributions. Since Du is Lipschitz, the distributional hessian $D^2 u$ coincides with the standard hessian which exists almost everywhere. We conclude that the Lipschitz constant of Du is equal to C. ∎

The previous results can be used to show that a given function is not semiconcave.

Example 3.3.9 We have observed in Example 1.1.4 that the function $u(x) = |x|$ is not semiconcave with a linear modulus in any neighborhood of 0. To see that it is not semiconcave with any modulus, it suffices to observe that u is convex; were u also semiconcave, it would be differentiable by Theorem 3.3.7.

It is well known that the superdifferential of a concave function has a monotonicity property. Semiconcave functions satisfy this property in a weaker form, as shown by the next result.

Proposition 3.3.10 *Let $u : A \to \mathbb{R}$ be a semiconcave function with modulus ω. Given $x, y \in A$ such that $[x, y] \subset A$ we have, for any $p \in D^+ u(x)$ and $q \in D^+ u(y)$,*

$$\langle q - p, y - x \rangle \leq 2|y - x| \omega(|y - x|). \tag{3.24}$$

Proof — Applying Proposition 3.3.1 twice, we find

$$-\langle q, x - y \rangle - |y - x| \omega(|y - x|) \leq u(y) - u(x) \leq \langle p, y - x \rangle + |y - x| \omega(|y - x|).$$

The conclusion follows. ∎

Property (3.24) characterizes in fact semiconcave functions, as we show below.

Proposition 3.3.11 *Let $u : A \to \mathbb{R}^n$ be a Lipschitz continuous function and let $\partial_c u$ denote its Clarke gradient. Suppose that there exists a nondecreasing upper semicontinuous function $\omega : [0, \infty[\to [0, \infty[$ such that $\omega(0) = 0$ and such that*

$$\langle q - p, y - x \rangle \leq |y - x| \omega(|y - x|) \tag{3.25}$$

for all x, y such that $[x, y] \subset A$, $p \in \partial_c u(x)$ and $q \in \partial_c u(y)$. Then u is semiconcave in A with modulus ω.

Proof — We can proceed exactly as in the proof of Proposition 2.1.2, using the nonsmooth version of the mean value theorem (Theorem 3.1.13).   ∎

Before giving further results, let us recall a standard definition from convex analysis.

Definition 3.3.12 *Given a convex set $C \subset \mathbb{R}^n$, we say that $x \in C$ is an* extreme point *of C if x cannot be expressed as the convex combination of $x_0, x_1 \in C$, with $x_0, x_1 \neq x$. The set of all the extreme points of C is denoted by* Ext(C).

Formula (3.20) trivially implies that, for a semiconcave function,

$$\text{Ext}(D^+ u(x)) \subset D^* u(x), \quad \forall x \in A. \tag{3.26}$$

One may wonder if such an inclusion is in fact an equality. This is not the case, as the following example shows.

Example 3.3.13 Let $u : \mathbb{R}^2 \to \mathbb{R}$ be the concave function

$$u(x, y) = -\sqrt{x^2 + y^4}.$$

Then u is differentiable on $\mathbb{R}^2 \setminus \{(0, 0)\}$, whereas

$$D^+ u(0, 0) = [-1, 1] \times \{0\}.$$

We can check this by computing the directional derivatives

$$\partial u((0, 0), (\theta_1, \theta_2)) = -|\theta_1|, \quad \forall (\theta_1, \theta_2) \in \mathbb{R}^2,$$

and recalling Proposition 3.1.5(a). Therefore, $\text{Ext}(D^+ u(0, 0)) = \{(-1, 0), (1, 0)\}$. On the other hand, we claim that $D^* u(0, 0) = [-1, 1] \times \{0\}$. Indeed, we first observe that

$$Du(x, y) = -\left(\frac{x}{\sqrt{x^2 + y^4}}, \frac{2y^3}{\sqrt{x^2 + y^4}} \right), \quad (x, y) \in \mathbb{R}^2 \setminus \{(0, 0)\}.$$

We know from (3.26) that $\{(-1, 0), (1, 0)\} \subset D^* u(0, 0)$, so it suffices to show that vectors of the form $(r, 0)$ with $|r| < 1$ belong to $D^* u(0, 0)$. For such an r, set $a = -r/\sqrt{1 - r^2}$ and consider the parabola $(x, y) = (at^2, t)$. We have

$$\lim_{t \to 0} Du(at^2, t) = \left(-\frac{a}{\sqrt{a^2 + 1}}, 0 \right) = (r, 0).$$

This proves that $D^*u(0, 0) = [-1, 1] \times \{0\}$.   ∎

Let us recall some terminology from convex analysis.

Definition 3.3.14 *Let $u : A \to \mathbb{R}$ be a Lipschitz function. Given $x \in A$, $\theta \in \mathbb{R}^n \setminus \{0\}$ we set*

$$D^+u(x, \theta) = \{p \in D^+u(x) : p \cdot \theta \le q \cdot \theta, \ \forall q \in D^+u(x)\}.$$

This set is called the exposed face *of $D^+u(x)$ in direction θ. If $D^+u(x, \theta) = \{p\}$, then we say that θ is an* exposed direction *and p is an* exposed point *of $D^+u(x)$. We denote the set of exposed points of $D^+u(x)$ by $Eu(x)$.*

Then we have the following property, which can be regarded as a refinement of Proposition 3.3.4–(a).

Proposition 3.3.15 *Let $u : A \to \mathbb{R}$ be semiconcave and let $x \in A$, $\bar{p} \in \mathbb{R}^n$ be given. Suppose that there exist two sequences $\{x_k\} \subset (A \setminus \{x\})$, $\{p_k\} \subset \mathbb{R}^n$ such that*

$$\lim_{k \to \infty} x_k = x, \quad \lim_{k \to \infty} \frac{x_k - x}{|x_k - x|} = \theta, \quad p_k \in D^+u(x_k), \quad \lim_{k \to \infty} p_k = \bar{p} \quad (3.27)$$

for some unit vector θ. Then $\bar{p} \in D^+u(x, \theta)$.

Proof — From Proposition 3.3.4(a) it follows that $\bar{p} \in D^+u(x)$. Moreover, by Proposition 3.3.1 we have

$$\frac{u(x_k) - u(x)}{|x_k - x|} \ge \langle p_k, \frac{x_k - x}{|x_k - x|} \rangle - \omega(|x - x_k|), \quad \forall k \ge 1.$$

Thus we have

$$\partial^+u(x, \theta) \ge \limsup_{k \to \infty} \frac{u(x_k) - u(x)}{|x_k - x|} \ge \bar{p} \cdot \theta.$$

On the other hand, any vector $p \in D^+u(x)$ satisfies the opposite inequality by Proposition 3.1.5(a). We conclude that

$$\bar{p} \cdot \theta = \partial^+u(x, \theta) = \min_{p \in D^+u(x)} p \cdot \theta.$$ ∎

Let us prove a technical result which is useful in the analysis of evolution problems where one considers functions u depending on $(t, x) \in]0, T[\times A$, for some $T > 0$ and $A \subset \mathbb{R}^n$. For these functions, it is natural to consider the generalized partial differentials with respect to x as follows:

$$\nabla^+u(t, x) := \left\{ \eta \in \mathbb{R}^n : \limsup_{h \to 0} \frac{u(t, x + h) - u(t, x) - \langle \eta, h \rangle}{|h|} \le 0 \right\}, \quad (3.28)$$

and

$$\nabla^* u(t, x) = \left\{ p \in \mathbb{R}^n : x_i \to x, \ \nabla u(t, x_i) \to p \right\}. \tag{3.29}$$

Observe that (3.20) immediately implies that if $u \in SC(\,]0, T[\,\times A)$, then

$$\nabla^+ u(t, x) = \mathrm{co}\, \nabla^* u(t, x). \tag{3.30}$$

Let $\Pi_x : \mathbb{R}^{n+1} \to \mathbb{R}^n$ be the projection onto the x-space, i.e., $\Pi_x(t, x) = x$. The next result explains the relationship between $\nabla^+ u(t, x)$ and $D^+ u(t, x)$.

Lemma 3.3.16 *For any $u \in SC(\,]0, T[\,\times A)$, we have that*

$$\Pi_x D^+ u(t, x) = \nabla^+ u(t, x) \qquad \forall (t, x) \in \,]0, T[\,\times U. \tag{3.31}$$

Proof — The inclusion $\Pi_x D^+ u(t, x) \subset \nabla^+ u(t, x)$ is a simple consequence of the definition. Let us prove that $\nabla^+ u(t, x) \subset \Pi_x D^+ u(t, x)$. First, we observe that it suffices to show that

$$\nabla^* u(t, x) \subset \Pi_x D^* u(t, x). \tag{3.32}$$

To prove this inclusion, fix $p^x := \lim_{i \to \infty} \nabla u(t, x_i) \in \nabla^* u(t, x)$. By the upper semicontinuity of $D^+ u$ we can find, for any $i \in \mathbb{N}$, a point $\{(t_i, y_i)\}$ where u is differentiable and such that

$$|(t, x_i) - (t_i, y_i)| < \frac{1}{i}, \qquad Du(t_i, y_i) \in D^+ u(t, x_i) + \frac{1}{i} B_1.$$

The last property implies that $|\nabla u(t_i, y_i) - \nabla u(t, x_i)| \le 1/i$. Thus, we deduce that $(t_i, x_i) \to (t, x)$ and $\nabla u(t_i, y_i) \to p^x$, which implies that $p^x \in \Pi_x D^* u(t, x)$. ∎

We conclude the section by giving an equivalent characterization of the set $D^* u$ for a semiconcave function.

Proposition 3.3.17 *If $u : A \to \mathbb{R}$ is semiconcave then, for any $x \in A$,*

$$D^* u(x) = \left\{ \lim_{i \to \infty} p_i : p_i \in D^+ u(x_i), \ x_i \to x, \ \mathrm{diam}(D^+ u(x_i)) \to 0 \right\}. \tag{3.33}$$

Proof — Let us denote by $D^\sharp u(x)$ the right-hand side of (3.33). Since $D^+ u$ reduces to the gradient at any differentiability point of u, we have that $D^* u(x) \subset D^\sharp u(x)$. Now, to show the reverse inclusion, let us fix a point $p = \lim_{i \to \infty} p_i$, with $p_i \in D^+ u(x_i)$ and $x_i \in A$ as in (3.33). Then, for all $i \in \mathbb{N}$, let us pick any $p_i^* \in D^* u(x_i)$. Recalling the definition of $D^* u$, we can find a point $x_i^* \in A$, at which u is differentiable, such that

$$|x_i - x_i^*| + |Du(x_i^*) - p_i^*| \le \frac{1}{i}.$$

Then, $x_i^* \to x$ and

$$|Du(x_i^*) - p| \le |Du(x_i^*) - p_i^*| + |p_i^* - p_i| + |p_i - p|$$
$$\le \frac{1}{i} + \mathrm{diam}(D^+ u(x_i)) + |p_i - p| \to 0,$$

as $i \to \infty$. Hence, $p \in D^* u(x)$. ∎

3.4 Marginal functions

A function $u : A \to \mathbb{R}$ is called a *marginal function* if it can be written in the form

$$u(x) = \inf_{s \in S} F(s, x), \tag{3.34}$$

where S is some topological space and the function $F : S \times A \to \mathbb{R}$ depends smoothly on x. Functions of this kind appear often in the literature, sometimes with different names (see e.g., the *lower C^k-functions* in [123]).

Under suitable regularity assumptions for F, a marginal function is semiconcave. For instance, Corollary 2.1.6 immediately implies the following.

Proposition 3.4.1 *Let $A \subset \mathbb{R}^n$ be open and let $S \subset \mathbb{R}^m$ be compact. If $F = F(s, x)$ is continuous in $C(S \times A)$ together with its partial derivatives $D_x F$, then the function u defined in (3.34) belongs to $\mathrm{SC}_{loc}(A)$. If $D^2_{xx} F$ also exists and is continuous in $S \times A$, then $u \in \mathrm{SCL}_{loc}(A)$.*

We now show that the converse also holds.

Theorem 3.4.2 *Let $u : A \to \mathbb{R}$ be a semiconcave function. Then u can be locally written as the minimum of functions of class C^1. More precisely, for any $K \subset A$ compact, there exists a compact set $S \subset \mathbb{R}^{2n}$ and a continuous function $F : S \times K \to \mathbb{R}$ such that $F(s, \cdot)$ is C^1 for any $s \in S$, the gradients $D_x F(s, \cdot)$ are equicontinuous, and*

$$u(x) = \min_{s \in S} F(s, x), \qquad \forall x \in K. \tag{3.35}$$

If the modulus of semiconcavity of u is linear, then F can be chosen such that $F(s, \cdot)$ is C^2 for any s, with uniformly bounded C^2 norm.

Proof — Let ω be the modulus of semiconcavity of u and let ω_1 be a function such that $\omega_1(0) = 0$, that $\omega_1(r) \geq \omega(r)$ and that the function $x \to |x|\omega_1(|x|)$ belongs to $C^1(\mathbb{R}^n)$. The existence of such an ω_1 has been proved in Lemma 3.1.8. If ω is linear we simply take $\omega_1 \equiv \omega$.

Let us set $S = \{(y, p) : y \in K, p \in D^+u(y)\}$. By Proposition 3.3.4(a) and the local Lipschitz continuity of u, S is a compact set. Then we define

$$F(y, p, x) = u(y) + \langle p, x - y \rangle + |y - x|\omega_1(|y - x|).$$

Then F has the required regularity properties. In addition $F(y, p, x) \geq u(x)$ for all $(y, p, x) \in S \times K$ by Proposition 3.3.1. On the other hand, if $x \in K$, then $D^+u(x)$ is nonempty and so there exists at least a vector p such that $(x, p) \in S$. Since $F(x, p, x) = u(x)$, we obtain (3.35). ∎

If u is semiconcave with a linear modulus, then it admits another representation as the infimum of regular functions by a procedure that is very similar to the Legendre transformation.

Proposition 3.4.3 *Let* $u : \mathbb{R}^n \to \mathbb{R}$ *be semiconcave with linear modulus and semi-concavity constant C. For any given* $K > C$, *define*

$$\tilde{u}(y) = \min_{x \in \mathbb{R}^n} \left[\frac{K}{2} |x - y|^2 - u(x) \right], \qquad y \in \mathbb{R}^n. \qquad (3.36)$$

Then

$$u(x) = \min_{y \in \mathbb{R}^n} \left[\frac{K}{2} |x - y|^2 - \tilde{u}(y) \right], \qquad x \in \mathbb{R}^n.$$

Proof — The result is a direct consequence of the involutive character of the Legendre transform of convex functions (see Appendix A. 2). Observe that, by Proposition 1.1.3(c), the function $v(x) = \frac{K}{2} |x|^2 - u(x)$ is convex and superlinear. This shows in particular that the minimum in the definition of $\tilde{u}$ exists. Let us denote by v^* the Legendre transform of v. Then we have

$$v^*(p) = \max_{x \in \mathbb{R}^n} [x \cdot p - v(x)] = \max_{x \in \mathbb{R}^n} \left[x \cdot p - \frac{K}{2} |x|^2 + u(x) \right]$$

$$= \max_{x \in \mathbb{R}^n} \left[-\frac{K}{2} \left| x - \frac{p}{K} \right|^2 + \frac{|p|^2}{2K} + u(x) \right]$$

$$= -\min_{x \in \mathbb{R}^n} \left[\frac{K}{2} \left| x - \frac{p}{K} \right|^2 - u(x) \right] + \frac{|p|^2}{2K}$$

$$= -\tilde{u} \left(\frac{p}{K} \right) + \frac{|p|^2}{2K}.$$

Recalling Theorem A. 2.3–(c), we obtain

$$u(x) = \frac{K}{2} |x|^2 - v(x) = \frac{K}{2} |x|^2 - \max_{p \in \mathbb{R}^n} \left[x \cdot p - v^*(p) \right]$$

$$= \frac{K}{2} |x|^2 + \min_{p \in \mathbb{R}^n} \left[-x \cdot p - \tilde{u} \left(\frac{p}{K} \right) + \frac{|p|^2}{2K} \right]$$

$$= \min_{p \in \mathbb{R}^n} \left[\frac{K}{2} \left| \frac{p}{K} - x \right|^2 - \tilde{u} \left(\frac{p}{K} \right) \right]$$

$$= \min_{y \in \mathbb{R}^n} \left[\frac{K}{2} |y - x|^2 - \tilde{u}(y) \right]. \qquad \blacksquare$$

The function $\tilde{u}$ defined in (3.36) is called *semiconcave conjugate* of u. The above result can be generalized to cases where u is not defined in the whole space (see [79, Theorem 4.6]).

When we have a representation of the form (3.34) for a semiconcave function, it is interesting to relate the generalized differentials of u with the partial derivative $D_x F$.

Theorem 3.4.4 *Let* $A \subset \mathbb{R}^n$ *be open and let* $S \subset \mathbb{R}^m$ *be compact. Let the function* $F = F(s, x)$ *be continuous in* $S \times A$ *together with its partial derivative* $D_x F$, *and let us define* $u(x) = \min_{s \in S} F(s, x)$. *Given* $x \in A$, *let us set*

$$M(x) = \{ s \in S : u(x) = F(s, x) \}, \qquad Y(x) = \{ D_x F(s, x) : s \in M(x) \}.$$

Then, for any $x \in A$

$$D^+u(x) = \text{co}\, Y(x) \qquad (3.37)$$

$$D^-u(x) = \begin{cases} \{p\} & \text{if } Y(x) = \{p\} \\ \emptyset & \text{if } Y(x) \text{ is not a singleton.} \end{cases} \qquad (3.38)$$

Proof — From Proposition 3.4.1 we know that u is locally semiconcave in A and so we can apply the results of the previous sections. We first observe that if $s \in M(x)$, then $F(s, \cdot)$ touches u from above at the point x, and therefore $D_x F(s, x) \in D^+u(x)$. This shows that $Y(x)$ is contained in $D^+u(x)$; since $D^+u(x)$ is convex, the same holds for the convex hull of $Y(x)$.

Let us now prove that $D^+u(x)$ is contained in the convex hull of $Y(x)$. In view of (3.20), it is enough to prove that $D^*u(x) \subset Y(x)$. Let us therefore pick $p \in D^*u(x)$. Let $\{x_n\}$ be a sequence of points of differentiability of u such that $x_n \to x$ and $Du(x_n) \to p$. Let $s_n \in S$ be such that $u(x_n) = F(s_n, x_n)$. By the first part of the proof, we have $D_x F(s_n, x_n) = Du(x_n)$. By possibly extracting a subsequence, we can assume that $s_n \to \bar{s}$ for some $\bar{s} \in S$. But then we have

$$u(x) = \lim u(x_n) = \lim F(s_n, x_n) = F(\bar{s}, x),$$

$$p = \lim Du(x_n) = \lim D_x F(s_n, x_n) = D_x F(\bar{s}, x).$$

This shows that $p \in Y(x)$ and concludes the proof of (3.37). To obtain (3.38), it suffices to recall Proposition 3.3.4(d) and that $D^-u(x)$ is necessarily empty if $D^+u(x)$ contains more than one element. ∎

As an application of the above result, we can provide a description of the semidifferentials of the distance function d_S from a nonempty closed set $S \subset \mathbb{R}^n$. We denote by $\text{proj}_S(x)$ the set of closest points in S to x, i.e.,

$$\text{proj}_S(x) = \{y \in S : d_S(x) = |x - y|\} \qquad x \in \mathbb{R}^n.$$

Corollary 3.4.5 *Let S be a nonempty closed subset of $\mathbb{R}^n$. Then the following properties hold.*

(i) d_S is differentiable at $x \notin S$ if and only if $\text{proj}_S(x)$ is a singleton and in this case

$$Dd_S(x) = \frac{x - y}{|x - y|}$$

where y is the unique element of $\text{proj}_S(x)$.
(ii) If $\text{proj}_S(x)$ is not a singleton, then we have

$$D^+d_S(x) = \text{co}\left\{ \frac{x - y}{|x - y|} : y \in \text{proj}_S(x) \right\} = \frac{x - \text{co}\,(\text{proj}_S(x))}{d_S(x)} \qquad (3.39)$$

while $D^-d_S(x) = \emptyset$.

*(iii) For any $x \notin S$ and any $y \in \text{proj}_S(x)$, d_S is differentiable along the segment
   $]x, y[$.*
(iv) For any $x \notin S$

$$D^* d_S(x) = \left\{ \frac{x - y}{|x - y|} : y \in \text{proj}_S(x) \right\}. \tag{3.40}$$

Proof — We have $d_S(x) = \min_{y \in S} F(x, y)$ where $F(x, y) = |x - y|$. Observe that
for any fixed $y \in S$, F is a smooth function of x on the complement of S and has
derivative $D_x F(x, y) = (x - y)/|x - y|$. Parts (i) and (ii) are then an immediate
consequence of Theorem 3.4.4 observing that

$$M(x) = \text{proj}_S(x), \qquad Y(x) = \left\{ \frac{x - y}{|x - y|} : y \in \text{proj}_S(x) \right\}.$$

To obtain (iii) we observe that if $y \in \text{proj}_S(x)$, then $\text{proj}_S(z)$ reduces to the singleton
$\{y\}$ for all $z \in]x, y[$. Thus the assertion follows from (i).
   It remains to check (iv). In the proof of Theorem 3.4.4 we have observed that
$D^* u(x) \subset Y(x)$ for a general marginal function u. Thus, it suffices to show that
in the case of the distance function we have $Y(x) \subset D^* d_S(x)$ as well. Let us take
any $p \in Y(x)$; then $p = (x - y)/|x - y|$ for some $y \in \text{proj}_S(x)$. By part (iii),
d_S is differentiable at any $z \in]x, y[$ and satisfies $D d_S(z) = p$; this shows that
$p \in D^* d_S(x)$.   ∎

3.5 Inf-convolutions

Given $g : \mathbb{R}^n \to \mathbb{R}$ and $\varepsilon > 0$, the functions

$$x \to \inf_{y \in \mathbb{R}^n} \left(g(y) + \frac{|x - y|^2}{2\varepsilon} \right) \qquad x \to \sup_{y \in \mathbb{R}^n} \left(g(y) - \frac{|x - y|^2}{2\varepsilon} \right)$$

are called inf- and sup-convolutions of g respectively, due to the formal analogy with
the usual convolution. They have been used in various contexts as a way to approximate g; one example is the uniqueness theory for viscosity solutions of Hamilton–
Jacobi equations. In some cases it is useful to consider more general expressions,
where the quadratic term above is replaced by some other coercive function. In this
section we analyze such general convolutions, showing that their regularity properties are strictly related with the properties of semiconcave functions studied in the
previous sections.

Definition 3.5.1 *Let $g \in C(\mathbb{R}^n)$ satisfy*

$$|g(x)| \le K(1 + |x|) \tag{3.41}$$

for some $K > 0$ and let $\phi \in C(\mathbb{R}^n)$ be such that

$$\lim_{|q|\to+\infty} \frac{\phi(q)}{|q|} = +\infty. \tag{3.42}$$

The inf-convolution *of g with kernel ϕ is the function*

$$g_\phi(x) = \inf_{y\in\mathbb{R}^n} [g(y) + \phi(x - y)], \tag{3.43}$$

while the sup-convolution *of g with kernel ϕ is defined by*

$$g^\phi(x) = \sup_{y\in\mathbb{R}^n} [g(y) - \phi(x - y)]. \tag{3.44}$$

We observe that the function u given by Hopf's formula (1.10) is an inf-convolution with respect to the x variable for any fixed t. In addition, inf-convolutions are a particular case of the marginal functions introduced in the previous section.

We give below some regularity properties of the inf-convolutions. The corresponding statements about the sup-convolutions are easily obtained observing that $g^\phi = -((-g)_\phi)$.

Lemma 3.5.2 *Let g and ϕ be as in Definition 3.5.1. Then the infimum in formula* (3.43) *is a minimum. In addition, for any $r > 0$ there exists $R > 0$ such that, if $|x| \leq r$, then any y at which the minimum in* (3.43) *is attained satisfies $|y| < R$.*

Proof — Given $r > 0$, let $R > \max\{1, 2r\}$ be such that

$$\frac{\phi(q)}{|q|} \geq 3K, \qquad \forall q : |q| \geq R - r,$$

where K is the constant which appears in (3.41). Such an R exists by assumption (3.42); in addition, we can choose it large enough to have

$$KR > \phi(0) + 3Kr + \max_{x\in\overline{B}_r} g(x).$$

Given any x, y with $|x| < r$ and $|y| > R$ we have

$$|y - x| \geq |y| - |x| > R - r,$$

and so we can estimate

$$\begin{aligned}
g(y) + \phi(x - y) &\geq -K(1 + |y|) + \frac{\phi(x - y)}{|x - y|}|x - y| \\
&> -2K|y| + 3K(|y| - |x|) \\
&> K|y| - 3K|x| > KR - 3Kr \\
&> \phi(0) + g(x).
\end{aligned}$$

This shows that the function $y \to g(y) + \phi(x - y)$ attains at $y = x$ a value which is smaller than any value assumed for $|y| > R$. Therefore,

$$g(y) + \phi(x - y) > \min_{z\in\overline{B}_R} [g(z) + \phi(x - z)], \qquad \forall y : |y| > R,$$

and this proves the lemma. ∎

Theorem 3.5.3 *Let g and ϕ be as in Definition 3.5.1. Then the following properties hold.*

(i) If either ϕ or g is semiconcave with modulus ω, then the same holds for g_ϕ.
(ii) If g is Lipschitz continuous with constant L_g, then so is g_ϕ.
(iii) If g and ϕ are both convex, then so is g_ϕ.
(iv) Suppose that ϕ is uniformly convex, that is, that there exists $C > 0$ such that $\phi(x) - \frac{C}{2}|x|^2$ is convex. Suppose also that g is semiconvex with constant B for some $B < C$. Then g_ϕ is semiconvex with constant $BC(C - B)^{-1}$.

Proof — If ϕ is semiconcave, then g_ϕ is semiconcave with the same modulus as a consequence of Proposition 2.1.5. If g is semiconcave we obtain the same conclusion observing that

$$g_\phi(x) = \inf_{y \in \mathbb{R}^n} [g(y) + \phi(x - y)] = \inf_{z \in \mathbb{R}^n} [g(x - z) + \phi(z)].$$

To prove (ii) we can take any $x, h \in \mathbb{R}^n$ and estimate

$$
\begin{aligned}
g_\phi(x + h) &= \min_{y \in \mathbb{R}^n} [g(y) + \phi(x + h - y)] \\
&= \min_{z \in \mathbb{R}^n} [g(z + h) + \phi(x - z)] \\
&\geq -L_g|h| + \min_{z \in \mathbb{R}^n} [g(z) + \phi(x - z)] \\
&= g_\phi(x) - L_g|h|.
\end{aligned}
$$

We now turn to (iii) and (iv). We take $x_1, x_2 \in \mathbb{R}^n$, and denote by y_i the points where the infimum in (3.43) with $x = x_i$ is attained, for $i = 1, 2$. We find by definition

$$
\begin{aligned}
& g_\phi(x_1) + g_\phi(x_2) - 2g_\phi\left(\frac{x_1 + x_2}{2}\right) \\
&\geq \phi(x_1 - y_1) + \phi(x_2 - y_2) - 2\phi\left(\frac{x_1 + x_2}{2} - \frac{y_1 + y_2}{2}\right) \\
& + g(y_1) + g(y_2) - 2g\left(\frac{y_1 + y_2}{2}\right).
\end{aligned}
\tag{3.45}
$$

In case (iii) the above quantity is nonnegative, and this shows that g_ϕ is convex. In case (iv) we have, using our assumption for ϕ, that

$$
\begin{aligned}
& \phi(x_1 - y_1) + \phi(x_2 - y_2) - 2\phi\left(\frac{x_1 + x_2}{2} - \frac{y_1 + y_2}{2}\right) \\
&\geq \frac{C}{2}\left(|x_1 - y_1|^2 + |x_2 - y_2|^2 - 2\left|\frac{x_1 + x_2}{2} - \frac{y_1 + y_2}{2}\right|^2\right) \\
&= \frac{C}{4}(|x_1 - x_2|^2 + |y_1 - y_2|^2 - 2\langle x_2 - x_1, y_2 - y_1\rangle),
\end{aligned}
$$

$$\geq \frac{C}{4} \left(|x_1 - x_2|^2 + |y_1 - y_2|^2 - \frac{C}{C-B}|x_1 - x_2|^2 - \frac{C-B}{C}|y_1 - y_2|^2 \right)$$

$$= -\frac{BC}{C-B} \left| \frac{x_1 - x_2}{2} \right|^2 + B \left| \frac{y_1 - y_2}{2} \right|^2 .$$

On the other hand, since g is semiconvex with constant B, we have

$$g(y_1) + g(y_2) - 2g \left(\frac{y_1 + y_2}{2} \right) \geq -B \left| \frac{y_1 - y_2}{2} \right|^2 ,$$

and so we deduce from (3.45) that

$$g_\phi(x_1) + g_\phi(x_2) - 2g_\phi \left(\frac{x_1 + x_2}{2} \right) \geq -\frac{BC}{C-B} \left| \frac{x_1 - x_2}{2} \right|^2 ,$$

which proves (v).   ∎

From general results for marginal functions we can deduce the following description of the semidifferentials of an inf-convolution.

Proposition 3.5.4 *Let g, ϕ be as in Definition 3.5.1, and suppose in addition that ϕ is of class C^1. Given $\bar{x} \in \mathbb{R}^n$, let us denote by $Y(\bar{x})$ the set of the points where the minimum in (3.43) with $x = \bar{x}$ is attained. Then g_ϕ is differentiable at $\bar{x}$ if and only if $Y(\bar{x})$ is a singleton. More precisely, if $Y(\bar{x}) = \{\bar{y}\}$ we have $Dg_\phi = D\phi(\bar{y} - \bar{x})$, while if $Y(\bar{x})$ contains more than one point we have*

$$D^+ g_\phi(\bar{x}) = \mathrm{co}\{D\phi(y - \bar{x}) \,:\, y \in Y(\bar{x})\}, \qquad D^- g_\phi(\bar{x}) = \emptyset.$$

Proof — The result is a consequence of Theorem 3.4.4, after recalling that, by Lemma 3.5.2, inf-convolutions can be locally written as an infimum over a compact ball B_R.   ∎

We now give an example of how inf- and sup-convolutions can be used to approximate a given function g. We restrict ourselves to the case when g is Lipschitz continuous, but we point out that such convolutions have good approximating properties also in more general cases, for instance when g is just continuous or semicontinuous, or when g is a function defined on a Hilbert space.

Definition 3.5.5 *Given $g \in \mathrm{Lip}\,(\mathbb{R}^n)$, $\phi \in C(\mathbb{R}^n)$ satisfying (3.42) and $\varepsilon > 0$ we define*

$$g_\varepsilon(x) = \min_{y \in \mathbb{R}^n} \left[g(y) + \varepsilon\phi \left(\frac{x - y}{\varepsilon} \right) \right], \qquad x \in \mathbb{R}^n, \tag{3.46}$$

$$g^\varepsilon(x) = \max_{y \in \mathbb{R}^n} \left[g(y) - \varepsilon\phi \left(\frac{x - y}{\varepsilon} \right) \right], \qquad x \in \mathbb{R}^n. \tag{3.47}$$

Remark 3.5.6 If ϕ is convex, the above definitions are related to the theory of Hamilton–Jacobi equations. To see this, let us denote by H the Legendre transform of ϕ. Then, recalling Hopf's formula, $g_\varepsilon = u(\varepsilon, \cdot)$, where u is the solution of the equation $u_t + H(\nabla u) = 0$ with initial value $u(0, \cdot) = g$. Analogously, g^ε is the solution of $u_t - H(\nabla u) = 0$, with initial value g, evaluated at time $t = \varepsilon$.

In what follows we will also consider compositions of the form $(g^\varepsilon)_\delta$, $(g_\varepsilon)^\delta$, which are called *sup-inf-* and *inf-sup-convolutions*. Let us observe a preliminary property.

Lemma 3.5.7 *Let us denote by $M_\varepsilon(x)$ and $M^\varepsilon(x)$ the set of points where the minimum in (3.46) and the maximum in (3.47), respectively, is attained. Then there exists $R > 0$, depending only on ϕ and on the Lipschitz constant of g, such that any point $y \in M_\varepsilon(x) \cup M^\varepsilon(x)$ satisfies $|y - x| < \varepsilon R$.*

Proof — Let us denote by L_g the Lipschitz constant of g and let R be such that $(\phi(x) - \phi(0))/|x| > L_g$ for $|x| > R$. Such an R exists by assumption (3.42). Given $x \in \mathbb{R}^n$, suppose that $y \in M_\varepsilon(x)$. Then we have

$$g(y) + \varepsilon\phi\left(\frac{x - y}{\varepsilon}\right) \le g(x) + \varepsilon\phi(0)$$
$$\le g(y) + L_g|x - y| + \varepsilon\phi(0),$$

which implies

$$L_g\frac{|x - y|}{\varepsilon} \ge \phi\left(\frac{x - y}{\varepsilon}\right) - \phi(0).$$

By definition of R, we obtain that $|x - y| \le \varepsilon R$. The proof for $M^\varepsilon(x)$ is analogous. ∎

Theorem 3.5.8 *Let $g \in \mathrm{Lip}\,(\mathbb{R}^n)$, and let $\phi \in C(\mathbb{R}^n)$ satisfy (3.42). Then we have, with the notation of Definition 3.5.5,*

$$\lim_{\varepsilon \to 0^+} (g^\varepsilon)(x) = \lim_{\varepsilon \to 0^+} (g_\varepsilon)(x) = g(x), \qquad x \in \mathbb{R}^n \text{ uniformly} \tag{3.48}$$

$$\lim_{\varepsilon,\delta \to 0^+} (g^\varepsilon)_\delta(x) = \lim_{\varepsilon,\delta \to 0^+} (g_\varepsilon)^\delta(x) = g(x), \qquad x \in \mathbb{R}^n \text{ uniformly}. \tag{3.49}$$

Suppose in addition that ϕ is convex and belongs to $C^2(\mathbb{R}^n)$. Let $R > 0$ be as in the previous lemma and let $\lambda, \Lambda > 0$ be such that

$$\lambda I \le D^2\phi(q) \le \Lambda I, \qquad \forall q \in B_R. \tag{3.50}$$

Then, for any $\varepsilon > 0$, g_ε (resp. g^ε) is semiconcave (resp. semiconvex) with semiconcavity constant Λ/ε. In addition, if $\delta < \Lambda^{-1}\lambda\varepsilon$, the functions $(g^\varepsilon)_\delta$ and $(g_\varepsilon)^\delta$ are of class $C^{1,1}$.

Proof — For given $x \in \mathbb{R}^n$ and $\varepsilon > 0$, let us take $y_\varepsilon \in M_\varepsilon(x)$. Recalling the definition of g_ε and Lemma 3.5.7 we find that

$$\varepsilon\phi(0) \geq g_\varepsilon(x) - g(x) \geq g(y_\varepsilon) - g(x) + \varepsilon \min \phi$$
$$\geq -L_g|y_\varepsilon - x| + \varepsilon \min \phi \geq -\varepsilon(L_g R - \min \phi),$$

and therefore, setting $M = \max\{|\phi(0)|, |L_g R - \min\phi|\}$, we have that $|g(x) - g_\varepsilon(x)| \leq \varepsilon M$ for any x. Similarly, we find $|g(x) - g^\varepsilon(x)| \leq \varepsilon M$. Since g_ε has the same Lipschitz constant of g, we also find that

$$|g(x) - (g_\varepsilon)^\delta(x)| \leq |g(x) - g_\varepsilon(x)| + |g_\varepsilon(x) - (g_\varepsilon)^\delta(x)| \leq (\varepsilon + \delta)M,$$

and similarly for $(g^\varepsilon)_\delta$. This proves the first part of the statement.

The assertion about the semiconcavity of g_ε and the semiconvexity of g^ε follows from Proposition 1.1.3, Theorem 3.5.3(i), and Lemma 3.5.7.

Let us now consider the inf-sup convolution $(g_\varepsilon)^\delta$. From the previous property we know that $(g_\varepsilon)^\delta$ is semiconvex with constant Λ/δ and that g_ε is semiconcave with modulus Λ/ε. We now apply Theorem 3.5.3–(iv) with $\phi(x)$ replaced by $\delta\phi(x/\delta)$, and g replaced by $-(g_\varepsilon)$. Our assumption on δ implies that the hypotheses of the theorem are satisfied with $C = \lambda/\delta$ and $B = \Lambda/\varepsilon$. Thus we obtain that $(g_\varepsilon)^\delta = (-(g_\varepsilon))_\delta$ is semiconvex with linear modulus. Then, by Corollary 3.3.8, we conclude that $(g_\varepsilon)^\delta$ is of class $C^{1,1}$. ∎

Remark 3.5.9 If we take the standard kernel $\phi(x) = |x|^2/2$, then (3.50) holds with $\lambda = \Lambda = 1$, and so $(g_\varepsilon)^\delta$ is smooth whenever $0 < \delta < \varepsilon$. One may wonder whether the same holds for a general smooth convex superlinear ϕ. As observed in Remark 3.5.6, in this case inf- and sup-convolutions have an interpretation in terms of Hamilton-Jacobi equations. The smoothness property conjectured here would show an interesting connection between nondifferentiability of solutions and irreversibility in time. This property has been proved in [23] in the one-dimensional case, while it remains a conjecture when $n > 1$.

3.6 Proximal analysis and semiconcavity

We introduce now another generalized gradient whose definition relies on the existence of a parabola touching from above or below the function under consideration.

Definition 3.6.1 *Let $u : \Omega \to \mathbb{R}$, with $\Omega \subset \mathbb{R}^n$ open. We say that $p \in \mathbb{R}^n$ is a proximal subgradient (resp. proximal supergradient) of u at $x \in \Omega$ if there exists $\sigma > 0$ such that*

$$u(y) \geq u(x) + p \cdot (y - x) - \sigma|x - y|^2$$
$$(resp.\ u(y) \leq u(x) + p \cdot (y - x) + \sigma|x - y|^2)$$

for all y in a neighborhood of x.

Clearly, a proximal subgradient (supergradient) belongs to $D^- u(x)$ (resp. to $D^+ u(x)$) but the converse is not true in general. The two notions coincide, however, for a semiconvex (semiconcave) function with a linear modulus.

Proposition 3.6.2 *Let $u \in \text{SCL}(\Omega)$, with $\Omega \subset \mathbb{R}^n$ open. Then p is a proximal supergradient of u at $x \in \Omega$ if and only if $p \in D^+ u(x)$.*

Proof — The assertion is a direct consequence of Proposition 3.3.1.   ∎

Definition 3.6.3 *Let $S \subset \mathbb{R}^n$. A vector $v \in \mathbb{R}^n$ is called a* perpendicular *(or* proximal normal*) to S at a point $\bar{x} \in \partial S$ if, for some $\lambda > 0$,*

$$d_S(\bar{x} + \lambda v) = \lambda |v|, \tag{3.51}$$

that is, the open ball with center $\bar{x} + \lambda v$ and radius $\lambda |v|$ does not intersect S. We denote by $N_S(\bar{x})$ the set of perpendiculars to S at $\bar{x}$.

Remark 3.6.4 It is immediate from the definition that $N_S(\bar{x})$ is a cone. It coincides with the half-line parallel to the standard outer normal if ∂S is a C^2 manifold. Property (3.51) is equivalent to saying that $\bar{x}$ is a projection of $\bar{x} + \lambda v$ onto S. It is easy to see that if (3.51) holds, then for any $\alpha \in]0, \lambda[$ the closed ball with center $\bar{x} + \alpha v$ and radius $\alpha |v|$ intersects S only at $\bar{x}$, that is, $\bar{x}$ is the unique projection of $\bar{x} + \alpha v$ onto S.

Definition 3.6.5 *A nonempty set $S \subset \mathbb{R}^n$ is called* proximally smooth *with radius $r > 0$ if all unit vectors $v \in N_S(\bar{x})$ are such that $B_r(\bar{x} + rv) \cap S = \emptyset$.*

Sets of the above form, which admit different equivalent definitions, have been widely studied in the literature (see e.g., the *sets with positive reach* in [75]). It turns out that these sets are exactly the ones possessing a neighborhood where the distance function is differentiable (see Theorems 3.6.7, 3.6.8). We call such sets "proximally smooth," following [56] where the equivalence between various definitions is proved.

By Proposition 3.3.1, the graph of a semiconcave function with linear modulus can be touched from above at each point by a paraboloid with fixed curvature. Thus it is natural to expect that semiconcavity is related to the proximal smoothness of the subgraph. The next result shows that this is the case.

Theorem 3.6.6 *Let $u \in \text{Lip}(\Omega)$, with $\Omega \subset \mathbb{R}^n$ open, convex and bounded. Then u is semiconcave with a linear modulus if and only if its subgraph is proximally smooth.*

Proof — See [56, Theorem 5.2].   ∎

The Lipschitz continuity assumption in the above statement is essential. Roughly speaking, the correspondence between tangent spheres to a graph and tangent paraboloids breaks down if the graph becomes arbitrarily steep. In fact, one can check that the function $u(x) = -\sqrt{|x|}$, $x \in \mathbb{R}$, has a proximally smooth subgraph but is not locally Lipschitz (and thus it is not semiconcave).

The proximal smoothness property is a possible way to characterize the sets S possessing a neighborhood N such that the distance function d_S is differentiable in $N \setminus S$. More precisely, for a given $r > 0$, let us set

$$S_r = \{x \; : \; 0 < d_S(x) < r\}.$$

Then we have the following result.

Theorem 3.6.7 *Let a closed set S be proximally smooth with radius r. Then $d_S \in C^{1,1}_{loc}(S_r)$.*

Proof — From Proposition 2.2.2-(ii) we know that, for a general set S, d_S is locally semiconcave with a linear modulus on the complement of S. It suffices to show that under the hypothesis of proximal smoothness, d_S is semiconvex with a linear modulus on S_r; then the $C^{1,1}$ regularity will follow from Theorem 3.3.7. To this end, let us set $K(r) = \{x \; : \; d_S(x) \geq r\}$. The proximal smoothness of S easily implies that $K(r)$ is nonempty. We claim that

$$d_S(y) + d_{K(r)}(y) = r, \qquad \forall y \in S_r. \tag{3.52}$$

To see this, take $y \in S_r$, and let $\bar{x}$ and $\hat{x}$ denote a projection of y onto S and $K(r)$ respectively. Then we have

$$d_S(y) + d_{K(r)}(y) = |\bar{x} - y| + |\hat{x} - y| \geq |\bar{x} - \hat{x}| \geq r. \tag{3.53}$$

On the other hand, since $\bar{x}$ is a projection of y onto S, we have $y - \bar{x} \in N_S(\bar{x})$. Let us set

$$x^* = \bar{x} + r \frac{y - \bar{x}}{|y - \bar{x}|}.$$

By the proximal smoothness assumption, the ball $B_r(x^*)$ does not intersect S. Hence $d_S(x^*) \geq r$; in fact, $d_S(x^*) = r$, observing that $|x^* - \bar{x}| = r$. Since the three points $\bar{x}$, y and x^* are on the same line, we have

$$|x^* - y| = |x^* - \bar{x}| - |y - \bar{x}| = r - d_S(y).$$

Recalling that $x^* \in K(r)$ we find that

$$d_{K(r)}(y) \leq |x^* - y| = r - d_S(y),$$

which, together with (3.53), proves (3.52). We know from Proposition 2.2.2–(ii) that $d_{K(r)}$ is locally semiconcave with a linear modulus on the complement of $K(r)$, and so $d_S = r - d_{K(r)}$ is locally semiconvex on S_r. This proves the theorem. ∎

The converse implication is also true (see [56, Theorem 4.1]).

Theorem 3.6.8 *If the distance function to a closed set S is differentiable in S_r for some $r > 0$, then S is proximally smooth with radius r.*

For other characterizations of proximally smooth sets (also in infinite dimensional spaces) see e.g., [56, 117] and the references therein.

Bibliographical notes

The notion of superdifferential is classical in convex analysis. Subsequently, several kinds of generalized gradients have been introduced in the framework of nonsmooth analysis, see e.g., [54, 55]. The importance of the Fréchet semidifferentials for the study of Hamilton–Jacobi equations was pointed out in [65].

The results of Sections 3.2 and 3.3 are well known from convex analysis in the case of concave functions. The proofs given here for general semiconcave functions follow [45, 46, 32].

Properties of pointwise infima of functions have been widely studied in the literature (see e.g., Section 2.8 in [54] or Section II.2 in [20] and the references therein). In our exposition we have emphasized the relation with the semiconcavity property, showing how some results about marginal functions can be derived from the general properties of semiconcave functions. Let us also recall that the equivalence between the monotonicity properties of the Clarke differential and the property of being the infimum of smooth functions has been analyzed by Spingarn [132] and Rockafellar [123].

The duality representation of Proposition 3.4.3 was derived by Fleming and McEneaney in [79] and applied in the analysis of the so-called filtering problem for nonlinear systems.

A detailed treatment of inf–convolutions with a quadratic kernel in Hilbert spaces is given in [105]. These convolutions have been applied to obtain uniqueness results for viscosity solutions of Hamilton–Jacobi equations (see [98], [66], [81]). Convolutions with more general kernels have been considered in [96].

The notion of perpendicular, or proximal normal, of Section 3.6 already appears in [54]; it has played an important role in the recent developments of nonsmooth analysis [55]. Sets possessing a neighborhood where the distance function is differentiable have been studied by many authors during the last decades, see the references in [56, 117]. Let us also mention that the regularity of the tubular neighborhoods of sets has been investigated in [82] which exploits the semiconcavity properties of the distance function.

4

Singularities of Semiconcave Functions

By a *singular point*, or *singularity*, of a semiconcave function u we mean a point where u is not differentiable. This chapter is devoted to the analysis of the set of all singular points for u, which is called *singular set* and is denoted here by $\Sigma(u)$. As we have already remarked, the singular set of a semiconcave function has zero measure by Rademacher's theorem. However, we will see that much more detailed properties can be proved.

In Section 4.1 we study the rectifiability properties of the singular set. We divide the singular points according to the dimension of the superdifferential of u denoting by $\Sigma^k(u)$ the set of points x such that $D^+u(x)$ has dimension k. Then we show that $\Sigma^k(u)$ is countably $(n-k)$-rectifiable for all integers $k = 1, \ldots, n$. In particular, the whole singular set $\Sigma(u)$ is countably $(n-1)$-rectifiable.

Sections 4.2 and 4.3 are devoted to the *propagation of singularities* for semiconcave functions: given a singular point x_0, we look for conditions ensuring that x_0 belongs to a connected component of dimension $\nu \geq 1$ of the singular set. We study first the propagation along Lipschitz arcs and then along Lipschitz manifolds of higher dimension. In general we find that a sufficient condition for the propagation of singularities from x_0 is that the inclusion $D^*u(x_0) \subset \partial D^+u(x_0)$ (see Proposition 3.3.4-(b)) is strict.

As an application of the previous analysis, we study in Section 4.4 some properties of the distance function from a closed set S. Using our propagation results, we show that the distance function has no isolated singularities except for the special case when the singularity is the center of a spherical connected component of the complement of S. In general, we show that a point x_0 which is singular for the distance function belongs to a connected set of singular points whose Hausdorff dimension is at least $n - k$, with $k = \dim(D^+d(x_0))$.

4.1 Rectifiability of the singular sets

Throughout this chapter $\Omega \subset \mathbb{R}^n$ is an open set and $u : \Omega \to \mathbb{R}$ is a semiconcave function. We denote by $\Sigma(u)$ the set of points of Ω where u is not differentiable and we call it the *singular set* of u. In the following we use some notions from measure theory, like Hausdorff measures and rectifiable sets, which are recalled in Appendix A. 3.

We know from Theorem 2.3.1–(ii) that Du is a function with bounded variation if u is semiconcave with a linear modulus. For functions of bounded variation one can introduce the *jump set*, whose rectifiability properties have been widely studied (see Appendix A. 6). We now show that the jump set of Du coincides with the singular set $\Sigma(u)$. To this purpose we need two preliminary results. The first one is a lemma about approximate limits (see Definition A. 6.2).

Lemma 4.1.1 *Let* $w \in L^1(A)$, *with* $A \subset \mathbb{R}^n$ *open, let* $\bar{x} \in A$, *and let* ap $\lim_{x \to \bar{x}} w(x) = \bar{w}$. *Then for any* $\theta \in \mathbb{R}^n$ *with* $|\theta| = 1$ *we can find a sequence* $\{x_k\} \subset A$ *such that*

$$x_k \to \bar{x}, \quad \frac{x_k - \bar{x}}{|x_k - \bar{x}|} \to \theta, \quad w(x_k) \to \bar{w} \qquad as \ k \to \infty.$$

Proof — For any $k \in \mathbb{N}$, let us define

$$A_k = \left\{ x \in A \setminus \{\bar{x}\} : \ \left| \frac{x - \bar{x}}{|x - \bar{x}|} - \theta \right| < \frac{1}{k} \right\}.$$

Any such set A_k is the intersection of A with an open cone of vertex $\bar{x}$. Therefore

$$\lim_{\rho \to 0^+} \frac{\text{meas} (B_\rho(\bar{x}) \cap A_k)}{\rho^n} > 0.$$

By the definition of approximate limit we have

$$\lim_{\rho \to 0^+} \frac{\text{meas} (\{x \in B_\rho(\bar{x}) \cap A_k \ : \ |w(x) - \bar{w}| > 1/k\})}{\rho^n} = 0.$$

Comparing the above relations we see that the set

$$\{x \in B_\rho(\bar{x}) \cap A_k \ : \ |w(x) - \bar{w}| \leq 1/k\}$$

is nonempty if ρ is small enough. Thus, we can find $x_k \in A_k$ such that $|w(x_k) - \bar{w}| \leq 1/k$, $|x_k - \bar{x}| \leq 1/k$. Repeating this construction for all k we obtain a sequence $\{x_k\}$ with the desired properties. ∎

Next we give a result showing, roughly speaking, that for the gradient of a semiconcave function the notions of limit and of approximate limit coincide.

Theorem 4.1.2 *Let* $u : \Omega \to \mathbb{R}$ *be a semiconcave function, with* Ω *open. Then the following properties are equivalent for a point* $y \in \Omega$:

(i) ap $\lim_{y \to x} Du(y) = p_1$;

(ii) $\lim_{\substack{y \to x \\ y \notin \Sigma(u)}} Du(y) = p_1$;

(iii) u is differentiable at x and $Du(x) = p_1$.

Proof — Property (iii) implies (ii) thanks to Proposition 3.3.4(a), while property (ii) implies (i) because the set $\Sigma(u)$ has zero measure. Therefore it suffices to prove that property (i) implies (iii).

Let us suppose that $\operatorname{ap\,lim}_{y \to x} Du(y) = p_1$. Then $p_1 \in D^*u(x)$. If we show that $D^*u(x)$ contains no other elements, then we obtain (iii), thanks to property (3.20) and to Proposition 3.3.4(d). We argue by contradiction and suppose that there is $p_2 \in D^*u(x)$, $p_2 \neq p_1$. By the previous lemma, we can find a sequence $\{y_k\} \subset A$ such that u is differentiable at y_k and such that

$$y_k \to x, \quad \frac{y_k - x}{|y_k - x|} \to \frac{p_1 - p_2}{|p_1 - p_2|}, \quad Du(y_k) \to p_1 \qquad \text{as } k \to \infty.$$

Then, by Proposition 3.3.10,

$$\langle Du(y_k) - p_2, y_k - x \rangle \leq 2\omega(|y_k - x|)|y_k - x|.$$

Dividing by $|y_k - x|$ and letting $k \to \infty$ we obtain $|p_1 - p_2| \leq 0$, which is a contradiction. ∎

The above results allow us to obtain a first rectifiability result for the singular set of a semiconcave function with linear modulus.

Proposition 4.1.3 *Suppose that $u : \Omega \to \mathbb{R}$ is semiconcave with a linear modulus. Then $\Sigma(u)$ coincides with the jump set S_{Du} of the gradient Du considered as a function of $BV_{loc}(\Omega, \mathbb{R}^n)$ and is a countably $\mathcal{H}^{n-1}$-rectifiable set.*

Proof — It is a direct consequence of Theorem 2.3.1-(ii), Theorem 4.1.2 and of general results on BV functions (Theorem A. 6.5). ∎

Using completely independent techniques, we will now obtain a rectifiability result which is more detailed than the one above and which holds for semiconcave functions with arbitrary modulus. Our strategy will be to derive first a rectifiability criterion (Theorem 4.1.6) and then show that this criterion can be applied to $\Sigma(u)$.

In the following we call *k-planes* the k-dimensional subspaces of $\mathbb{R}^n$. Given a k-plane Π, we denote by $\Pi^\perp$ its orthogonal complement. Given $x \in \mathbb{R}^n$, we denote by $\pi(x)$ and $\pi^\perp(x)$ the orthogonal projections of x onto Π and $\Pi^\perp$ respectively. If we have several k-planes $\Pi_1, \ldots, \Pi_r$, we will denote by $\pi_i(x), \pi_i^\perp(x)$ the projections onto $\Pi_i, \Pi_i^\perp$ for any $i = 1, \ldots, r$

Let $A \subset \Pi$, where $\Pi \subset \mathbb{R}^n$ is a k-plane and let $\phi : A \to \Pi^\perp$ be a Lipschitz continuous map. Then the graph

$$\Gamma_\phi := \{x \in \mathbb{R}^n \,:\, \pi(x) \in A, \, \pi^\perp(x) = \phi(\pi(x))\} \tag{4.1}$$

is clearly a k-rectifiable set (see Definition A. 3.4).

Definition 4.1.4 *Let* $\Pi \subset \mathbb{R}^n$ *be a k-plane and let* $M > 0$. *The set*

$$K_M(\Pi) = \{x \in \mathbb{R}^n \ : \ |\pi^\perp(x)| \le M|\pi(x)|\}$$

is called a cone of axis Π *and width* M.

Proposition 4.1.5 *Given a set* $\Gamma \subset \mathbb{R}^n$, *a k-plane* Π *and* $M > 0$, *the two following properties are equivalent.*

(i) *There exists* $A \subset \Pi$ *and a Lipschitz function* $\phi : A \to \Pi^\perp$ *with* $\mathrm{Lip}\,(\phi) \le M$
such that Γ *coincides with the graph* Γ_ϕ *defined in* (4.1).
(ii) *For any* $x \in \Gamma$, *we have* $\Gamma \subset x + K_M(\Pi)$.

Proof — Suppose that (i) holds and let x, x' be any two points of Γ. Then we have

$$|\pi^\perp(x' - x)| = |\phi(\pi(x')) - \phi(\pi(x))| \le M|\pi(x' - x)|,$$

which means that $x' \in x + K_M(\Pi)$. Therefore, (ii) holds.

Conversely, suppose that (ii) is satisfied. Then we have, for any $x, x' \in \Gamma$,

$$|\pi^\perp(x' - x)| \le M|\pi(x' - x)|.$$

Thus, if $x, x' \in \Gamma$ and $\pi(x) = \pi(x')$, then $x = x'$. It follows that for any $y \in \pi(\Gamma)$, there exists a unique $z \in \Pi^\perp$ such that $y + z \in \Gamma$. Setting $A = \pi(\Gamma)$ and $z = \phi(y)$ we have that ϕ is Lipschitz continuous with Lipschitz constant M and that $\Gamma = \Gamma_\phi$. ∎

We now give our rectifiability criterion.

Theorem 4.1.6 *Let* $S \subset \mathbb{R}^n$ *have the following property: for all* $x \in S$ *there exist constants* $\rho_x > 0$, $M_x \ge 0$ *and a k-plane* Π_x *such that*

$$S \cap B_{\rho_x}(x) \subset x + K_{M_x}(\Pi_x). \tag{4.2}$$

Then S is countably k-rectifiable.

Proof — Let us set $\alpha = \sup_{x \in S} M_x$. It is not restrictive to assume that α is finite, since otherwise we can split S in a countable union of subsets with this property.

Let us now set, for any positive integer i,

$$S_i = \left\{x \in S \ : \ \rho_x > \frac{1}{i}\right\},$$

and let us prove the rectifiability of S_i. Let us fix $\delta > 0$ small enough to have

$$\delta(\alpha + 1) < 1, \qquad \frac{\alpha + \delta(\alpha + 1)}{1 - \delta(\alpha + 1)} \le 2\alpha. \tag{4.3}$$

Let us now choose a family $\{\Pi_1, \dots, \Pi_N\}$ of k-planes with the property that given any other k-plane Π, we have

$$\min_{1 \leq j \leq N} \|\pi_j - \pi\| < \delta. \tag{4.4}$$

In this formula we have denoted by $\|\pi_j - \pi\|$ the operator norm of the linear function $\pi_j - \pi$. To see that such a family $\{\Pi_1, \ldots, \Pi_N\}$ exists, it suffices to observe that any projection π satisfies $\|\pi\| = 1$, and thus the set of all projections is a bounded subset of $\mathcal{L}(\mathbb{R}^n, \mathbb{R}^n)$.

Let us define, for $j = 1, \ldots, N$,

$$S_{ij} = \{x \in S_i \ : \ \|\pi_x - \pi_j\| < \delta\}.$$

We claim that any set $T \subset S_{ij}$ with diameter less than $1/i$ is contained in a k-dimensional Lipschitz graph with constant not greater than 2α. This proves the theorem, since by (4.4) the sets S_{ij} cover S_i.

To prove our claim, let us take $x, x' \in T$. Then, using the definition of S_{ij} and assumption (4.2) we obtain

$$
\begin{aligned}
|\pi_j^\perp(x - x')| &= |x - x' - \pi_j(x - x')| \\
&= |\pi_x^\perp(x - x') + \pi_x(x - x') - \pi_j(x - x')| \\
&\leq |\pi_x^\perp(x - x')| + \delta|x - x'| \leq \alpha|\pi_x(x - x')| + \delta|x - x'| \\
&\leq \alpha|\pi_j(x - x')| + \delta(\alpha + 1)|x - x'| \\
&\leq [\alpha + \delta(\alpha + 1)]|\pi_j(x - x')| + \delta(\alpha + 1)|\pi_j^\perp(x - x')|.
\end{aligned}
$$

Thus, by inequality (4.3),

$$|\pi_j^\perp(x - x')| \leq 2\alpha|\pi_j(x - x')|$$

and so $T \subset x + K_{2\alpha}(\Pi_j)$. Our claim follows from Proposition 4.1.5. ∎

We now show that the above criterion can be applied to the analysis of the singular set of a semiconcave function. Let us first recall a definition from nonsmooth analysis.

Definition 4.1.7 *Let $S \subset \mathbb{R}^n$ and $x \in \overline{S}$ be given. The* contingent cone *(or Bouligand's tangent cone) to S at x is the set*

$$T(x, S) = \left\{ \lim_{i \to \infty} \frac{x_i - x}{t_i} \ : \ x_i \in S, x_i \to x, t_i \in \mathbb{R}_+, t_i \downarrow 0 \right\}.$$

The vector space generated by $T(x, S)$ is called the tangent space *to S at x and is denoted by $\mathrm{Tan}(x, S)$.*

Remark 4.1.8 It is easily checked that $T(x, S)$ is a cone and that $\mathrm{Tan}(x, S)$ is in general strictly larger than $T(x, S)$. For instance, if

$$S = \{(x, y) \in \mathbb{R}^2 \ : \ xy = 0\},$$

then $T((0, 0), S) = S$, while $\mathrm{Tan}((0, 0), S) = \mathbb{R}^2$.

Corollary 4.1.9 *Let $S \subset \mathbb{R}^n$ be a set such that*

$$\dim \mathrm{Tan}(x, S) \leq k, \qquad \forall x \in S$$

for a given integer $k \in [0, n]$. Then S is countably k-rectifiable.

Proof — Let us check that the hypotheses of Theorem 4.1.6 are satisfied. Given $x \in S$, let Π_x be a k-plane containing $\mathrm{Tan}(x, S)$. We claim that

$$\forall M > 0 \, \exists \rho > 0 \, : \, S \cap B_\rho(x) \subset x + K_M(\Pi_x). \tag{4.5}$$

We prove this by contradiction. Suppose that (4.5) is not valid. Then there exists $M > 0$ and a sequence $\{x_i\} \subset S$ converging to x and such that

$$|\pi_x^\perp(x_i - x)| > M|\pi_x(x_i - x)|, \qquad \forall i. \tag{4.6}$$

By passing to a subsequence, we can assume that

$$\frac{x_i - x}{|x_i - x|} \to \theta$$

for some unit vector θ. By definition, θ belongs to $T(x, S)$ and therefore to Π_x. This implies that

$$\lim_{i \to \infty} \pi_x^\perp \left(\frac{x_i - x}{|x_i - x|} \right) = 0.$$

On the other hand, by (4.6),

$$\left| \pi_x^\perp \left(\frac{x_i - x}{|x_i - x|} \right) \right| > M \left| \pi_x \left(\frac{x_i - x}{|x_i - x|} \right) \right| \to M \text{ as } i \to \infty.$$

The contradiction shows that our claim (4.5) is valid, and so we can apply Theorem 4.1.6. ∎

We now turn back to the analysis of the singular set $\Sigma(u)$ of a semiconcave function $u : \Omega \to \mathbb{R}^n$. To obtain a fine description, it is convenient to introduce a hierarchy of subsets of $\Sigma(u)$ according to the dimension of the superdifferential.

Definition 4.1.10 *The* **magnitude** *of a point $x \in \Omega$ (with respect to u) is the integer*

$$\kappa(x) = \dim D^+ u(x).$$

Given an integer $k \in \{0, \ldots, n\}$ we set

$$\Sigma^k(u) = \{x \in \Omega \, : \, \kappa(x) = k\}.$$

Let us also introduce the following sets. Given $\rho > 0$, we denote by $\Sigma_\rho^k(u)$ the set of all $x \in \Omega$ such that $D^+u(x)$ contains a k-dimensional sphere of radius ρ. By well-known properties of convex sets (see Proposition A. 1.10), we have $\Sigma^k(u) \subset \cup_{\rho > 0} \Sigma_\rho^k(u)$. On the other hand, a point $x \in \Sigma_\rho^k(u)$ does not necessarily belong to $\Sigma^k(u)$, since $D^+u(x)$ may have dimension greater than k.

Proposition 4.1.11 *If u is semiconcave in Ω, then $\Sigma_\rho^k(u)$ is closed for any $\rho > 0$ and $k = 0, \ldots, n$.*

Proof — Let $x \in \Omega$ be the limit of a sequence $\{x_i\} \subset \Sigma_\rho^k(u)$. Then, for any i we can find a k-dimensional ball B_ρ^i of radius ρ and contained in $D^+u(x_i)$. These spheres are all contained in a bounded set in $\mathbb{R}^n$ by the local Lipschitz continuity of u. Passing to a subsequence and using the upper semicontinuity of D^+u (see Proposition 3.3.4(a)) we obtain that $D^+u(x)$ contains a k-dimensional ball sphere B_ρ of radius ρ whose points are limit of points of the spheres B_ρ^i. This proves that $x \in \Sigma_\rho^k(u)$. ∎

Theorem 4.1.12 *If u is semiconcave in Ω, then*

$$\text{Tan}(x, \Sigma_\rho^k(u)) \subset [D^+u(x)]^\perp, \quad \forall x \in \Sigma_\rho^k(u) \cap \Sigma^k(u) \tag{4.7}$$

for any $\rho > 0$ and $k = 0, \ldots, n$.

Proof — It suffices to prove the desired inclusion for the tangent cone $T(x, \Sigma_\rho^k(u))$. Let us take any nonzero vector $\theta \in T(x, \Sigma_\rho^k(u))$. Then there exist $x_i \in \Sigma_\rho^k(u)$ and t_i such that as $i \to \infty$,

$$x_i \to x, \qquad t_i \downarrow 0, \qquad \frac{x_i - x}{t_i} \to \theta.$$

As in the proof of Proposition 4.1.11, we can find a k-dimensional ball $B_\rho \subset D^+u(x)$ whose elements can all be approximated by elements of $D^+u(x_i)$.

Let p, p' be two arbitrary elements of B_ρ and let $p_i \in D^+u(x_i)$ be such that $p_i \to p'$ as $i \to \infty$. From Proposition 3.3.10 we know that

$$\langle p_i - p, x_i - x \rangle \leq 2|x_i - x|\omega(|x_i - x|)$$

where ω is a modulus of semiconcavity for u. Letting $i \to \infty$ we obtain that $\langle p' - p, \theta \rangle \leq 0$. Exchanging the role of p and p' we find that $\langle p' - p, \theta \rangle = 0$ for all $p, p' \in B_\rho$. Therefore the k-plane containing B_ρ is orthogonal to θ. Since $D^+u(x)$ is k-dimensional by assumption and contains B_ρ, we conclude that $D^+u(x)$ is also orthogonal to θ. ∎

Corollary 4.1.13 *If u is semiconcave in Ω, then the set $\Sigma^k(u)$ is countably $(n - k)$-rectifiable for any integer $k = 0, \ldots, n$.*

Proof — From Theorem 4.1.12 we obtain that

$$\dim\left[\text{Tan}\left(x, \Sigma_\rho^k(u)\right)\right] \leq \dim\left[D^+u(x)\right]^\perp = n - k$$

for all $x \in \Sigma_\rho^k(u) \cap \Sigma^k(u)$. We can apply Corollary 4.1.9 to conclude that $\Sigma_\rho^k(u) \cap \Sigma^k(u)$ is countably $(n - k)$-rectifiable. Therefore, $\Sigma^k(u)$ is also countably $(n - k)$-rectifiable. ∎

4.2 Propagation along Lipschitz arcs

Let u be a semiconcave function in an open domain $\Omega \subset \mathbb{R}^n$. The rectifiability properties of $\Sigma(u)$, obtained in the previous section, can be regarded as "upper bounds" for $\Sigma(u)$. From now on, we shall study the singular set of u trying to obtain "lower bounds" for such a set. In the rest of the chapter, we restrict our attention to semiconcave functions with a linear modulus.

Given a point $x_0 \in \Sigma(u)$, we are interested in conditions ensuring the existence of other singular points approaching x_0. The following example explains the nature of such conditions.

Example 4.2.1 The functions

$$u_1(x_1, x_2) = -\sqrt{x_1^2 + x_2^2}, \qquad u_2(x_1, x_2) = -|x_1| - |x_2|$$

are concave in $\mathbb{R}^2$, and $(0, 0)$ is a singular point for both of them. Moreover, $(0, 0)$ is the only singularity for u_1 while

$$\Sigma(u_2) = \{(x_1, x_2) : x_1 x_2 = 0\}.$$

So, $(0, 0)$ is the intersection point of two singular lines of u_2. Notice that $(0, 0)$ has magnitude 2 with respect to both functions as

$$D^+ u_1(0, 0) = \{(p_1, p_2) : p_1^2 + p_2^2 \leq 1\}$$
$$D^+ u_2(0, 0) = \{(p_1, p_2) : |p_1| \leq 1, \ |p_2| \leq 1\}.$$

The different structure of $\Sigma(u_1)$ and $\Sigma(u_2)$ in a neighborhood of x_0 is captured by the reachable gradients. In fact,

$$D^* u_1(0, 0) = \{(p_1, p_2) : p_1^2 + p_2^2 = 1\} = \partial D^+ u_1(0, 0)$$
$$D^* u_2(0, 0) = \{(p_1, p_2) : |p_1| = 1, \ |p_2| = 1\} \neq \partial D^+ u_2(0, 0).$$

In other words, the inclusion $D^* u(x) \subset \partial D^+ u(x)$ (see Proposition 3.3.4(b)) is an equality for u_1 and a proper inclusion for u_2.   ∎

The above example suggests that a sufficient condition to exclude that x_0 is an isolated point of $\Sigma(u)$ should be that $D^* u(x_0)$ fails to cover the whole boundary of $D^+ u(x_0)$. As we shall see, such a condition implies a much stronger property, namely that x_0 is the initial point of a Lipschitz singular arc.

In the following we call an *arc* a continuous map $\mathbf{x} : [0, \rho] \to \mathbb{R}^n$, $\rho > 0$. We shall say that the arc $\mathbf{x}$ is *singular* for u if the support of $\mathbf{x}$ is contained in Ω and $\mathbf{x}(s) \in \Sigma(u)$ for every $s \in [0, \rho]$. The following result describes the "arc structure" of the singular set $\Sigma(u)$.

Theorem 4.2.2 *Let $x_0 \in \Omega$ be a singular point of a function $u \in \mathrm{SCL}(\Omega)$. Suppose that*

$$\partial D^+ u(x_0) \setminus D^* u(x_0) \neq \emptyset. \tag{4.8}$$

Then there exist a Lipschitz singular arc $\mathbf{x} : [0, \rho] \to \mathbb{R}^n$ for u, with $\mathbf{x}(0) = x_0$, and a positive number δ such that

$$\lim_{s \to 0^+} \frac{\mathbf{x}(s) - x_0}{s} \neq 0 \tag{4.9}$$

$$\mathrm{diam}\Big(D^+ u(\mathbf{x}(s))\Big) \geq \delta \qquad \forall s \in [0, \rho]. \tag{4.10}$$

Moreover, $\mathbf{x}(s) \neq x_0$ for any $s \in \,]0, \rho]$.

Observe that Theorem 4.2.2 gives no information on the magnitude of $\mathbf{x}(s)$ as a singular point. However, by Corollary 4.1.13, we can exclude that the support of $\mathbf{x}$ consists entirely of singular points of magnitude n.

Remark 4.2.3 We note that condition (4.8) is equivalent to the existence of two vectors, $p_0 \in \mathbb{R}^n$ and $q \in \mathbb{R}^n \setminus \{0\}$, such that

$$p_0 \in D^+ u(x_0) \setminus D^* u(x_0) \tag{4.11}$$

$$\langle q, p - p_0 \rangle \geq 0 \qquad \forall\, p \in D^+ u(x_0). \tag{4.12}$$

Indeed, (4.11) and (4.12) imply that $p_0 \in \partial D^+ u(x_0) \setminus D^* u(x_0)$. Conversely, if (4.8) holds true, then (4.11) is trivially satisfied by any vector $p_0 \in \partial D^+ u(x_0) \setminus D^* u(x_0)$, and (4.12) follows taking $-q$ in the normal cone to the convex set $D^+ u(x_0)$ at p_0 (see Definition A. 1.12).

Remark 4.2.4 It is easy to see that the support of the singular arc $\mathbf{x}$, given by Theorem 4.2.2, is a connected set of Hausdorff dimension 1. Indeed, from the Lipschitz continuity of $\mathbf{x}$ it follows that the support of $\mathbf{x}$ is 1-rectifiable, while property (4.9) implies that the 1-dimensional Hausdorff measure of $\mathbf{x}([0, T])$ is positive.

The idea of the proof of Theorem 4.2.2 relies on a simple geometric argument: intuition suggests that the distance between the graph of u and a transverse plane through $(x_0, u(x_0))$ should be maximized along the singular arc we expect to find. We will be able to construct such a plane using the vector $p_0 - q$, where p_0 and q are chosen as in Remark 4.2.3. Indeed, condition (4.12) implies that $p_0 - q \notin D^+ u(x_0)$, and so the graph of

$$x \mapsto u(x_0) + \langle p_0 - q, x - x_0 \rangle \qquad x \in \mathbb{R}^n \tag{4.13}$$

is transverse to the graph of u. We single out this step of the proof in the next lemma, as such a technique applies to any point x_0 of the domain of a semiconcave function, not necessarily singular. For $x_0 \in \Sigma(u)$, we will then show that the arc we construct in this way is singular for u.

Lemma 4.2.5 *Let u be semiconcave in $\overline{B}_R(x_0) \subset \Omega$ with semiconcavity constant C. Fix $p_0 \in \partial D^+u(x_0)$ and let $q \in \mathbb{R}^n \setminus \{0\}$ be such that*

$$\langle q, p - p_0 \rangle \geq 0 \qquad \forall p \in D^+u(x_0). \tag{4.14}$$

Define

$$\sigma = \min\left\{\frac{R}{4|q|}, \frac{1}{2C}\right\}. \tag{4.15}$$

Then there exists a Lipschitz arc $\mathbf{x} : [0, \sigma] \to B_R(x_0)$, with $\mathbf{x}(0) = x_0$, such that

$$0 < |\mathbf{x}(s) - x_0| < 4|q|s \qquad \forall s \in \,]0, \sigma] \tag{4.16}$$

$$\lim_{s \downarrow 0} \frac{\mathbf{x}(s) - x_0}{s} = q \tag{4.17}$$

$$\mathbf{p}(s) := p_0 + \frac{\mathbf{x}(s) - x_0}{s} - q \in D^+u(\mathbf{x}(s)) \qquad \forall s \in \,]0, \sigma]. \tag{4.18}$$

Moreover,

$$\mathrm{Lip}\,(\mathbf{x}) \leq 4\,L + 2\,|q| \tag{4.19}$$

where L is a Lipschitz constant for u in $B_R(x_0)$.

Proof — Let us define, for any $s \in \,]0, 1/C[\,$,

$$\phi_s(x) = u(x) - u(x_0) - \langle p_0 - q, x - x_0 \rangle - \frac{1}{2s}|x - x_0|^2 \qquad x \in \overline{B}_R(x_0).$$

Notice that ϕ_s is the difference between the affine function in (4.13) and the function $x \mapsto u(x) - \frac{1}{2s}|x - x_0|^2$, which is a concave perturbation of u.

Being strictly concave, ϕ_s has a unique maximum point in $\overline{B}_R(x_0)$, which we term x_s. For technical reasons to be clarified in the sequel, we restrict our attention to the interval $0 \leq s \leq \sigma$, where σ is the number given by (4.15). Let us define $\mathbf{x}$ by

$$\mathbf{x}(s) = \begin{cases} x_0 & \text{if } s = 0 \\ x_s & \text{if } s \in \,]0, \sigma]. \end{cases}$$

We now proceed to show that $\mathbf{x}$ possesses all the required properties.

First, we claim that $\mathbf{x}$ satisfies estimate (4.16). Indeed, by property (3.18) satisfied by the elements of D^+u, we have that

$$\phi_s(x) \leq \langle q, x - x_0 \rangle + \left(\frac{C}{2} - \frac{1}{2s}\right)|x - x_0|^2 \tag{4.20}$$

for any $x \in \overline{B}_R(x_0)$. Moreover, $p_0 - q \notin D^+u(x_0)$ in view of condition (4.14). Since this fact implies that there are points in $\overline{B}_R(x_0)$ at which ϕ_s is positive, we conclude that $\phi_s(\mathbf{x}(s)) > 0$. The last estimate and (4.20) yield

$$0 < |\mathbf{x}(s) - x_0| < \frac{2s|q|}{1 - Cs} \qquad \forall s \in \ ]0, \sigma] \tag{4.21}$$

and so (4.16) follows from (4.15).

Second, we proceed to check (4.18). For this purpose we note that, on account of estimate (4.21), the choice of σ forces $\mathbf{x}(s) \in B_R(x_0)$ for any $s \in [0, \sigma]$. Hence, $\mathbf{x}(s)$ is also a local maximum point of ϕ_s. This implies that

$$0 \in D^+\phi_s(\mathbf{x}(s)) = D^+u(\mathbf{x}(s)) - p_0 + q - \frac{\mathbf{x}(s) - x_0}{s}$$

for any $s \in \ ]0, \sigma]$. Clearly, the last inclusion can be recast in the desired form (4.18).

Third, to prove (4.17), we show that $\lim_{s \to 0} \mathbf{p}(s) = p_0$, where $\mathbf{p}$ is defined in (4.18). Let $\overline{p} = \lim_{k \to \infty} \mathbf{p}(s_k)$ for some sequence $s_k \downarrow 0$. Then, taking the scalar product of both sides of the identity

$$\mathbf{p}(s_k) - p_0 + q = \frac{\mathbf{x}(s_k) - x_0}{s_k},$$

with $\mathbf{p}(s_k) - p_0$ and recalling property (3.24), we obtain

$$|\mathbf{p}(s_k) - p_0|^2 + \langle q, \mathbf{p}(s_k) - p_0 \rangle \tag{4.22}$$
$$= \frac{1}{s_k} \langle \mathbf{p}(s_k) - p_0, \mathbf{x}(s_k) - x_0 \rangle \leq \frac{C}{s_k} |\mathbf{x}(s_k) - x_0|^2.$$

Now, observe that the right-hand side above tends to 0 as $k \to \infty$, in view of (4.21). Moreover, $\overline{p} \in D^+u(x_0)$ since D^+u is upper semicontinuous; so $\langle q, \overline{p} - p_0 \rangle \geq 0$ by assumption (4.14). Therefore, (4.22) yields $|\overline{p} - p_0|^2 \leq 0$ in the limit as $k \to \infty$. This proves that $\overline{p} = p_0$ as required.

Finally, let us derive the Lipschitz estimate (4.19). Let $r, s \in [0, \sigma]$. Using (4.18) to evaluate $\mathbf{x}(s)$ and $\mathbf{x}(r)$ one can easily compute that

$$\mathbf{x}(s) - \mathbf{x}(r) = s[\mathbf{p}(s) - \mathbf{p}(r)] + (s - r)[\mathbf{p}(r) - p_0 + q]. \tag{4.23}$$

Now, taking the scalar product of both sides of the above equality with $\mathbf{x}(s) - \mathbf{x}(r)$, and recalling (3.24), we obtain

$$|\mathbf{x}(s) - \mathbf{x}(r)|^2 \leq Cs|\mathbf{x}(s) - \mathbf{x}(r)|^2 + |s - r|\,|\mathbf{x}(s) - \mathbf{x}(r)|\,|\mathbf{p}(r) - p_0 + q|.$$

Hence, for any $r, s \in [0, \sigma]$,

$$(1 - Cs)|\mathbf{x}(s) - \mathbf{x}(r)| \leq |\mathbf{p}(r) - p_0 + q|\,|s - r| \leq (2L + |q|)\,|s - r|,$$

because L provides a bound for D^+u in $B_R(x_0)$. This completes the proof. ∎

Proof of Theorem 4.2.2 — To begin, let us fix a radius $R > 0$ such that $\overline{B}_R(x_0) \subset \Omega$. We recall that, as noted in Remark 4.2.3, the geometric assumption that $\partial D^+u(x_0) \backslash D^*u(x_0)$ be nonempty is equivalent to the existence of vectors $p_0 \in \mathbb{R}^n$ and $q \in \mathbb{R}^n \backslash \{0\}$ satisfying

$$p_0 \in D^+u(x_0) \setminus D^*u(x_0) \quad \& \quad \langle q, p - p_0 \rangle \geq 0 \quad \forall p \in D^+u(x_0). \tag{4.24}$$

Applying Lemma 4.2.5 to such a pair, we can construct two arcs, $\mathbf{x}$ and $\mathbf{p}$, that enjoy properties (4.17) and (4.18). Moreover, the same lemma ensures that $\mathbf{x}$ is Lipschitz continuous and that $\mathbf{x}(s) \neq x_0$ for any $s \in \,]0, \sigma]$.

Therefore, it remains to show that the restriction of $\mathbf{x}$ to a suitable subinterval $[0, \rho]$ is singular for u, and that the diameter of $D^+u(\mathbf{x}(s))$ is bounded away from 0 for all $s \in [0, \rho]$. In fact, it suffices to check the latter point. Let us argue by contradiction: suppose that a sequence $s_k \downarrow 0$ exists such that $\mathrm{diam}(D^+u(\mathbf{x}(s_k))) \to 0$ as $k \to \infty$. Then, by Proposition 3.3.17, $p_0 = \lim_{k \to \infty} \mathbf{p}(s_k)$ belongs to $D^*u(x_0)$, in contradiction with (4.24). This proves the theorem. $\blacksquare$

4.3 Singular sets of higher dimension

In the previous section we proved that the singularities of a function $u \in \mathrm{SCL}\,(\Omega)$ propagate along a Lipschitz image of an interval $[0, \rho]$, from any point $x_0 \in \Omega$ at which $D^*u(x_0)$ fails to cover the whole boundary of $D^+u(x_0)$. Such a result describes a sort of basic structure for the propagation of singularities of a semiconcave function. Our next step will be to give a more detailed propagation result in the case where the singular set has dimension greater than 1. Let us consider a simple example.

Example 4.3.1 It is easy to check that the singular set of the concave function

$$u(x) = -|x_3| \qquad x = (x_1, x_2, x_3) \in \mathbb{R}^3$$

is given by the coordinate plane

$$\Sigma(u) = \{x \in \mathbb{R}^3 : x_3 = 0\}.$$

Moreover,

$$D^*u(0) = \{(0, 0, 1), (0, 0, -1)\} \qquad D^+u(0) = \{(0, 0, p_3) : |p_3| \leq 1\}.$$

Therefore, one can apply Theorem 4.2.2 with $x_0 = 0$ and $p_0 = 0$, but this procedure only gives a Lipschitz singular arc starting at 0, whereas our function has a 2-dimensional singular set. Actually, a more careful application of Lemma 4.2.5 suggests that a singular arc for u should correspond to any vector $q \neq 0$ satisfying (4.14). Moreover, such a correspondence should be 1-to-1 in light of (4.17). Since (4.14) is satisfied by all vectors $q \in \mathbb{R}^3$ such that $q_1^2 + q_2^2 = 1$, $q_3 = 0$, one can imagine constructing the whole singular plane $x_3 = 0$ in this way. $\blacksquare$

The next result generalizes Theorem 4.2.2, showing the propagation of singularities along a ν-dimensional set. The integer $\nu \geq 1$ is given by the number of linearly independent directions of the normal cone to the superdifferential of u at the initial singular point, as conjectured in the above example.

In the following we write $N_S(x)$ to denote the normal cone to a convex set S at a point $x \in S$. Moreover, we use the notation

$$A \ni x \to x_0$$

to mean that $x \in A$ and $x \to x_0$. For a family $\{A_i\}_{i \in \mathbb{N}}$ of subsets of $\mathbb{R}^n$, we write

$$A_i \ni x_i \to x$$

to denote a sequence $x_i \in A_i$ converging to x.

Theorem 4.3.2 *Let $x_0 \in \Omega$ be a singular point of a function $u \in$ SCL (Ω). Suppose that*

$$\partial D^+ u(x_0) \setminus D^* u(x_0) \neq \emptyset$$

and, having fixed a point $p_0 \in \partial D^+ u(x_0) \setminus D^* u(x_0)$, define

$$\nu := \dim N_{D^+ u(x_0)}(p_0).$$

Then a number $\rho > 0$ and a Lipschitz map $\mathbf{f} : N_{D^+ u(x_0)}(p_0) \cap B_\rho \to \Sigma(u)$ exist such that

$$\mathbf{f}(q) = x_0 - q + |q|\mathbf{h}(q)$$
$$\text{with} \quad \mathbf{h}(q) \to 0 \quad \text{as} \quad N_{D^+ u(x_0)} \cap B_\rho \ni q \to 0 \tag{4.25}$$

$$\liminf_{r \to 0^+} r^{-\nu} \mathcal{H}^\nu \left(\mathbf{f}(N_{D^+ u(x_0)}(p_0) \cap B_\rho) \cap B_r(x_0) \right) > 0. \tag{4.26}$$

Moreover,

$$\mathrm{diam}\left(D^+ u(\mathbf{f}(q)) \right) \geq \delta \qquad \forall q \in N_{D^+ u(x_0)}(p_0) \cap B_\rho \tag{4.27}$$

for some $\delta > 0$.

Remark 4.3.3 As one can easily realize, Theorem 4.3.2 is an extension of Theorem 4.2.2. The property

$$\mathbf{f}(q) \neq x_0 \quad \forall q \in N_{D^+ u(x_0)}(p_0) \cap B_\rho \setminus \{0\},$$

though absent from the statement of Theorem 4.3.2, is valid in the present case as well. In fact, it can also be derived from (4.25), possibly restricting the domain of $\mathbf{f}$. The reason for treating the one-dimensional case first is that the Hausdorff estimate for the density of $\Sigma(u)$ is immediate if $\nu = 1$, and so the main ideas of the approach can be more easily understood.

As in the previous section, we first prove a preliminary result.

Lemma 4.3.4 *Let u be semiconcave in $\overline{B}_R(x_0) \subset \Omega$ with semiconcavity constant C. Fix $p_0 \in \partial D^+ u(x_0)$ and define*

$$\sigma = \min\left\{\frac{1}{2C}, \frac{R}{4}\right\}. \tag{4.28}$$

Then a Lipschitz map $\mathbf{f} : N_{D^+u(x_0)} \cap B_\sigma \to \mathbb{R}^n$ exists such that

$$\mathbf{f}(q) = x_0 - q + |q|\mathbf{h}(q)$$
$$\text{with} \quad \mathbf{h}(q) \to 0 \quad \text{as} \quad N_{D^+u(x_0)} \cap B_\sigma \ni q \to 0 \tag{4.29}$$

$$p_0 + \mathbf{h}(q) \in D^+ u(\mathbf{f}(q)) \qquad \forall q \in N_{D^+u(x_0)}(p_0) \cap B_\sigma. \tag{4.30}$$

Moreover,

$$\mathrm{Lip}\,(\mathbf{f}) \le 4L + 2 \tag{4.31}$$

where L is a Lipschitz constant for u in $B_R(x_0)$.

Proof — Let us set, for brevity,

$$N := N_{D^+u(x_0)}(p_0).$$

We proceed as in the proof of Lemma 4.2.5 and consider, for any $q \in N \setminus \{0\}$, the function

$$\phi_q(x) = u(x) - u(x_0) - \left\langle p_0 + \frac{q}{|q|}, x - x_0 \right\rangle - \frac{1}{2|q|}|x - x_0|^2 \qquad x \in \overline{B}_R(x_0)$$

and the point x_q given by the relation

$$\phi_q(x_q) = \max_{x \in \overline{B}_R(x_0)} \phi_q(x).$$

Then, we define σ as in (4.28) and $\mathbf{f}$ by

$$\mathbf{f}(q) = \begin{cases} x_0 & \text{if } q = 0 \\ x_q & \text{if } q \in N \cap B_\sigma \setminus \{0\}. \end{cases}$$

Arguing as in the proof of Lemma 4.2.5, we obtain

$$0 < |\mathbf{f}(q) - x_0| < \frac{2|q|}{1 - C|q|} \qquad \forall q \in N \cap B_\sigma \setminus \{0\} \tag{4.32}$$

and

$$p_0 + \frac{\mathbf{f}(q) - x_0 + q}{|q|} \in D^+ u(\mathbf{f}(q)) \qquad \forall q \in N \cap B_\sigma. \tag{4.33}$$

Hence, denoting by $\mathbf{h}(q)$ the quotient in (4.33), assertion (4.30) follows.

To prove (4.29) we must show that

$$\mathbf{h}(q) \to 0 \qquad \text{as} \qquad N \cap B_\sigma \ni q \to 0. \tag{4.34}$$

For this purpose, let $\{q_i\}$ be an arbitrary sequence in $N \cap B_\sigma \setminus \{0\}$ such that $q_i \to 0$. Since $\mathbf{h}$ is bounded, we can extract a subsequence (still termed $\{q_i\}$) such that $\lim_{i \to \infty} \mathbf{h}(q_i)$ exists and

$$\lim_{i \to \infty} \frac{q_i}{|q_i|} = \overline{q}$$

for some $\overline{q} \in N$ satisfying $|\overline{q}| = 1$. We claim that $\lim_{i \to \infty} \mathbf{h}(q_i) = 0$, which in turn implies (4.34). Indeed, let us set

$$\overline{p} := p_0 + \lim_{i \to \infty} \mathbf{h}(q_i)$$

and observe that $\overline{p} \in D^+ u(x_0)$ as $D^+ u$ is upper semicontinuous and $\mathbf{f}$ is continuous at 0. Taking the scalar product of both sides of the identity

$$\mathbf{h}(q_i) - \frac{q_i}{|q_i|} = \frac{\mathbf{f}(q_i) - x_0}{|q_i|}$$

with $\mathbf{h}(q_i)$ and applying inequality (3.24), we deduce that

$$|\mathbf{h}(q_i)|^2 - \left\langle \frac{q_i}{|q_i|}, \mathbf{h}(q_i) \right\rangle = \frac{1}{|q_i|} \langle \mathbf{f}(q_i) - x_0, \mathbf{h}(q_i) \rangle$$

$$\leq \frac{C}{|q_i|} |\mathbf{f}(q_i) - x_0|^2 \leq \frac{C}{|q_i|} \left(\frac{2|q_i|}{1 - C|q_i|} \right)^2$$

where the last estimate follows from (4.32). In the limit as $i \to \infty$, the above inequality yields

$$|\overline{p} - p_0|^2 - \langle \overline{q}, \overline{p} - p_0 \rangle \leq 0.$$

Hence, recalling that $\overline{q} \in N$, we conclude that $\overline{p} = p_0$, which proves our claim and implies (4.29). Finally, the Lipschitz estimate (4.31) can be obtained by the same reasoning as in the proof of Lemma 4.2.5. ∎

Proof of Theorem 4.3.2 — Keeping the abbreviated notation $N = N_{D^+ u(x_0)}(p_0)$ as in the proof of Lemma 4.3.4, let us denote by $L(N)$ the linear subspace generated by N, and by $\pi : \mathbb{R}^n \to L(N)$ the orthogonal projection of $\mathbb{R}^n$ onto $L(N)$.

Having fixed $R > 0$ so that $\overline{B}_R(x_0) \subset \Omega$, let $\mathbf{f} : N \cap B_\sigma \to \mathbb{R}^n$ be the map given by Lemma 4.3.4. Arguing by contradiction—as in the proof of Theorem 4.2.2—the reader can easily show that a suitable restriction of $\mathbf{f}$ to $N \cap B_\rho$, $0 < \rho \leq \sigma$, satisfies (4.27). In particular, $\mathbf{f}(q) \in \Sigma(u)$ for any $q \in N \cap B_\rho$. Moreover, (4.25) is an immediate consequence of (4.29).

Therefore, the only point of the conclusion that needs to be demonstrated is estimate (4.26) for the ν-dimensional Hausdorff density of the singular set $\mathbf{f}(N \cap B_\rho)$, or, since $\mathcal{H}^\nu$ is translation invariant,

$$\liminf_{r \to 0^+} r^{-\nu} \mathcal{H}^\nu \left([\mathbf{f}(N \cap B_\rho) - x_0] \cap B_r \right) > 0.$$

We note that the above inequality can be deduced from the lower bound

$$\liminf_{r \to 0^+} r^{-\nu} \mathcal{H}^\nu \left(\pi([\mathbf{f}(N \cap B_\rho) - x_0] \cap B_r) \right) > 0 \qquad (4.35)$$

since

$$\mathcal{H}^\nu \left([\mathbf{f}(N \cap B_\rho) - x_0] \cap B_r \right) \geq \mathcal{H}^\nu \left(\pi([\mathbf{f}(N \cap B_\rho) - x_0] \cap B_r) \right).$$

Now, using the shorter notation $F(q) = \pi(x_0 - \mathbf{f}(q))$, we observe that (4.35) can in turn be derived from the lower bound

$$\liminf_{r \to 0^+} r^{-\nu} \mathcal{H}^\nu (F(N \cap B_r)) > 0. \qquad (4.36)$$

Indeed, for any sufficiently small r, say $0 < r \leq M\rho$ where $M := \operatorname{Lip}(\mathbf{f})$, we have that

$$F(N \cap B_{r/M}) = \pi \left[x_0 - \mathbf{f}(N \cap B_{r/M}) \right] \subset \pi \left[\left(x_0 - \mathbf{f}(N \cap B_\rho) \right) \cap B_r \right].$$

The rest of our reasoning will therefore be devoted to the proof of (4.36). To begin, we note that in view of (4.25) the map $F : N \cap B_\rho \to L(N)$ introduced above can be also represented as

$$F(q) = q + H(q)$$

where $H(q)/|q| \to 0$ as $N \cap B_\rho \ni q \to 0$. Consequently, the function

$$\lambda(r) := \sup\{|H(q)| : q \in N \cap B_r\}$$

satisfies

$$\lim_{r \downarrow 0} \frac{\lambda(r)}{r} = 0. \qquad (4.37)$$

Let us define

$$N_r := \left\{ q \in N \cap B_r : d_{\Gamma_r}(q) \geq \lambda(r) \right\} \qquad 0 < r \leq \rho, \qquad (4.38)$$

where Γ_r denotes the relative boundary of $N \cap B_r$ and d_{Γ_r} the distance from Γ_r. Using the limit (4.37), it is easy to check that $N_r \neq \emptyset$ provided r is sufficiently small, say $0 < r < \rho_0 \leq \rho$. We claim that

$$N_r \subset F(N \cap B_r) \qquad \forall r \in (0, \rho_0). \qquad (4.39)$$

Indeed, having fixed $y \in N_r$, let us rewrite the equation $F(q) = y$ as

$$q = y - H(q) =: H_y(q).$$

Now, observe that $\overline{B}_{\lambda(r)}(y) \subset N \cap B_r$ and that the continuous map $H_y(q)$ satisfies

$$H_y\left(\overline{B}_{\lambda(r)}(y)\right) \subset y - H(\overline{N \cap B_r}) \subset y + \overline{B}_{\lambda(r)} = \overline{B}_{\lambda(r)}(y).$$

Therefore, applying Brouwer's fixed point theorem, we conclude that $q = H_y(q)$ has a solution $q \in \overline{B}_{\lambda(r)}(y)$. So, our claim (4.39) follows.

Our next step is to obtain the lower bound

$$\liminf_{r \to 0^+} \frac{\mathcal{H}^\nu(N_r)}{r^\nu} > 0 \tag{4.40}$$

for the density of the set N_r introduced in (4.38). To verify the above estimate let $\overline{q}$ be a point in the relative interior of N, with $|\overline{q}| = 1$. Then, using the notation $B_\alpha := B_\alpha \cap L(N)$ to denote ν-dimensional balls, we have that

$$\overline{q} + \widehat{B}_{2\alpha} \subset N$$

for some $\alpha \in \,]0, 1/2]$. On account of (4.37), there exists $r_0 \in \,]0, \rho_0[$ such that

$$\lambda(r) \le \frac{r\alpha}{2} \qquad \forall r \in \,]0, r_0]. \tag{4.41}$$

Hence,

$$\frac{r}{2}\left(\overline{q} + \widehat{B}_{2\alpha}\right) \subset N \cap B_r \qquad \forall r \in \,]0, r_0[. \tag{4.42}$$

Now, combining (4.41), (4.42) and the definition of N_r, we discover

$$\frac{r}{2}\left(\overline{q} + \widehat{B}_\alpha\right) \subset N_r \quad \forall r \in \,]0, r_0[.$$

Estimate (4.40) is an immediate consequence of the last inclusion.

Finally, to complete the proof, it suffices to observe that (4.36) follows from (4.39) and (4.40). ∎

Remark 4.3.5 Though clear from (4.26), we explicitly note that Theorem 4.3.2 ensures that in a neighborhood of x_0 $\Sigma(u)$ covers a rectifiable set of Hausdorff dimension ν. Moreover, from formula (4.25) it follows that the set $\mathbf{f}(N_{D^+u(x_0)}(p_0) \cap B_\rho) \subset \Sigma(u)$ possesses a tangent space at x_0 whenever the normal cone $N_{D^+u(x_0)}(p_0)$ is actually a vector space. This happens, for instance, when x_0 is a singular point of magnitude $k(x_0) < n$ and one can find a point p_0 in the relative interior of $D^+u(x_0)$, but not in $D^*u(x_0)$. Then, it is easy to check that $N_{D^+u(x_0)}(p_0)$ is a vector space of dimension $\nu = n - k(x_0)$.

4.4 Application to the distance function

In this section we examine some properties of the singular set of the distance function d_S associated to a nonempty closed subset S of $\mathbb{R}^n$. As in Section 3.4, we denote by $\text{proj}_S(x)$ the set of closest points in S to x, i.e.,

$$\text{proj}_S(x) = \{y \in S \ : \ d_S(x) = |x - y|\} \qquad x \in \mathbb{R}^n.$$

Our first result characterizes the isolated singularities of d_S.

Theorem 4.4.1 *Let S be a nonempty closed subset of $\mathbb{R}^n$ and $x \notin S$ a singular point of d_S. Then the following properties are equivalent:*

(a) x is an isolated point of $\Sigma(d_S)$.
(b) $\partial D^+ d_S(x) = D^* d_S(x)$.
(c) $\text{proj}_S(x) = \partial B_r(x)$ where $r := d_S(x)$.

Proof — The implication (a)$\Rightarrow$(b) is an immediate corollary of the propagation result of Section 4.2. Indeed, if $\partial D^+ d_S(x) \setminus D^* d_S(x)$ is nonempty, then Theorem 4.2.2 ensures the existence of a nonconstant singular arc with initial point x. In particular, x could not be isolated.

Let us now show that (b) implies (c). First, we claim that, if (b) holds, then x must be a singular point of magnitude $\kappa(x) = n$, i.e., $\dim D^+ d_S(x) = n$. For suppose the strict inequality $\kappa(x) < n$ is verified. Then, the whole superdifferential would be made of boundary points and so, owing to (b), $D^+ d_S(x) = D^* d_S(x)$. Therefore, $D^+ d_S(x) \subset \partial B_1$ as all reachable gradients of d_S are unit vectors. But the last inclusion contradicts the fact that $D^+ d_S(x)$ is a convex set of dimension at least 1. Our claim is thus proved.

Now, we use the fact that $D^+ d_S(x)$ is an n-dimensional convex set with

$$\partial D^+ d_S(x) = D^* d_S(x) \subset \partial B_1$$

to conclude that $D^+ d_S(x) = \overline{B}_1$ and $D^* d_S(x) = \partial B_1$. Then, we invoke formula (3.40) to discover

$$\text{proj}_S(x) = x - d_S(x) D^* d_S(x) = \partial B_r(x), \tag{4.43}$$

which proves (c).

Finally, let us show that (c) implies (a). From Corollary 3.4.5 (iii) we know that d_S is differentiable along each segment $]x, y[$ with $y \in \text{proj}_S(x) = \partial B_r(x)$. So, $d_S \in C^1(B_r(x) \setminus \{x\})$ and the proof is complete. $\blacksquare$

In other words, the previous result shows that a point x_0 is an isolated singularity for the distance function from a set S only if there exists an open sphere B centered at x_0, such that $B \cap S = \emptyset$ and $\partial B \subset S$. In particular, if S is a simply connected set in $\mathbb{R}^2$, or a set in $\mathbb{R}^n$ with trivial $n - 1$ homotopy group, then the distance from S has no isolated singularities in the complement of S.

We now apply the theory of Section 4.3 to obtain a result on the propagation of singularities for the distance function. In the following, we use the shorter notation

$$P_S(x) = \text{co}[\text{proj}_S(x)].$$

In view of (3.39),

$$P_S(x) = x - d_S(x)D^+d_S(x). \tag{4.44}$$

Theorem 4.4.2 *Let S be a nonempty closed subset of $\mathbb{R}^n$ and let $x \notin S$ be a non-isolated singular point of d_S. Then $P_S(x)$ has an exposed face of dimension at least 1. Moreover, if y is in the relative interior of an exposed face E of $P_S(x)$ satisfying $\dim E \geq 1$, then $\dim N_{P_S(x)}(y)$ is a lower bound for the Hausdorff dimension of the connected component of x in $\Sigma(d_S)$.*

Proof — Since x is a non-isolated singular point of d_S, Theorem 4.4.1 ensures that

$$\partial D^+d_S(x) \setminus D^*d_S(x) \neq \emptyset.$$

Now, simple arguments of convex analysis show that if $p \in \partial D^+d_S(x) \setminus D^*d_S(x)$, then p belongs to some exposed face V of $D^+d_S(x)$, with $\dim V \geq 1$. Then, recalling (4.44) we conclude that $x - d_S(x)V$ is an exposed face of $P_S(x)$.

By similar arguments we have that if y is in the relative interior of an exposed face E of $P_S(x)$ with $\dim E \geq 1$, then

$$p := \frac{x - y}{d_S(x)} \in \partial D^+d_S(x) \setminus D^*d_S(x)$$

and $N_{P_S(x)}(y) = -N_{D^+d_S(x)}(p)$. Then, to complete the proof, it suffices to apply Theorem 4.3.2 to the distance function. ∎

The next result immediately follows from the previous one since, due to (4.44), the dimension of $P_S(x)$ coincides with $\kappa(x)$, the magnitude of x.

Corollary 4.4.3 *Let S be a nonempty closed subset of $\mathbb{R}^n$ and $x \notin S$ a singular point of d_S. Then $n - \kappa(x)$ is a lower bound for Hausdorff dimension of the connected component of x in $\Sigma(d_S)$.*

A typical situation that is covered by Corollary 4.4.3 is when $\text{proj}_S(x) = \{y_0, \ldots, y_k\}$ for some $k \in \{1, \ldots, n-1\}$ provided that the vectors $y_1 - y_0, \ldots, y_k - y_0$ are linearly independent. In this case the connected component of x in $\Sigma(d_S)$ has dimension $\geq n - k$.

Bibliographical notes

Apart from earlier contributions for specific problems as in [70], to our knowledge the first general results about the rectifiability of the singular sets of concave functions are due to Zajíček [144, 145] and Veselý [137, 138]. Similar properties were later extended to semiconcave functions in [10]. Related results on the singularities of concave and semiconcave functions can be found in [4, 9, 15, 60, 61, 135].

The propagation of singularities was first studied in [45] for semiconcave solutions to Hamilton–Jacobi–Bellman equations, and then in [13] for general semiconvex functions. In the first reference, singularities were shown to propagate just along a sequence of points. In the second one, on the contrary, conditions were given to derive estimates for the Hausdorff dimension of the singular set in a neighborhood of a point $x_0 \in \Sigma(u)$. The importance of a condition of type (4.11) was initially pointed out in [13].

The results of Sections 4.2, 4.3 and 4.4 are taken from [5]. An earlier version of the technique used in the proof of Theorem 4.2.2 was given in [4]. Some extensions of these results to semiconcave functions with arbitrary modulus can be found in [1].

In [37], the structural analysis of [5] has been used to derive a representation formula for a sandpile growth model, as well as to provide a simplified proof of the Lipschitz regularity result in [97, 107] in dimension two.

From the point of view of best approximation theory, the singularities of the distance function d_S have been investigated by many authors. In [24], the set $\Sigma(d_S)$ is described proving that, in the case of $n = 2$, the non-isolated singularities of the metric projection propagate along Lipschitz arcs. Such a result is extended to Hilbert spaces in [140]. Property (c) in Theorem 4.4.1 was observed by Motzkin [113] in the case of $n = 2$; it was later extended to Hilbert spaces in [140]. The propagation of the non-isolated singularities of d_S along Lipschitz arcs has been established in both Euclidean and Hilbert spaces; see [24] and [139, 140] respectively.

Let us finally mention here the recent papers [119, 120, 121], where the author studies the stabilization problem in control theory by the introduction of suitable semiconcave Lyapunov functions, and an important role is played by the properties of the singular set of a semiconcave function.

5

Hamilton–Jacobi Equations

Hamilton–Jacobi equations are nonlinear first order equations which have been first introduced in classical mechanics, but find application in many other fields of mathematics. Our interest in these equations lies mainly in the connection with calculus of variations and optimal control. We have seen in Chapter 1 how the dynamic programming approach leads to the analysis of a Hamilton–Jacobi equation and other examples will be considered in the remainder of the book. However, our point of view in this chapter will be to study Hamilton–Jacobi equations for their intrinsic interest without referring to specific applications.

We begin by giving, in Section 5.1, a fairly general exposition of the method of characteristics. This method allows us to construct smooth solutions of first order equations, and in general can be applied only locally. However, this method is interesting also for the study of solutions that are not smooth. As we will see in the following, characteristic curves (or suitable generalizations) often play an important role for generalized solutions and are related to the optimal trajectories of the associated control problem.

In Section 5.2 we recall the basic definitions and results from the theory of viscosity solutions for Hamilton–Jacobi equations. In this theory solutions are defined by means of inequalities satisfied by the generalized differentials or by test functions. With such a definition it is possible to obtain existence and uniqueness theorems under quite general hypotheses. In addition, in most cases where the equation is associated to a control problem, the viscosity solution coincides with the value function of the problem. Although this section is meant to be a collection of results whose proof can be found in specialized monographs, we have included the proofs of some simple statements in order to give to the reader the flavor of the techniques of the theory.

In Section 5.3 we analyze the relation between semiconcavity and the viscosity property. Roughly speaking, it turns out that the two properties are equivalent when the hamiltonian is a convex function of the gradient of the solution. However, it is also possible to obtain semiconcavity results under different assumptions on the hamiltonian.

In the remaining sections we analyze the singular set of semiconcave functions that are solutions of a Hamilton–Jacobi equation. We show that for such functions the results of the previous chapter can be improved; in particular, we give sufficient conditions for the propagation of singularities which are easier to verify than for a general semiconcave function. In addition, we can prove the propagation of singularities along *generalized characteristics*, i.e., arcs which solve a differential inclusion that generalizes the equation satisfied by the characteristic curves in the smooth case.

5.1 Method of characteristics

In Section 1.5 we have introduced the method of characteristics to construct a local classical solution of the Cauchy problem for equations of the form $\partial_t u + H(\nabla u) = 0$. We now show how this method can be extended to study general first order equations.

As a first step, let us show how the procedure of Section 1.5 can be generalized to Cauchy problems where the hamiltonian depends also on t, x. Let us consider the problem

$$\partial_t u(t, x) + H(t, x, \nabla u(t, x)) = 0, \qquad (t, x) \in [0, \infty[\times \mathbb{R}^n \qquad (5.1)$$

$$u(0, x) = u_0(x), \qquad x \in \mathbb{R}^n, \qquad (5.2)$$

with H and u_0 of class C^2.

Suppose, first, we have a solution $u \in C^2([0, T] \times \mathbb{R}^n)$ of the above problem. Given $z \in \mathbb{R}^n$, we call *characteristic curve* associated to u starting from z the curve $t \to (t, X(t; z))$, where $X(\cdot; z)$ solves

$$\dot{X} = H_p(t, X, \nabla u(t, X)), \qquad X(0) = z.$$

Here and in the following the dot denotes differentiation with respect to t. Now, if we set

$$U(t; z) = u(t, X(t; z)), \qquad P(t; z) = \nabla u(t, X(t; z)).$$

we find that

$$\dot{U} = u_t(t, X) + \nabla u(t, X) \cdot \dot{X} = -H(t, X, P) + P \cdot H_p(t, X, P),$$

$$\dot{P} = \nabla u_t(t, X) + \nabla^2 u(t, X) H_p(t, X, \nabla u(t, X)).$$

Taking into account that

$$0 = \nabla(u_t(t, x) + H(t, x, \nabla u(t, x)))$$
$$= \nabla u_t(t, x) + H_x(t, x, \nabla u(t, x)) + \nabla^2 u(t, x) H_p(t, x, \nabla u(t, x))$$

we obtain that

$$\dot{P} = -H_x(t, X, \nabla u(t, X)) = -H_x(t, X, P).$$

Therefore, the pair X, P solves the ordinary differential problem

$$\begin{cases} \dot{X} = H_p(t, X, P) \\ \dot{P} = -H_x(t, X, P) \end{cases} \qquad \begin{cases} X(0) = z \\ P(0) = \nabla u_0(z), \end{cases} \qquad (5.3)$$

while U satisfies

$$\dot{U} = -H(t, X, P) + P \cdot H_p(t, X, P), \qquad U(0; z) = u_0(z). \qquad (5.4)$$

This shows that X, P and U are uniquely determined by the initial value u_0. As in Section 1.5, the above arguments suggest that one can obtain a solution to the partial differential equation (5.1) by solving the characteristic system (5.3), provided the map $z \to X(t; z)$ is invertible. We have the following result, which can be proved by a reasoning similar to the one of Theorem 1.5.3.

Theorem 5.1.1 *For any $z \in \mathbb{R}^n$, let $X(t; z)$, $P(t; z)$ denote the solution of problem (5.3) and let $U(t; z)$ be defined by (5.4). Suppose that there exists $T^* > 0$ such that*

(i) the maximal interval of existence of the solution to (5.3) contains $[0, T^*[$ for all $z \in \mathbb{R}^n$;

(ii) the map $z \to X(t; z)$ is invertible with C^1 inverse $x \to Z(t; x)$ for all $t \in [0, T^*[$.

Then there exists a unique solution $u \in C^2([0, T^*[\times \mathbb{R}^n)$ of problem (5.1), which is given by

$$u(t, x) = U(t; Z(t; x)), \qquad (t, x) \in [0, T^*[\times \mathbb{R}^n. \qquad (5.5)$$

Remark 5.1.2 In the cases where no global smooth solution exists, it is natural to wonder whether one can still use the method of characteristics to obtain a function defined globally, which is a solution in some generalized sense. To be more precise, let us consider again the maps X, U, P defined above. In general, the map $X(t; \cdot)$ is not injective if t is larger than some critical time T^*; for larger times, no smooth solution exists, and formula (5.5) is not well defined. However, we can still define a multivalued function in this way:

$$\mathcal{U}(t, x) = \{U(t; z) : z \text{ such that } X(t; z) = x\}.$$

Then we may look for a suitable selection of $\mathcal{U}$, that is, a function $u(t, x)$ such that $u(t, x) \in \mathcal{U}(t, x)$ for all t, x, and check whether $u(t, x)$ is a generalized solution of the equation in some sense. It turns out that this can be done when the hamiltonian is either convex or concave with respect to p (and the derivatives of H, u_0 satisfy suitable growth restrictions, which will not be specified here). For instance, it can be shown that, in situations where the hamiltonian H is convex with respect to p, the minimal selection

$$u(t, x) = \min\{U(t; z) : z \text{ such that } X(t; z) = x\} \qquad (5.6)$$

yields the unique semiconcave function satisfying the equation almost everywhere. We have already proved this result (see Proposition 1.5.9) in the case where $H = H(p)$ and we will see another case in Chapter 6. A symmetric property holds for hamiltonians that are concave in p: in this case the maximal selection of $\mathcal{U}$ yields the unique semiconvex almost everywhere solution of the equation. Let us point out, however, that these results are usually obtained "a posteriori," as in Proposition 1.5.9. Rather than studying directly the function u defined in (5.6), one defines a "good" notion of generalized solution using semiconcavity (or the more general theory of viscosity solutions, see the next section). Then one obtains a representation formula of the solution as the value function of an optimization problem (such as the Hopf formula of Chapter 1). Finally, one proves that the generalized solution coincides with u given in (5.6) by showing that the extremal arcs of the optimization problem coincide with the characteristic curves associated to the Hamilton–Jacobi equation. Trying to derive a general theory of weak solutions using only the method of characteristics would not be convenient in general; however, the constructive nature of this method provides a valuable tool for the study of weak solutions, once one knows that there is the relationship described above.   ∎

Let us now consider a more general situation. Let $\Omega \subset \mathbb{R}^n$ be an open set and let $\Gamma \subset \Omega$ be an $(n - 1)$–dimensional surface of class C^2 without boundary. We consider the problem

$$
\begin{cases}
H(x, u(x), Du(x)) = 0, & x \in \Omega \\
u(x) = g(x) & x \in \Gamma,
\end{cases}
\tag{5.7}
$$

where $H \in C^2(\Omega \times \mathbb{R} \times \mathbb{R}^n)$ and $g \in C^2(\Omega)$. Our arguments can be easily adapted to the case of a Dirichlet problem, corresponding to $\Gamma = \partial\Omega$, under the assumptions that H, g are C^2 in $\overline{\Omega}$.

Given a point $z \in \Gamma$, let us denote by $\nu(z)$ the unit normal to Γ at z. If $u \in C^1(\Omega)$ satisfies $u = g$ on Γ, then it is known from elementary calculus that $Du(z) = Dg(z) + \lambda\nu(z)$ for some $\lambda = \lambda(z) \in \mathbb{R}$. If in addition u solves equation (5.7), then λ must belong to the set

$$
\Lambda(z) = \{\lambda \in \mathbb{R} : H(z, g(z), Dg(z) + \lambda\nu(z)) = 0\}.
\tag{5.8}
$$

As in the evolutionary case, we first perform some computations assuming that we already have a solution $u \in C^2(\Omega)$ of the problem (5.7). Given $z \in \Gamma$, we define $X(\cdot; z)$ to be the solution of

$$
\dot{X} = H_p(X, u(X), Du(X)), \qquad X(0; z) = z
\tag{5.9}
$$

and we set

$$
U(s; z) = u(X(s; z)), \qquad P(s; z) = \nabla u(X(s; z)).
\tag{5.10}
$$

We then check whether the triple (X, U, P) satisfies an ordinary differential system with initial conditions determined by g. By a straightforward computation we obtain the following.

Lemma 5.1.3 *Let* $u \in C^2(\Omega)$ *be a solution of* (5.7). *Given* $z \in \Gamma$, *let* X, U, P *be defined as in* (5.9), (5.10). *Then there exists* $\lambda \in \Lambda(z)$ *such that* X, U, P *satisfy*

$$
\begin{cases}
\dot{X} = H_p(X, U, P) \\
\dot{P} = -H_x(X, U, P) - H_u(X, U, P)P \\
\dot{U} = P \cdot H_p(X, U, P)
\end{cases}
\tag{5.11}
$$

with initial conditions

$$
X(0) = z, \quad U(0) = g(z), \quad P(0) = Dg(z) + \lambda \nu(z). \tag{5.12}
$$

We now no longer assume a priori the existence of a solution u, and try to construct one by using system (5.11), called the *characteristic system*. Before going further let us give the following definition. Given $z_0 \in \Gamma$, we say that the vector $p_0 \in \mathbb{R}^n$ is *characteristic* for problem (5.7) at z_0 if

$$
H_p(z_0, g(z_0), p_0) \cdot \nu(z_0) = 0.
$$

Let us denote by $\Lambda^*(z_0)$ the set of all $\lambda \in \Lambda(z_0)$ such that $Dg(z_0) + \lambda \nu(z_0)$ is not characteristic at z_0. It is easy to see that $\Lambda^*(z_0)$ is a discrete set.

Recall that H_p is the speed of the characteristic curves X introduced above. Therefore, a vector p_0 is characteristic if and only if the curve X obtained solving (5.11) with initial conditions $X(0) = z_0$, $P(0) = p_0$, $U(0) = g(z_0)$ is tangent to Γ at z_0. Such a behavior is not good for our purposes because we want the characteristic curves to cover a neighborhood of Γ. Therefore, the choice of λ in the initial conditions (5.12) will be restricted to the elements of $\Lambda^*(z_0)$.

Lemma 5.1.4 *Let* $z_0 \in \Gamma$ *and let* $\lambda_0 \in \Lambda^*(z_0)$. *Then there exists a neighborhood of* z_0 *(in the relative topology of* Γ) $\Gamma' \subset \Gamma$, *and a unique* C^1 *map* $\mu : \Gamma' \to \mathbb{R}$ *such that*

$$
\mu(z_0) = \lambda_0, \quad \mu(z) \in \Lambda^*(z) \quad \forall z \in \Gamma'. \tag{5.13}
$$

Moreover, if we take $\lambda = \mu(z)$ *in* (5.12) *and denote by* $X(t; z)$, $U(t; z)$, $P(t; z)$ *the corresponding solution of* (5.11)–(5.12), *then the map* $(t, z) \to X(t; z)$ *is a* C^1-*diffeomorphism between a neighborhood of* $(0, z_0) \in \mathbb{R} \times \Gamma'$ *and a neighborhood of* $z_0 \in \mathbb{R}^n$.

Proof — Let us introduce a parametrization of Γ near z_0. More precisely, let $A \subset \mathbb{R}^{n-1}$ be an open set, let $\phi : A \to \mathbb{R}^n$ be a map of class C^2 such that $\phi(A) \subset \Gamma$, $D\phi(y)$ has rank $n - 1$ for all $y \in A$ and $\phi(y_0) = z_0$ for some $y_0 \in A$. Let us define, for $(y, \lambda) \in A \times \mathbb{R}$,

$$
F(y, \lambda) = H(\phi(y), g(\phi(y)), Dg(\phi(y)) + \lambda \nu(\phi(y))).
$$

If $\lambda_0 \in \Lambda^*(z_0)$, then $F(y_0, \lambda_0) = 0$ and $F_\lambda(y_0, \lambda_0) \neq 0$. By the implicit function theorem, there is a unique function $\lambda = f(y)$ of class C^1 defined in a neighborhood of y_0 such that $f(y_0) = \lambda_0$ and $F(y, f(y)) = 0$. If we set $\mu(z) = f(\phi^{-1}(z))$ for $z \in \Gamma$ close to z_0 we have that $\mu(z) \in \Lambda(z)$; if we take our neighborhoods sufficiently small we also have, by continuity, that $Dg(z) + \mu(z)\nu(z)$ is noncharacteristic.

The statement about X will follow from the inverse function theorem if we show that the jacobian of the function $\tilde{X}(t, y) := X(t, \phi(y))$ is nonsingular at $(0, y_0)$. Now, since $\tilde{X}(0, y) \equiv \phi(y)$, the partial derivatives of $\tilde{X}$ with respect to the y components at the point $(0, y_0)$ are linearly independent vectors generating the tangent space to Γ at z_0. On the other hand, the vector $\partial_t \tilde{X}(0, y_0) = H_p(z_0, g(z_0), Dg(z_0) + \lambda_0 \nu(z_0))$ does not belong to this space, by the assumption that $\lambda_0 \in \Lambda^*(z_0)$. Thus, the jacobian of $\tilde{X}$ is nonsingular and the lemma is proved. ∎

Theorem 5.1.5 *Given $z_0 \in \Gamma$ and $\lambda_0 \in \Lambda^*(z_0)$, we can choose $R > 0$ such that there exists a unique $u \in C^2$ which solves problem (5.7) in $B_R(z_0)$ and satisfies $Du(z_0) = Dg(z_0) + \lambda_0 \nu(z_0)$.*

Proof — As in the previous lemma, we define $X(t; z)$, $U(t; z)$, $P(t; z)$ solving system (5.11)–(5.12) with $\lambda = \mu(z)$ in (5.12). By the lemma, $X(t; z)$ is locally invertible; therefore we can define in a unique way, for x near z_0, two functions $\tau(x)$ and $\zeta(x)$ such that $X(\tau(x); \zeta(x)) = x$. We then set $u(t, x) = U(\tau(x); \zeta(x))$; we claim that u satisfies the required properties.

First we observe that $H(X(t; z), U(t; z), P(t; z)) \equiv 0$. In fact, recalling equations (5.11) and writing for simplicity $U = U(t, z)$, $H_p = H_p(X, U, Z)$, etc., we find

$$\frac{d}{dt} H(X(t; z), U(t; z), P(t; z)) = H_x \dot{X} + H_u \dot{U} + H_p \dot{P}$$
$$= H_x \cdot H_p + H_u P \cdot H_p - H_p \cdot (H_x + H_u P) = 0,$$

which implies

$$H(X(t; z), U(t; z), P(t; z)) = H(X(0; z), U(0; z), P(0; z))$$
$$= H(z, g(z), Dg(z) + \mu(z)\nu(z)) = 0.$$

Let us now denote by $\phi : A \to \mathbb{R}^n$ a local parametrization of Γ. It is convenient to change variables and set

$$\bar{X}(t, y) = X(t; \phi(y)), \qquad \bar{U}(t, y) = U(t; \phi(y)), \qquad \bar{P}(t, y) = P(t; \phi(y)),$$

where $y \in A \subset \mathbb{R}^{n-1}$ and $t \in \,]-r, r[$ with $r > 0$ suitably small. We want to compute the derivatives of $\bar{U}$. To this purpose we introduce, for any fixed $y \in A$ and $i = 1, \ldots, n-1$, the function

$$v(t) = \frac{\partial \bar{U}}{\partial y_i}(t, y) - \bar{P}(t, y) \cdot \frac{\partial \bar{X}}{\partial y_i}(t, y).$$

Then we have

$$v(0) = \frac{\partial}{\partial y_i} g(\phi(y)) - [Dg(\phi(y)) + \mu(\phi(y))v(\phi(y))] \cdot \frac{\partial \phi}{\partial y_i}(y)$$

$$= -\mu(\phi(y))v(\phi(y)) \cdot \frac{\partial \phi}{\partial y_i}(y) = 0.$$

In addition we have, writing for simplicity H_p instead of $H_p(\bar{X}(t, y), \ldots)$, etc.,

$$\dot{v}(t) = \frac{\partial \bar{P}}{\partial y_i} \cdot H_p + \bar{P} \cdot \frac{\partial}{\partial y_i} H_p + (H_x + H_u \bar{P}) \frac{\partial \bar{X}}{\partial y_i} - \bar{P} \cdot \frac{\partial}{\partial y_i} H_p$$

$$= H_p \cdot \frac{\partial \bar{P}}{\partial y_i} + (H_x + H_u \bar{P}) \cdot \frac{\partial \bar{X}}{\partial y_i}.$$

Now, differentiating the identity $H(\bar{X}, \bar{U}, \bar{P}) \equiv 0$, we obtain

$$H_x \cdot \frac{\partial \bar{X}}{\partial y_i} + H_u \frac{\partial \bar{U}}{\partial y_i} + H_p \cdot \frac{\partial \bar{P}}{\partial y_i} = 0.$$

Therefore

$$\dot{v}(t) = H_u \left(\bar{P} \cdot \frac{\partial \bar{X}}{\partial y_i} - \frac{\partial \bar{U}}{\partial y_i} \right) = -H_u v(t).$$

Since $v(0) = 0$, this implies that $v(t) = 0$ for all t. Thus, we conclude that

$$\frac{\partial \bar{U}}{\partial y_i}(t, y) = \bar{P}(t, y) \cdot \frac{\partial \bar{X}}{\partial y_i}(t, y).$$

We are now ready to compute the derivatives of $u(x)$. If we set $\eta(x) = \phi^{-1}(\zeta(x))$, it follows from our definitions that $u(x) = \bar{U}(\tau(x), \eta(x))$, and that $\bar{X}(\tau(x), \eta(x)) = x$ for all x in a neighborhood of z_0. Thus, we find

$$\frac{\partial u}{\partial x_j}(x) = \frac{\partial \bar{U}}{\partial t} \frac{\partial \tau}{\partial x_j}(x) + \sum_{i=1}^{n-1} \frac{\partial \bar{U}}{\partial y_i} \frac{\partial \eta_i}{\partial x_j}(x)$$

$$= \bar{P} \cdot H_p \frac{\partial \tau}{\partial x_j}(x) + \sum_{i=1}^{n-1} \bar{P} \cdot \frac{\partial \bar{X}}{\partial y_i} \frac{\partial \eta_i}{\partial x_j}(x)$$

$$= \bar{P} \cdot \left(\frac{\partial \bar{X}}{\partial t} \frac{\partial \tau}{\partial x_j}(x) + \sum_{i=1}^{n-1} \frac{\partial \bar{X}}{\partial y_i} \frac{\partial \eta_i}{\partial x_j}(x) \right)$$

$$= \bar{P} \cdot \frac{\partial}{\partial x_j} \bar{X}(\tau(x), \eta(x)) = P \cdot \frac{\partial}{\partial x_j} x = P_j,$$

where P_j is the j-th component of P. Therefore,

$$Du(x) = P(\tau(x), \zeta(x)).$$

This implies that u is of class C^2. In addition

$$H(x, u(x), Du(x)) = H(X(\tau(x); \zeta(x)), U(\tau(x); \zeta(x)), P(\tau(x); \zeta(x))) = 0,$$

and so u solves the equation. The fact that u is the unique solution in a neighborhood of z_0 follows from Lemma 5.1.3 and from the uniqueness of the function $\mu(\cdot)$ given by Lemma 5.1.4.   ∎

It is clear from the above statement that neither local existence nor local unique-ness of a classical solution to problem (5.7) are ensured. In fact, if $\Lambda(z_0)$ is empty, then no smooth solution of (5.7) exists in a neighborhood of z_0. On the other hand, if $\Lambda^*(z_0)$ contains more than one element λ_0, we find different classical solutions corresponding to each choice of λ_0. Finally, nothing can be said in general about the existence and uniqueness of solutions corresponding to values of $\lambda \in \Lambda(z_0) \setminus \Lambda^*(z_0)$.

Applying the previous theorem along the whole surface Γ we easily obtain the following result.

Corollary 5.1.6 *Let $\mu \in C(\Gamma)$ be such that $\mu(z) \in \Lambda^*(z)$ for all $z \in \Gamma$. Then there exists a neighborhood $\mathcal{N}$ of Γ and a unique $u \in C^2(\mathcal{N})$ which solves problem (5.7) and satisfies $Du(z) = Dg(z) + \mu(z)\nu(z)$ for all $z \in \Gamma$.*

As in the evolutionary case, the applicability of the method of characteristics is guaranteed only in a neighborhood of Γ and we cannot expect in general the exis-tence of a smooth solution in the whole set Ω. Observe that in this stationary case we have the additional hypothesis that $\Lambda^*(z)$ be nonempty; this can be interpreted as a compatibility condition on the problem data.

Example 5.1.7 Let us consider the *eikonal equation*

$$|Du(x)|^2 - 1 = 0, \qquad x \in \mathbb{R}^n$$

with data

$$u(z) = 0 \text{ for all } z \in \Gamma := \{z \in \mathbb{R}^n \ : \ |z| = 1\}.$$

Then

$$\Lambda(z) = \Lambda^*(z) = \{-1, +1\}$$

for all $z \in \Gamma$. The two corresponding solutions to the characteristic system are

$$X(t; z) = (1 \pm 2t)z, \qquad U(t; z) = 2t, \qquad P(t; z) = \pm z.$$

We can invert X as long as $1 \pm 2t > 0$ and find

$$z(x) = \frac{x}{|x|}, \qquad t(x) = \pm \frac{|x| - 1}{2}.$$

Thus we obtain the two solutions

$$u(x) = U(t(x); z(x)) = 2t(x) = \pm(|x| - 1).$$

Observe that the two solutions are smooth everywhere except at the origin, which is the point where the characteristic lines intersect. Notice also that the distance func-tion from Γ is given by $d_\Gamma(x) = ||x| - 1|$; thus, both solutions coincide with d_Γ multiplied by a minus sign either inside or outside the unit ball.   ∎

5.2 Viscosity solutions

Let $\Omega \subset \mathbb{R}^n$ be an open set and let $H \in C(\Omega \times \mathbb{R} \times \mathbb{R}^n)$. Let us again consider the general nonlinear first order equation

$$H(x, u, Du) = 0, \qquad x \in \Omega \subset \mathbb{R}^n, \tag{5.14}$$

in the unknown $u : \Omega \to \mathbb{R}$. As usual, evolution equations can be recast in this form by considering time as an additional space variable.

As we have already mentioned, when one considers boundary value problems or Cauchy problems for equations of the above form, one finds that in general no global smooth solutions exist even if the data are smooth. On the other hand, the property of being a Lipschitz continuous function satisfying the equation almost everywhere is usually too weak to have uniqueness results. Therefore, a crucial step in the analysis is to give a notion of a generalized solution such that global existence and uniqueness results can be obtained. In Chapter 1 we have seen a class of problems which are well posed in the class of semiconcave solutions. Here we present the notion of a viscosity solution, which has a much wider range of applicability.

Definition 5.2.1 *A function* $u \in C(\Omega)$ *is called a* viscosity subsolution *of equation* (5.14) *if, for any* $x \in \Omega$, *it satisfies*

$$H(x, u(x), p) \le 0, \qquad \forall p \in D^+ u(x). \tag{5.15}$$

Similarly, we say that u *is a* viscosity supersolution *of equation* (5.14) *if, for any* $x \in \Omega$, *we have*

$$H(x, u(x), p) \ge 0, \qquad \forall p \in D^- u(x). \tag{5.16}$$

If u *satisfies both of the above properties, it is called a* viscosity solution *of equation* (5.14).

Observe that, by virtue of Proposition 3.1.7, condition (5.15) (resp. (5.16)) can be restated in an equivalent way by requiring

$$H(x, u(x), D\phi(x)) \le 0 \qquad (\text{resp. } H(x, u(x), D\phi(x)) \ge 0) \tag{5.17}$$

for any $\phi \in C^1(\Omega)$ such that $u - \phi$ has a local maximum (resp. minimum) at x.

We see that if u is differentiable everywhere the notion of a viscosity solution coincides with the classical one since we have at any point $D^+ u(x) = D^- u(x) = \{Du(x)\}$. On the other hand, if u is not differentiable everywhere, the definition of a viscosity solution includes additional requirements at the points of nondifferentiability. The reason for taking inequalities (5.15)–(5.16) as the definition of solution might not be clear at first sight, as well as the relation with the semiconcavity property considered in Chapter 1. However, we will see that with this definition one can obtain existence and uniqueness results for many classes of Hamilton–Jacobi equations, and that the viscosity solution usually coincides with the one which is relevant for the applications, like the value function in optimal control. The relationship with semiconcavity will be examined in detail in the next section.

Example 5.2.2 Consider the equation $(u'(x))^2 - 1 = 0$ in the interval $]-1, 1[$ with boundary data $u(-1) = u(1) = 0$. We first observe that no classical solution exists: if $u \in C^1([-1, 1])$ satisfies $u(-1) = u(1) = 0$, then we have $u'(x_0) = 0$ at some $x_0 \in]-1, 1[$ and so the equation does not hold at x_0. On the other hand, we can check that $u(x) = 1 - |x|$ is a viscosity solution of the equation. The equation is clearly satisfied for $x \neq 0$ where u is differentiable. At $x = 0$ we have $D^- u(0) = \emptyset$ and $D^+ u(0) = [-1, 1]$; thus, the requirement in (5.16) is empty, while (5.15) holds since $p^2 - 1 \leq 0$ for all $p \in D^+ u(0)$. In the following (see Theorem 5.2.10) we will see a uniqueness result which ensures that u is the only viscosity solution of our boundary value problem.

If we consider the function $v(x) = -u(x) = |x| - 1$, we find by a similar analysis that v is a subsolution but not a supersolution. Observe that the situation changes if we consider equation $1 - (u'(x))^2 = 0$; then v is a solution, while u is only a supersolution. This shows that viscosity solutions are not invariant after changing the sign of the whole equation; this unusual feature can be better understood in the light of results such as Theorem 5.2.3 below. ∎

Let us mention that the notion of a viscosity solution can be generalized to cases where u, or H, or both, are discontinuous, or to second order equations whose principal part is degenerate elliptic. These topics are beyond the scope of the present work.

In this section we summarize some of the main results about viscosity solutions. For the reader who is not familiar with this theory, we first prove some statements under special hypotheses in order to present some of the basic techniques. Then we give, without proof, more general existence and uniqueness results. For a comprehensive exposition of the theory, the reader may consult [110, 66, 22, 20, 64, 71] and the references therein.

We first present an existence result based on the vanishing viscosity method.

Theorem 5.2.3 *Let $\{\varepsilon_k\} \subset]0, +\infty[$ be a sequence converging to 0 and let $u_k \in C^2(\Omega)$ be a solution of*

$$H(x, u_k(x), Du_k(x)) = \varepsilon_k \Delta u_k(x), \qquad x \in \Omega. \tag{5.18}$$

Suppose in addition that the sequence u_k converges to some function u locally uniformly in Ω. Then u is a viscosity solution of equation (5.14).

Proof — It is convenient to use the equivalent formulation (5.17) of the definition of a viscosity solution. We prove only that u is a subsolution since the other part is completely analogous. Let $\phi \in C^1(\Omega)$ be such that $u - \phi$ has a maximum at some point $x_0 \in \Omega$. We have to show that

$$H(x_0, u(x_0), D\phi(x_0)) \leq 0. \tag{5.19}$$

It suffices to consider the case where $u - \phi$ has a strict maximum. In fact, if the maximum is not strict, we can replace $\phi(x)$ by $\psi(x) = \phi(x) + |x - x_0|^2$. Then

$u - \psi$ has a strict maximum at x_0 and $D\phi(x_0) = D\psi(x_0)$; therefore, if ψ satisfies (5.19), the same holds for ϕ.

We first prove (5.19) under the additional assumption that ϕ is of class C^2. Then, since x_0 is a strict local maximum for $u - \phi$ and $u_k \to u$ locally uniformly, it is easily seen that there exists a sequence $\{x_k\}$ of points converging to x_0 such that $u_k - \phi$ has a local maximum at x_k. Then we have

$$Du_k(x_k) = D\phi_k(x_k), \qquad \Delta u_k(x_k) \le \Delta\phi(x_k).$$

Using the fact that u_k solves (5.18) we obtain

$$H(x_k, u(x_k), D\phi(x_k)) \le \varepsilon_k \Delta\phi(x_k),$$

and so inequality (5.19) follows as $k \to \infty$.

If ϕ is only of class C^1, we pick a sequence $\phi_k \in C^2(\Omega)$ such that $\phi_k \to \phi$, $D\phi_k \to D\phi$ locally uniformly in Ω. Then, since $u - \phi$ has a strict local maximum at x_0, there exists a sequence $\{y_k\}$ of points converging to x_0 such that $u - \phi_k$ has a local maximum at y_k. By the first part of the proof, we have that $H(y_k, u(y_k), D\phi_k(y_k)) \le 0$. Letting $k \to \infty$, we obtain the assertion. ∎

Remark 5.2.4 In the previous result one assumes the existence of a solution for the approximate equation (5.18). Such an equation is semilinear elliptic, and there are standard techniques (see [110]) which ensure, under general hypotheses, the existence of a solution, as well as suitable bounds yielding the uniform convergence of some sequence $\{u_k\}$. The above result shows how powerful the notion of viscosity solution is since the passage to the limit can be done under the assumption only that $\{u_k\}$ converges and without requiring explicitly any convergence of the derivatives.

As explained in Remark 2.2.7, the term $\varepsilon \Delta u$, with $\varepsilon \to 0^+$, is called *vanishing viscosity*. This kind of approximation explains why the viscosity solutions of a Hamilton–Jacobi change if we multiply the equation by -1; the approximating equation becomes $-H(x, u, Du) = \varepsilon \Delta u$, which does not have the same solutions as (5.18). The vanishing viscosity method played an important role in the beginning of the theory of viscosity solutions (see [67, 110]), and explains the name of these solutions. Since then several other approaches have been introduced to prove existence results, some of which will be mentioned later in this section. ∎

By an argument completely analogous to the proof of the previous theorem one can obtain the following stability result for viscosity solutions.

Theorem 5.2.5 *Let $u_k \in C(\Omega)$ be viscosity subsolutions (resp. supersolutions) of*

$$H_k(x, u_k, Du_k) = 0$$

where $H_k \in C(\Omega \times \mathbb{R} \times \mathbb{R}^n)$. Suppose that $u_k \to u$ and $H_k \to H$ locally uniformly. Then u is a viscosity subsolution (resp. supersolution) of

$$H(x, u, Du) = 0.$$

We now give a uniqueness statement for viscosity solutions of a problem on the whole space.

Theorem 5.2.6 *Let $u_1, u_2 \in C(\mathbb{R}^n)$ be viscosity solutions of*

$$u(x) + H(Du(x)) - n(x) = 0, \qquad x \in \mathbb{R}^n,$$

where H is continuous and n is uniformly continuous. Suppose in addition that $u_i(x) \to 0$ as $|x| \to +\infty$ for $i = 1, 2$. Then $u_1 \equiv u_2$.

Proof — We argue by contradiction and suppose that the difference $u_2 - u_1$ is not identically zero. Since $u_2(x) - u_1(x) \to 0$ as $|x| \to \infty$, it has either a positive maximum or a negative minimum. Let us suppose for instance that it attains a positive maximum m at some point $\bar{x}$. Let us also set $M = \sup |u_1| + \sup |u_2|$.

For given $\varepsilon > 0$, let us consider the function of $2n$ variables

$$\Phi_\varepsilon(x, y) = u_2(x) - u_1(y) - \frac{|x - y|^2}{2\varepsilon}, \qquad x, y \in \mathbb{R}^n.$$

It is easy to check that the assumption that $u_1(x) \to 0$ and $u_2(x) \to 0$ as $|x| \to \infty$ implies that $\limsup_{||(x,y)|| \to \infty} \Phi_\varepsilon(x, y) = 0$. Then, since $\Phi_\varepsilon(\bar{x}, \bar{x}) = m > 0$, we deduce that Φ_ε attains a maximum at some point $(x_\varepsilon, y_\varepsilon)$ and that $\Phi_\varepsilon(x_\varepsilon, y_\varepsilon) \geq m$. Therefore

$$|x_\varepsilon - y_\varepsilon|^2 = 2\varepsilon(u_2(x_\varepsilon) - u_1(y_\varepsilon) - \Phi_\varepsilon(x_\varepsilon, y_\varepsilon)) \leq 2\varepsilon(M - m). \qquad (5.20)$$

The property that $(x_\varepsilon, y_\varepsilon)$ is a maximum point for Φ_ε implies in particular that x_ε is a maximum point for the function $x \to u_2(x) - (|x - y_\varepsilon|^2)/(2\varepsilon)$. Hence, the superdifferential of this function at the point $x = x_\varepsilon$ contains 0, and this is equivalent to $(x_\varepsilon - y_\varepsilon)/\varepsilon \in D^+ u_2(x_\varepsilon)$. Using the definition of viscosity solution, we deduce that

$$u_2(x_\varepsilon) + H\left(\frac{x_\varepsilon - y_\varepsilon}{\varepsilon}\right) \leq n(x_\varepsilon).$$

Analogously, from the property that $y \to -u_1(y) - (|x_\varepsilon - y|^2)/(2\varepsilon)$ attains its maximum at y_ε we deduce that $(x_\varepsilon - y_\varepsilon)/\varepsilon \in D^- u_1(y_\varepsilon)$ and therefore

$$u_1(y_\varepsilon) + H\left(\frac{x_\varepsilon - y_\varepsilon}{\varepsilon}\right) \geq n(y_\varepsilon).$$

Hence

$$u_2(x_\varepsilon) - u_1(y_\varepsilon) \leq n(x_\varepsilon) - n(y_\varepsilon) \leq \omega(|y_\varepsilon - x_\varepsilon|),$$

where ω is the modulus of continuity of n. Taking into account (5.20) we obtain that

$$\limsup_{\varepsilon \to 0} [u_2(x_\varepsilon) - u_1(y_\varepsilon)] \leq 0.$$

On the other hand, we have that $u_2(x_\varepsilon) - u_1(y_\varepsilon) \geq \Phi_\varepsilon(x_\varepsilon, y_\varepsilon) \geq m > 0$ for every $\varepsilon > 0$. The contradiction shows that $u_1 \equiv u_2$. ∎

The above uniqueness result holds under less restrictive assumptions on the behavior of u_1, u_2 at infinity; the hypothesis that u_1, u_2 tend to 0 allows us to avoid some technicalities in the proof.

We now give some more general statements without proof. We begin by giving a comparison result for the stationary case, first in the case of a bounded domain, then in the whole space. It is possible to give many different versions of these results by changing the hypotheses on the hamiltonian and on the solutions; a comprehensive account can be found in the general references quoted earlier.

Theorem 5.2.7 *Let $\Omega \subset \mathbb{R}^n$ be open and bounded, and let $H \in C(\Omega \times \mathbb{R}^n)$. Suppose that u_1, $u_2 \in \mathrm{Lip}\,(\overline{\Omega})$ are a viscosity subsolution and supersolution, respectively, of*

$$u(x) + H(x, Du(x)) = 0, \qquad x \in \Omega. \tag{5.21}$$

If $u_1 \le u_2$ on $\partial\Omega$, then $u_1 \le u_2$ in Ω.

Proof – See e.g., [22, Theorem 2.4]. ∎

Theorem 5.2.8 *Let $H \in C(\mathbb{R}^n \times \mathbb{R}^n)$ and suppose that, for any $R > 0$, H is uniformly continuous in $\mathbb{R}^n \times B_R$. Let u_1, $u_2 \in \mathrm{Lip}\,(\mathbb{R}^n)$ be a viscosity subsolution and supersolution respectively of*

$$u(x) + H(x, Du(x)) = 0, \qquad x \in \mathbb{R}^n. \tag{5.22}$$

Then $u_1 \le u_2$ in $\mathbb{R}^n$.

Proof – See [22, Theorem 2.11]. ∎

The previous two theorems can be extended to hamiltonians of the form $H(x, u, Du)$ provided H is strictly increasing in the u argument. If this monotonicity assumption is violated, then the solution to the Dirichlet problem may no longer be unique, as shown by the following example (taken from [96]).

Example 5.2.9 The Dirichlet problem

$$\begin{cases} u'(x)^2 = x^2, & x \in \,]-1, 1[\\ u(-1) = u(1) = 0 \end{cases}$$

has infinitely many viscosity solutions, given by

$$u_c(x) = \frac{1}{2} \min\{(1 - x^2), (c + x^2)\}$$

for $c \in [-1, 1]$.

However, the strict monotonicity in u can be replaced by other assumptions, as shown in the next result.

Theorem 5.2.10 *Let $H \in C(\Omega \times \mathbb{R}^n)$ satisfy the following.*

(i) $H(x, \cdot)$ is convex for any $x \in \Omega$;
(ii) there exists $\phi \in C^1(\Omega) \cap C(\overline{\Omega})$ such that $\sup_{x \in \Omega} H(x, D\phi(x)) < 0$.

Let $u_1, u_2 \in \text{Lip}(\overline{\Omega})$ be a viscosity subsolution and supersolution, respectively, of

$$H(x, Du(x)) = 0, \qquad x \in \Omega.$$

If $u_1 \le u_2$ on $\partial\Omega$, then $u_1 \le u_2$ in Ω.

Proof — See [22, Theorem 2.7]. ■

Example 5.2.11 The above theorem can be applied to equation $|Du(x)|^2 - n(x) = 0$ if $n \in C(\overline{\Omega})$ is a positive function. In this case, in fact, hypothesis (ii) holds taking $\phi \equiv 0$. Let us remark that the same argument does not apply to Example 5.2.9 since the function on the right-hand side vanishes at $x = 0$.

We now turn to the evolutionary case and give a comparison result for the Cauchy problem.

Theorem 5.2.12 *Let $H \in C([0, T] \times \mathbb{R}^n \times \mathbb{R}^n)$ satisfy*

$$|H(t, x, p) - H(t, x, q)| \le K(|x| + 1)|p - q|, \qquad \forall t, x, p, q$$

for some $K > 0$. Suppose also that for all $R > 0$, there exists $m_R : [0, \infty[\to [0, \infty[$ continuous, nondecreasing, with $m_R(0) = 0$ and such that

$$|H(t, x, p) - H(t, y, p)| \le m_R(|x - y|) + m_R(|x - y||p|),$$

$$\forall x, y \in B_R, \ p \in \mathbb{R}^n, \ t \in [0, T].$$

Let $u_1, u_2 \in C([0, T] \times \mathbb{R}^n)$ be, respectively, a viscosity subsolution and supersolution of the equation

$$u_t + H(t, x, \nabla u(t, x)) = 0, \qquad (t, x) \in]0, T[\times \mathbb{R}^n. \tag{5.23}$$

Then

$$\sup_{\mathbb{R}^n \times [0,T]} (u_1 - u_2) \le \sup_{\mathbb{R}^n} (u_1(0, \cdot) - u_2(0, \cdot)).$$

Proof — See [20, Theorem III.3.15, Exercise III.3.6]. ■

Remark 5.2.13 The previous result can be generalized to hamiltonians depending also on u. While in the stationary case it was necessary to assume that H is monotone increasing with respect to u, in the evolutionary case it suffices to assume that there exists $\gamma \ge 0$ such that

$$H(t, x, u, p) - H(t, x, v, p) \ge -\gamma(u - v), \qquad \text{for all } u > v.$$

However, the dependence on u cannot be completely arbitrary. In particular, the case of hyperbolic conservation laws is not covered by the theory of viscosity solutions and has to be treated using other techniques. This is not surprising if one thinks of the connection described in Section 1.7, which suggests that solutions to conservation laws are less regular than the ones to Hamilton–Jacobi equations (roughly speaking, they have "one derivative less"). ∎

Let us now mention some existence results for viscosity solutions. We first consider the stationary case in the whole space.

Theorem 5.2.14 *Let* $H : \mathbb{R}^n \times \mathbb{R}^n$ *be uniformly continuous on* $\mathbb{R}^n \times B_R$ *for every* $R > 0$. *Suppose in addition that* $H(\cdot, 0)$ *is bounded on* $\mathbb{R}^n$ *and that*

$$\lim_{|p| \to \infty} \inf_{x \in \mathbb{R}^n} H(x, p) = +\infty.$$

Then there exists a viscosity solution of $u + H(x, Du) = 0$ *in* $\mathbb{R}^n$ *which belongs to* $W^{1,\infty}(\mathbb{R}^n)$.

Proof — See [22, Theorem 2.12]. ∎

Existence theorems like the previous one are usually proved by a Perron-type technique which was introduced by Ishii [94]. Such a procedure can be applied, roughly speaking, to all Hamilton–Jacobi equations where the comparison principle for viscosity solutions holds.

Another way to obtain existence for Hamilton–Jacobi equations is illustrated in the next result. In this approach one introduces a suitable problem in the calculus of variations and proves that the value function is the viscosity solution of the equation. The connection between Hamilton–Jacobi equations and problems in the calculus of variations (or optimal control) has already been observed in Chapter 1 in a special case, and other examples will be given in the next chapters. Observe that this connection is exploited here in a reversed way compared to the approach followed elsewhere in the book: here one starts from the equation and then introduces a problem in the calculus of variations in order to solve the equation. Such an approach to prove existence does not have such a wide applicability as the Perron-type technique mentioned above (in particular, it is only applicable when H is convex with respect to p), but it has the important advantage of yielding a representation formula for the solution.

Theorem 5.2.15 *Let the open set* $\Omega \subset \mathbb{R}^n$ *be bounded, smooth and connected, let* $H \in C(\overline{\Omega} \times \mathbb{R}^n)$ *and* $\phi \in C(\partial\Omega)$. *Consider the Dirichlet problem*

$$\begin{cases} u(x) + H(x, Du(x)) = 0, & x \in \Omega \\ u(x) = \phi(x), & x \in \partial\Omega. \end{cases} \tag{5.24}$$

Suppose that H *is convex with respect to* p *and satisfies*

$$\lim_{|p|\to\infty} \left(\inf_{x\in\overline{\Omega}} H(x,p) \right) = +\infty.$$

Let $L(x,q) = \sup_{q\in\mathbb{R}^n}[-p\cdot q - H(x,p)]$. Define, for $x \in \overline{\Omega}$, $y \in \partial\Omega$,

$$v(x,y) = \inf \left\{ \int_0^T e^{-t}L(\gamma(t),\dot{\gamma}(t))\,dt + e^{-T}\phi(y) \right\}, \qquad (5.25)$$

where the infimum is taken over all $T \geq 0$ and all arcs $\gamma \in \mathrm{Lip}\,([0,T])$ such that $\gamma(0) = x$, $\gamma(T) = y$ and $\gamma(t) \in \overline{\Omega}$ for all $t \in [0,T]$.

Then problem (5.24) possesses a viscosity solution $u \in \mathrm{Lip}\,(\overline{\Omega})$ if and only if we have $v(x,y) \geq \phi(x)$ for all x, $y \in \partial\Omega$; in this case, the solution is given by

$$u(x) = \inf_{y\in\partial\Omega} v(x,y), \qquad x \in \overline{\Omega}.$$

Proof — See [110, Theorem 5.4]. ∎

Remark 5.2.16 From the previous statement we see that the Dirichlet problem under consideration may not possess any solution $u \in \mathrm{Lip}\,(\overline{\Omega})$ for some boundary data. In fact, the boundary data must satisfy the compatibility condition $\phi(x) \leq v(x,y)$ for all x, $y \in \partial\Omega$. To see, at least intuitively, why some compatibility condition on the Dirichlet data is needed, let us consider a slightly different equation, namely the eikonal equation

$$|Du(x)|^2 = 1, \qquad x \in \Omega.$$

It can be proved that viscosity solutions to this equation are Lipschitz continuous with constant 1. Therefore, if we are looking for a viscosity solution assuming the boundary data in the classical sense, we have to restrict ourselves to data $\phi \in C(\partial\Omega)$ which are Lipschitz with constant 1. Such a requirement can be restated as

$$\phi(x) \leq \phi(y) + |x - y|, \ \text{for all } x, y \in \partial\Omega,$$

which has some formal analogy with the compatibility condition in the above theorem. An alternative way of characterizing admissible boundary data for the Dirichlet problem is requiring the existence of a sub- and supersolution to the equation assuming the data on the boundary. More general problems can be treated by the theory of discontinuous viscosity solutions (see e.g., [22], [20] and the references therein). ∎

5.3 Semiconcavity and viscosity

We now analyze the relation between the notions of a semiconcave solution and a viscosity solution to Hamilton–Jacobi equations. We will see that the two notions are strictly related when the hamiltonian is a convex function of Du. We begin with the following result.

Proposition 5.3.1 *Let u be a semiconcave function satisfying equation* (5.14) *almost everywhere. If H is convex in the third argument, then u is a viscosity solution of the equation.*

Proof — As a first step, we show that u satisfies the equation at all points of differentiability (our assumption is a priori slightly weaker). Let u be differentiable at some $x_0 \in \Omega$. Then there exists a sequence of points x_k converging to x_0 such that

(i) u is differentiable at x_k;
(ii) $H(x_k, u(x_k), Du(x_k)) = 0$;
(iii) $Du(x_k)$ has a limit for $k \to \infty$.

From Proposition 3.3.4(a) we deduce that the limit of $Du(x_k)$ is $Du(x_0)$. By the continuity of H, the equation holds at x_0 as well.

Let us now take an arbitrary $x \in \Omega$ and check that (5.15) is satisfied. We first observe that

$$H(x, u(x), p) = 0, \qquad \forall\, p \in D^* u(x). \tag{5.26}$$

This follows directly from the definition of $D^* u$, the continuity of H and the property that the equation holds at the points of differentiability. Since $D^+ u(x)$ is the convex hull of $D^* u(x)$ (see Theorem 3.3.6) and H is convex, inequality (5.15) follows.

Now let us check inequality (5.16). For a given $x \in \Omega$, suppose that $D^- u(x)$ contains some vector p. Then, by Proposition 3.1.5(c) and Proposition 3.3.4(c), u is differentiable at x and $Du(x) = p$. Thus, (5.16) holds as an equality by the first part of the proof. ∎

Remark 5.3.2 A more careful analysis shows that the convexity of H and the semiconcavity of u play independent roles in the viscosity property. In fact, in the previous proof, the deduction that u is a supersolution uses only semiconcavity and is valid also if H is not convex. On the other hand, it is possible to prove (see see [110, p. 96] or [20, Prop. II.5.1]) that, if H is convex, any Lipschitz continuous u (not necessarily semiconcave) satisfying the equation almost everywhere is a viscosity subsolution of the equation.

Remark 5.3.3 If we consider equations where the hamiltonian is not convex, then the viscosity solution is not semiconcave in general. Consider for instance equation

$$1 - u'(x)^2 = 0, \qquad x \in\,]-1, 1[$$

with boundary data $u(-1) = u(1) = 0$ (see Example 5.2.2). Then the function $u_1(x) = 1 - |x|$ is semiconcave and satisfies the equation almost everywhere, but is not a viscosity solution. On the other hand, the viscosity solution of the problem is the semiconvex function $u_2(x) = |x| - 1$. In general, when we consider equations with a concave hamiltonian, all results of this section hold with semiconcavity replaced by semiconvexity. When the hamiltonian is neither convex nor concave, the viscosity property has in general no relation with semiconcavity or semiconvexity.

Let us now consider the case when the hamiltonian is convex and the viscosity solution is semiconvex rather than semiconcave. This behavior is somehow unusual, as one can understand from the other results of this section, but there are some interesting cases where it occurs (see for instance Theorem 7.2.11 and 8.3.1). In such cases one can derive consequences on the regularity of the solution, as we show here.

Proposition 5.3.4 Let $u : \Omega \to \mathbb{R}$ be a semiconvex viscosity solution of equation (5.14). If H is convex with respect to the last argument, then u satisfies

$$H(x, u(x), p) = 0, \qquad x \in \Omega, \ p \in D^- u(x). \tag{5.27}$$

If, in addition, for all (x, u) the zero level set of $H(x, u, \cdot)$ does not contain line segments (e.g., if H is strictly convex in p) then $u \in C^1(\Omega)$.

Proof — Since u is a viscosity solution, it satisfies the equation in the classical sense at all points of differentiability. Then, as in the proof of Proposition 5.3.1, we find that (5.26) holds at any point $x \in \Omega$. We recall that, by Theorem 3.3.6, the subdifferential of a semiconvex function satisfies $D^- u = \text{co } D^* u$. Thus, the convexity of H implies

$$H(x, u(x), p) \le 0, \qquad x \in \Omega, \ p \in D^- u(x).$$

But the converse inequality holds as well, by the definition of a viscosity solution, and so we have proved (5.27).

If the set $\{p : H(x, u(x), p) = 0\}$ does not contain any segment, then by (5.27) $D^- u(x)$ reduces to a point for every $x \in \Omega$. Since u is semiconvex, Proposition 3.3.4(e) shows that $u \in C^1(\Omega)$. ∎

As we have seen in Proposition 5.3.1, the fact that a semiconcave solution is also a viscosity solution is a rather direct consequence of the definitions and of the properties of generalized differentials. The converse implication is also true in many cases where the hamiltonian is convex, as we will see, but it cannot be proved in such a direct way. This shows intuitively that semiconcavity is a stronger property than viscosity; the fact that viscosity solutions are in many cases semiconcave is nontrivial, and can be regarded as an additional regularity information. As long as one is concerned with existence and uniqueness issues, the notion of a viscosity solution is the most convenient one. On the other hand, if one is also interested in the structure of the singularities, the viscosity property yields little information, while semiconcavity allows one to obtain the finer results described in Chapter 4.

Let us therefore consider the issue of showing that a viscosity solution is semiconcave. Two possible approaches to the proof of this property are:

(I) Consider the elliptic or parabolic approximation to the equation, obtained by adding a term of the form $\varepsilon \Delta u$ with $\varepsilon > 0$. Use the maximum principle to derive a semiconcavity estimate on u independent of ε (as in Proposition 2.2.6). Pass to the limit as $\varepsilon \to 0$ to obtain the same estimate for the viscosity solution of the Hamilton–Jacobi equation.

(II) Employ techniques from control theory to obtain a representation formula for the viscosity solution as the infimum of a suitable family of functions. Use this formula to prove semiconcavity by arguments similar to those used in Section 1.6 for solutions given by Hopf's formula.

We record here, without proof, some semiconcavity results for viscosity solutions which can be obtained by the two approaches described above. In the following chapters we shall prove semiconcavity results obtained by approach (II) for Hamilton–Jacobi equations associated to problems in the calculus of variations and optimal control.

We first give semiconcavity statements for hamiltonians which do not depend on x, u.

Theorem 5.3.5 *Let $\Omega \subset \mathbb{R}^n$ be an open set.*

(i) *Let $u \in C(\Omega)$ be a viscosity solution of $u + H(Du) - n(x) = 0$, with H uniformly convex and $n \in \mathrm{SCL}_{loc}(\Omega)$. Then $u \in \mathrm{SCL}_{loc}(\Omega)$.*

(ii) *Let $u \in C(\,]0, T[\,\times\Omega)$ be a viscosity solution of $\partial_t u + H(t, \nabla u) - n(x) = 0$ in $\,]0, T[\,\times\Omega$, with H locally Lipschitz continuous, $H(t, \cdot)$ uniformly convex, $u(0, \cdot) \in \mathrm{Lip}\,(\Omega)$, $n \in \mathrm{SCL}_{loc}(\Omega)$. Then $u(t, \cdot) \in \mathrm{SCL}_{loc}(\Omega)$ for every $t \in \,]0, T[$.*

The above results are proved in [110, Prop. 3.3, Cor. 9.2] by the vanishing viscosity method (see also [102]). Observe that part (ii) of the statement, in the special case of $\Omega = \mathbb{R}^n$, $H = H(\nabla u)$, $n \equiv 0$, follows from Corollary 1.6.6. These results show that the convexity of the hamiltonian induces a gain of regularity, since the solution is semiconcave, while the data are only Lipschitz continuous.

Let us now quote a semiconcavity result for general hamiltonians depending on (x, u, p). Observe that in this case stronger regularity assumptions are required on H, in addition to convexity with respect to Du.

Theorem 5.3.6 *Let $u \in \mathrm{Lip}_{loc}(\Omega)$ be a viscosity solution of (5.14) and suppose that the hamiltonian $H(x, u, p)$ satisfies the following properties on any compact subset of $\Omega \times \mathbb{R} \times \mathbb{R}^n$:*

(i) *H is Lipschitz continuous and the distributional derivatives H_{xu}, H_{xp}, H_{up}, H_{pp} are bounded.*

(ii) *H is uniformly convex with respect to p.*

(iii) *H is semiconcave in x with a linear modulus independent of u, p.*

Then u is locally semiconcave with a linear modulus in Ω.

For the proof, see [110, Theorem 3.3]. The reader may find other semiconcavity results with convex hamiltonians in [102, 83].

More recently, semiconcavity results have been obtained under much milder regularity assumptions on H. We remark, however, that these results are based on the generalized estimate of Section 2.5 and that they no longer yield semiconcavity with a linear modulus. Let us first consider the stationary case.

Theorem 5.3.7 *Let $\Omega \subset \mathbb{R}^n$ be an open set and let $u \in \mathrm{Lip}_{loc}(\Omega)$ be a viscosity solution of $H(x, u, Du) = 0$ in Ω, where $H \in \mathrm{Lip}_{loc}(\Omega \times \mathbb{R} \times \mathbb{R}^n)$ is strictly convex in the last argument. Then u is locally semiconcave in Ω.*

Proof — Since our statement is of a local nature, it is not restrictive to consider the case when Ω is bounded and $u \in \mathrm{Lip}(\Omega)$. We can also assume that the Lipschitz continuity of H is uniform with respect to x and, by possibly redefining $H(x, u, p)$ for large p, that H has superlinear growth with respect to p. In addition, it is enough to consider the case when the equation satisfied by u is of the form

$$u + H(x, Du) = 0. \tag{5.28}$$

In fact, in the general case we can set $\tilde{H}(x, p) = -u(x) + H(x, u(x), p)$; then u is a viscosity solution of $u + \tilde{H}(x, Du) = 0$ with $\tilde{H}$ satisfying the same assumptions as H.

We will exploit the property that the viscosity solution of an equation of the form (5.28) can be represented as the infimum of a variational problem, as it is stated in Theorem 5.2.15. Let B be any open ball contained in Ω. Given $\bar{x} \in B$, we consider the following set of arcs starting from $\bar{x}$:

$$\mathcal{A}(\bar{x}) = \{\xi \in \mathrm{Lip}_{loc}([0, \infty[\, , \mathbb{R}^n) : \xi(0) = \bar{x}\}. \tag{5.29}$$

Given $\xi \in \mathcal{A}(\bar{x})$, we denote by t_ξ the exit time of ξ from $\overline{B}$, defined as

$$t_\xi = \inf\{t > 0 : \xi(t) \notin \overline{B}\} \le +\infty.$$

Let us introduce the function

$$L(x, q) = \max_{p \in \mathbb{R}^n}[-q \cdot p - H(x, p)].$$

Up to a minus sign in front of q, this is the Legendre transform of H with respect to the p variable. We use this definition in order to avoid some minus signs in the following formulas; the results of Appendix A. 2 can be adapted in a straightforward way. We then define the functional

$$J(\bar{x}, \xi) = \int_0^{t_\xi} e^{-t} L(\xi(t), \dot{\xi}(t)) dt + e^{-t_\xi} u(\xi(t_\xi)), \qquad \forall \xi \in \mathcal{A}(\bar{x}). \tag{5.30}$$

Then, by Theorem 5.2.15, we have

$$u(\bar{x}) = \inf_{\xi \in \mathcal{A}(\bar{x})} J(\bar{x}, \xi). \tag{5.31}$$

In addition, by standard arguments in the calculus of variations which exploit the superlinear growth of L (see e.g., Theorem 7.4.6), in taking the above infimum it is enough to consider the arcs in $\mathcal{A}(\bar{x})$ satisfying the restriction $|\dot{\xi}(t)| \le C$, where C is a constant depending only on H and on the Lipschitz constant of u in B.

For a given $\varepsilon > 0$, let us set $\Omega_\varepsilon = \{x \in \Omega : \text{dist}(x, \partial\Omega) > \varepsilon\}$. By Theorem A. 2.6–(c)–(d) we can find $K > 0$ such that

$$|L(x, q) - L(y, q)| \leq K|x - y|, \qquad \forall x, y \in \Omega_{\varepsilon/2}, \forall q \in B_{3C}. \qquad (5.32)$$

In addition, $L_q(x, q)$ exists and is continuous. Therefore, arguing as in Proposition 2.1.2, if we denote by $\omega(\cdot)$ the modulus of continuity of $L_q(x, q)$ in $\Omega_{\varepsilon/2} \times B_{3C}$, we have

$$\lambda L(x, q_1) + (1 - \lambda)L(x, q_2) - L(x, \lambda q_1 + (1 - \lambda)q_2) \qquad (5.33)$$
$$\leq \lambda(1 - \lambda)|q_2 - q_1|\omega(|q_2 - q_1|), \qquad x \in \Omega_{\varepsilon/2}, \, q_1, q_2 \in B_{3C}, \, \lambda \in [0, 1].$$

We now want to show that u satisfies the generalized one-sided estimate (2.20) in Ω_ε. Aiming at this, let us fix an arbitrary $x_0 \in \Omega_\varepsilon$ and $\delta \in \,]0, \varepsilon/8[$. We will use the representation formula (5.31) with $B = B_{4\delta}(x_0)$. Let us take $x, y \in B_1, \lambda \in [0, 1]$. Let ξ be an arbitrary arc in $\mathcal{A}(x_0 + \delta(\lambda x + (1 - \lambda)y))$ such that $|\dot{\xi}| \leq C$. Then $\xi(t) \in B_{2\delta}(x_0)$ for $t \in [0, \delta C^{-1}]$ and $\xi(t) \in B$ for $t \in [0, 3\delta C^{-1}]$. This implies, in particular, that $t_\xi > 3\delta C^{-1}$. Therefore, if we set

$$\xi_1(t) = \begin{cases} \xi(t) + (1 - \lambda)(\delta - Ct)(x - y) & \text{if } t \leq \delta C^{-1} \\ \xi(t) & \text{if } t > \delta C^{-1}, \end{cases}$$

$$\xi_2(t) = \begin{cases} \xi(t) - \lambda(\delta - Ct)(x - y) & \text{if } t \leq \delta C^{-1} \\ \xi(t) & \text{if } t > \delta C^{-1}, \end{cases}$$

we have that $\xi_1 \in \mathcal{A}(x_0 + \delta x), \xi_2 \in \mathcal{A}(x_0 + \delta y)$ and $t_{\xi_1} = t_{\xi_2} = t_\xi$. Using (5.32) and (5.33) we deduce, for $t \in [0, \delta C^{-1}]$ a.e.,

$$\lambda L(\xi_1(t), \dot{\xi}_1(t)) + (1 - \lambda)L(\xi_2(t), \dot{\xi}_2(t)) - L(\xi(t), \dot{\xi}(t))$$
$$= \lambda[L(\xi_1(t), \dot{\xi}_1(t)) - L(\xi(t), \dot{\xi}_1(t))]$$
$$+ (1 - \lambda)[L(\xi_2(t), \dot{\xi}_2(t)) - L(\xi(t), \dot{\xi}_2(t))]$$
$$+ \lambda L(\xi(t), \dot{\xi}_1(t)) + (1 - \lambda)L(\xi(t), \dot{\xi}_2(t))$$
$$- L(\xi(t), \lambda\dot{\xi}_1(t) + (1 - \lambda)\dot{\xi}_2(t))$$
$$\leq 2\lambda(1 - \lambda)K|x - y|(\delta - Ct) + \lambda(1 - \lambda)C|x - y|\omega(C|x - y|).$$

Therefore,

$$\lambda J(x_0 + \delta x, \xi_1) + (1 - \lambda)J(x_0 + \delta y, \xi_2) - J(x_0 + \delta(\lambda x + (1 - \lambda)y), \xi)$$
$$= \int_0^{\delta C^{-1}} e^{-t}[\lambda L(\xi_1(t), \dot{\xi}_1(t)) + (1 - \lambda)L(\xi_2(t), \dot{\xi}_2(t)) - L(\xi(t), \dot{\xi}(t))]dt$$
$$\leq \int_0^{\delta C^{-1}} [2\lambda(1 - \lambda)K|x - y|(\delta - Ct) + \lambda(1 - \lambda)C|x - y|\omega(C|x - y|)]dt$$
$$= \lambda(1 - \lambda)|x - y|\{K\delta^2 C^{-1} + \delta\omega(C|x - y|)\}.$$

Since ξ was arbitrary we can use (5.31) to obtain the following estimate for the rescaled function $u_{x_0,\delta}$ (defined in (2.19)):

$$\lambda u_{x_0,\delta}(x) + (1 - \lambda)u_{x_0,\delta}(y) - u_{x_0,\delta}(\lambda x + (1 - \lambda)y)$$
$$= \delta^{-1}[\lambda u(x_0 + \delta x) + (1 - \lambda)u(x_0 + \delta y) - u(x_0 + \delta(\lambda x + (1 - \lambda)y))]$$
$$\leq \lambda(1 - \lambda)|x - y|\{KC^{-1}\delta + \omega(C|x - y|)\}.$$

Then, u satisfies the generalized one-sided estimate (2.20) and is semiconcave in Ω_ε, by Theorem 2.5.2.   ∎

A similar result holds in the evolutionary case (see [46, Th. 3.2]).

Theorem 5.3.8 *Let $u \in \text{Lip}_{loc}(\,]0, T[\,\times\Omega)$ be a viscosity solution of*

$$\partial_t u + H(t, x, u, \nabla u) = 0 \quad \text{in} \quad]0, T[\,\times\Omega,$$

where $H \in \text{Lip}_{loc}(\,]0, T[\,\times\Omega \times \mathbb{R} \times \mathbb{R}^n)$ is strictly convex in the last group of variables. Then u is locally semiconcave in $]0, T[\,\times\Omega$.

There are some classes of equations, interesting for control problems, which are not covered by the above results, since $H = H(x, p)$ is convex in p, but it is neither strictly convex nor smooth. Also in some of these cases it is possible to obtain semiconcavity results; these are better formulated starting from the optimal control problem and will be considered in the next chapters (see e.g., Theorem 7.2.8).

Compared with the results seen until here, the semiconcavity result we present to conclude this section is particularly interesting. In fact, it uses neither of the approaches (I) or (II) outlined previously, but it is only based on the definition of viscosity solution. In addition, it allows for possibly nonconvex hamiltonians.

Theorem 5.3.9 *Let $u \in L^\infty(\mathbb{R}^n) \cap \text{Lip}(\mathbb{R}^n)$ be a viscosity solution of*

$$u(x) + H(x, Du(x)) = 0, \qquad x \in \mathbb{R}^n. \tag{5.34}$$

Suppose that H satisfies

$$|H(x, p) - H(x, q)| \leq \omega(|p - q|), \qquad \forall x, p, q \in \mathbb{R}^n \tag{5.35}$$

for some nondecreasing function $\omega : \mathbb{R}_+ \to \mathbb{R}_+$ such that $\omega(r) \to 0$ as $r \to 0$. Assume in addition that for some $C > 0$ and $L' > 2\,\text{Lip}\,(u)$,

$$H(x + h, p + Ch) - 2H(x, p) + H(x - h, p - Ch) \geq -C|h|^2 \tag{5.36}$$

holds for all $x, h \in \mathbb{R}^n$, $p \in B_{L'}$. Then u is semiconcave in $\mathbb{R}^n$ with linear modulus and constant C.

Proof — We consider the function $v : \mathbb{R}^{3n} \to \mathbb{R}$ defined by

$$v(x, y, z) = u(x) + u(z) - 2u(y).$$

Then v is a viscosity solution of

$$v(x, y, z) + \tilde{H}(x, y, z, D_x v, D_y v, D_z v) = 0, \qquad (x, y, z) \in \mathbb{R}^{3n}, \qquad (5.37)$$

where

$$\tilde{H}(x, y, z, p, q, r) = H(x, p) + H(z, r) - 2H(y, -q/2).$$

Let us also define

$$\Phi(x, y, z) = v(x, y, z) - \frac{1}{\varepsilon}|x + z - 2y|^2 - \delta(1 + |y|^2)^{1/2}$$
$$- \frac{C}{2}|x - y|^2 - \frac{C}{2}|z - y|^2,$$

where ε, δ are positive parameters and C is as in (5.36). For simplicity we use in the following the notation $\langle y \rangle = (1 + |y|^2)^{1/2}$. Since u is assumed bounded, we find that Φ attains a maximum over $\mathbb{R}^{3n}$ at some point (x_0, y_0, z_0) depending on ε, δ. In addition we have, setting $p_0 = \frac{2}{\varepsilon}(x_0 + z_0 - 2y_0)$,

$$D_x(v - \Phi)(x_0, y_0, z_0) = p_0 + C(x_0 - y_0)$$
$$D_y(v - \Phi)(x_0, y_0, z_0) = -2p_0 + \delta\langle y_0 \rangle^{-1} y_0 - C(x_0 - y_0) - C(z_0 - y_0)$$
$$D_z(v - \Phi)(x_0, y_0, z_0) = p_0 + C(z_0 - y_0).$$

Using the fact the v is a subsolution of (5.37) we find

$$v(x_0, y_0, z_0) \leq -H(x_0, p_0 + C(x_0 - y_0)) - H(z_0, p_0 + C(z_0 - y_0))$$
$$+ 2H(y_0, p_0 - \frac{\delta}{2}\langle y_0 \rangle^{-1} y_0 + \frac{C}{2}(x_0 + z_0 - 2y_0)). \qquad (5.38)$$

By the definition of (x_0, y_0, z_0) we have $\Phi(x_0, y_0, z_0) \geq \Phi(y + h, y, y - h)$ for any $y, h \in R^n$, which implies

$$u(y + h) + u(y - h) - 2u(y) \leq C|h|^2 + \delta\langle y \rangle + M^*, \qquad (5.39)$$

where we have set

$$M^* = v(x_0, y_0, z_0) - \frac{C}{2}|x_0 - y_0|^2 - \frac{C}{2}|z_0 - y_0|^2 - \delta\langle y_0 \rangle.$$

Taking into account assumption (5.35), and observing that $|y_0| \leq \langle y_0 \rangle$, we obtain from (5.38) that

$$M^* \leq -H(x_0, p_0 + C(x_0 - y_0)) - H(z_0, p_0 + C(z_0 - y_0))$$
$$+ 2H\left(y_0, p_0 + \frac{C}{2}(x_0 + z_0 - 2y_0)\right) + 2\omega\left(\frac{\delta}{2}\right) \qquad (5.40)$$
$$- \frac{C}{2}|x_0 - y_0|^2 - \frac{C}{2}|z_0 - y_0|^2 - \delta\langle y_0 \rangle.$$

Our aim now is to show that, if ε, δ are small, then the point y_0 is close to the midpoint of x_0 and z_0. This will enable us to estimate the right-hand side of (5.40) by using

(5.36). To this end, let us first observe that the inequality $\Phi(0, 0, 0) \leq \Phi(x_0, y_0, z_0)$ yields

$$\delta\langle y_0\rangle + \frac{C}{2}|x_0 - y_0|^2 - L|x_0 - y_0| + \frac{C}{2}|z_0 - y_0|^2 - L|z_0 - y_0| \leq \delta,$$

where we have denoted by L a Lipschitz constant of u. This inequality easily implies that there exists $R > 0$ independent of ε, δ such that

$$|x_0 - y_0| \leq R, \quad |z_0 - x_0| \leq R, \quad \delta\langle y_0\rangle \leq R. \tag{5.41}$$

On the other hand, $\Phi(x_0, \frac{1}{2}(x_0 + z_0), z_0) \leq \Phi(x_0, y_0, z_0)$ gives

$$\frac{1}{\varepsilon}|x_0 + z_0 - 2y_0|^2 \leq 2\left(u\left(\frac{x_0 + z_0}{2}\right) - u(y_0)\right) + \delta\langle\frac{x_0 + z_0}{2}\rangle - \delta\langle y_0\rangle$$

$$\leq L|x_0 + z_0 - 2y_0| + \frac{\delta}{2}|x_0 + z_0 - 2y_0|, \tag{5.42}$$

since the function $t \to \langle t\rangle$ has Lipschitz constant 1. Therefore, if δ is enough small,

$$|p_0| \leq 2L + \delta < L', \tag{5.43}$$

where L' is the constant appearing in assumption (5.36). Let us now keep δ fixed, take a sequence $\{\varepsilon_n\}$ converging to 0 and denote by $x_n, y_n, z_n, p_n, M_n^*$ the quantities corresponding to x_0, y_0, z_0, p_0, M^* above when $\varepsilon = \varepsilon_n$. The previous estimates (5.41) and (5.43) imply that, up to extracting a subsequence, we have

$$x_n - y_n \to h, \quad y_n \to \bar{y}$$
$$y_n - z_n \to h', \quad p_n \to \bar{p},$$

for some $h, h', \bar{y} \in \overline{B}_R, \bar{p} \in \overline{B}_{L'}$. In addition,

$$x_n + z_n - 2y_n = \frac{\varepsilon_n}{2}p_n \to 0,$$

and so $h = h'$. Thus, passing to the limit in (5.40) and recalling assumption (5.36),

$$\limsup_{n\to\infty} M_n^* \leq -H(\bar{y} + h, \bar{p} + Ch) - H(\bar{y} - h, \bar{p} - Ch)$$

$$+2H(\bar{y}, \bar{p}) - C|h|^2 + 2\omega(\delta/2) - \delta\langle\bar{y}\rangle$$

$$\leq 2\omega(\delta/2) - \delta\langle\bar{y}\rangle.$$

Then the theorem follows from (5.39) by letting first $n \to \infty$ and then $\delta \to 0$. ■

5.4 Propagation of singularities

We now turn to the analysis of the singular set of semiconcave solutions to Hamilton–Jacobi equations. In Chapter 4 we have obtained results which are valid for any semiconcave function; here we focus our attention on semiconcave functions which are solutions to Hamilton–Jacobi equations and we obtain stronger results on the propagation of singularities in this case. As in the previous chapter, our discussion is restricted to semiconcave functions with linear modulus (some results in the case of a general modulus can be found in [1]).

We consider again an equation of the form

$$F(x, u(x), Du(x)) = 0 \qquad \text{a.e. in} \quad \Omega. \tag{5.44}$$

Throughout the rest of this chapter, $F : \overline{\Omega} \times \mathbb{R} \times \mathbb{R}^n \to \mathbb{R}$ is a continuous function satisfying the following assumptions:

(A1) $p \mapsto F(x, u, p)$ is convex;
(A2) for any $(x, u) \in \Omega \times \mathbb{R}$ and any $p_0, p_1 \in \mathbb{R}^n$,

$$[p_0, p_1] \subset \{p \in \mathbb{R}^n : F(x, u, p) = 0\} \qquad \Longrightarrow \qquad p_0 = p_1.$$

Condition (A2) requires that the 0-level set $\{p \in \mathbb{R}^n : F(x, u, p) = 0\}$ contains no straight line. Clearly, such a property holds, in particular, if F is strictly convex with respect to p. Observe, however, that it also holds for functions like $F(x, u, p) = |p|$ which are not strictly convex.

Remark 5.4.1 In Proposition 5.3.1 we have seen that, under the convexity assumption (A1), any semiconcave function u which solves equation (5.44) almost everywhere is also a viscosity solution of the equation. Thus, u satisfies for all $x \in \Omega$

$$F(x, u(x), p) = 0 \qquad \forall p \in D^*u(x), \tag{5.45}$$

$$F(x, u(x), p) \le 0 \qquad \forall p \in D^+u(x). \tag{5.46}$$

The first result we prove is that condition (4.8) is necessary and sufficient for the propagation of a singularity at x_0, if u is a semiconcave solution of (5.44). Notice that, for a general semiconcave function, (4.8) is only a sufficient condition.

Theorem 5.4.2 *Let (A1) and (A2) be satisfied. Let $u \in \mathrm{SCL}_{loc}(\Omega)$ be a solution of (5.44) and let $x_0 \in \Sigma(u)$. Then, the following properties are equivalent:*

(i) *$\partial D^+u(x_0) \setminus D^*u(x_0) \ne \emptyset$;*
(ii) *there exist a sequence $x_i \in \Sigma(u) \setminus \{x_0\}$, converging to x_0, and a number $\delta > 0$ such that $\mathrm{diam}\, D^+u(x_i) \ge \delta$, for any $i \in \mathbb{N}$.*

Proof — Since the implication (i)⇒(ii) is part of the conclusion of Theorem 4.2.2, we only need to prove that (ii)⇒(i). For this purpose, let us note that, in view of (5.45) and assumption (A2), the set $D^*u(x)$ contains no straight lines for any $x \in \Omega$. Now, let $x_i \in \Sigma(u) \setminus \{x_0\}$ be as in (ii). Without loss of generality, we can suppose that

$$\frac{x_i - x_0}{|x_i - x_0|} \to \theta \quad \text{as} \quad i \to \infty$$

for some unit vector θ. Since $D^+u(x_i)$ is convex and diam $D^+u(x_i) \geq \delta$, we can find vectors $p_i^1, p_i^2 \in D^+u(x_i)$ such that, for any $i \in \mathbb{N}$,

$$[p_i^1, p_i^2] \subset D^+u(x_i) \quad \text{and} \quad |p_i^2 - p_i^1| = \delta.$$

Possibly extracting a subsequence, we can also suppose that

$$\lim_{i \to \infty} p_i^j = p^j \quad (j = 1, 2) \quad \text{and} \quad |p^2 - p^1| = \delta.$$

Then, by Proposition 3.3.15, the segment $[p^1, p^2]$ belongs to the exposed face $D^+u(x_0, \theta)$, which by definition is contained in $\partial D^+u(x_0)$. Hence, $[p^1, p^2] \subset \partial D^+u(x_0)$, and since $D^*u(x_0)$ contains no straight lines by assumption , we conclude that the set $D^*u(x_0)$ cannot cover $\partial D^+u(x_0)$. ∎

We can exploit further the information that u is the solution of a Hamilton–Jacobi equation by finding conditions which ensure that the crucial assumption (i) of the previous theorem is satisfied. For this purpose we prove a preliminary proposition. Let us set, for any $x \in \Omega$,

$$\arg \min_{p \in D^+u(x)} F(x, u(x), p) :=$$

$$= \left\{ \bar{p} \in D^+u(x) \, : \, F(x, u(x), \bar{p}) = \min_{p \in D^+u(x)} F(x, u(x), p) \right\}.$$

Proposition 5.4.3 *Assume* (A1) *and* (A2) *and let F be differentiable with respect to p. Let $u \in \mathrm{SCL}_{loc}(\Omega)$ be a solution of equation* (5.44) *and let $x_0 \in \Omega$. Then,*

$$\arg \min_{p \in D^+u(x_0)} F(x_0, u(x_0), p) \neq \emptyset \tag{5.47}$$

and, for any $p_0 \in \arg\min_{p \in D^+u(x_0)} F(x_0, u(x_0), p)$, we have that

$$\langle F_p(x_0, u(x_0), p_0), p - p_0 \rangle \geq 0 \qquad \forall p \in D^+u(x_0). \tag{5.48}$$

Moreover,

$$x_0 \in \Sigma(u) \qquad \Longleftrightarrow \qquad \min_{p \in D^+u(x_0)} F(x_0, u(x_0), p) < 0. \tag{5.49}$$

Furthermore, if x_0 is a singular point of u, then

$$\arg \min_{p \in D^+u(x_0)} F(x_0, u(x_0), p) \subset D^+u(x_0) \setminus D^*u(x_0) \tag{5.50}$$

$$F_p(x_0, u(x_0), p_0) \neq 0 \quad \Longrightarrow \quad p_0 \in \partial D^+u(x_0) \setminus D^*u(x_0) \tag{5.51}$$

where p_0 is any element of $\arg\min_{p \in D^+u(x_0)} F(x_0, u(x_0), p)$.

Proof — We note first that assertions (5.47) and (5.48) are a consequence of the compactness and of the convexity of the superdifferential.

Next, to prove (5.49), let us suppose that x_0 is singular. Then, $D^*u(x_0)$ contains at least two elements. Since $F(x_0, u(x_0), p) = 0$ for all $p \in D^*u(x_0)$, assumption (A2) implies that $\min_{p \in D^+u(x_0)} F(x_0, u(x_0), p) < 0$. The converse implication of (5.49) is trivial. Inclusion (5.50) can also be shown by similar arguments.

To prove (5.51), observe that $p_0 - t F_p(x_0, u(x_0), p_0) \notin D^+u(x_0)$ for every $t > 0$, in view of (5.48). So, $p_0 \in \partial D^+u(x_0)$ and the proof is complete. ∎

The next result on propagation of singularities is a direct consequence of Theorem 4.2.2 and Proposition 5.4.3.

Theorem 5.4.4 *Assume* (A1) *and* (A2), *and let F be differentiable with respect to p. Let $x_0 \in \Omega$ be a singular point of a solution $u \in \mathrm{SCL}_{loc}(\Omega)$ of equation* (5.44). *Suppose that*

$$F_p(x_0, u(x_0), p_0) \neq 0 \text{ for some } p_0 \in \arg \min_{p \in D^+u(x_0)} F(x_0, u(x_0), p). \qquad (5.52)$$

Then there exist a Lipschitz arc $\mathbf{x} : [0, \rho] \to \mathbb{R}^n$ for u, with $\mathbf{x}(0) = x_0$, and a positive number δ such that

$$\lim_{s \to 0^+} \frac{\mathbf{x}(s) - x_0}{s} \neq 0$$

$$\mathrm{diam}\left(D^+u(\mathbf{x}(s))\right) \geq \delta \qquad \forall s \in [0, \rho].$$

In particular, $\mathbf{x}(s) \in \Sigma(u)$ for any $s \in \,]0, \rho]$, and $\mathbf{x}(s) \neq x_0$ for $s > 0$ sufficiently small.

Our next result shows that condition (5.52) is "almost" necessary for the propagation of singularities; in fact, each time property (i) of Theorem 5.4.2 holds, (5.52) is satisfied up to a possible change of equation.

Theorem 5.4.5 *Assume* (A1) *and* (A2), *and let F be differentiable with respect to p. Let $x_0 \in \Omega$ be a singular point of a solution $u \in \mathrm{SCL}_{loc}(\Omega)$ of equation* (5.44). *Fix*

$$p_0 \in \arg \min_{p \in D^+u(x_0)} F(x_0, u(x_0), p).$$

If $p_0 \in \partial D^+u(x_0) \setminus D^*u(x_0)$, then, for any open set $\Omega' \subset\subset \Omega$, there exists a function $G : \Omega' \times \mathbb{R} \times \mathbb{R}^n \to \mathbb{R}$ such that

(a) $G(x, u(x), Du(x)) = 0$ a.e. in Ω';
(b) G is differentiable with respect to p and satisfies assumption (A2);
(c) $G(x_0, u(x_0), p_0) = \min_{p \in D^+u(x_0)} G(x_0, u(x_0), p)$;
(d) $G_p(x_0, u(x_0), p_0) \neq 0$.

Proof — Let $\Omega' \subset \Omega$ be a bounded open set and let L be a Lipschitz constant for u in Ω'. Suppose that

$$F_p(x_0, u(x_0), p_0) = 0;$$

otherwise one can take $G = F$. Let $q \neq 0$ be a normal vector to $D^+u(x_0)$ at p_0, i.e.,

$$\langle p - p_0, q \rangle \leq 0 \qquad \forall p \in D^+u(x_0). \tag{5.53}$$

Fix $c > 0$ such that

$$c + \langle q, p \rangle > 0 \qquad \forall |p| \leq L. \tag{5.54}$$

Notice that, in particular, (5.54) holds for every $p \in D^+u(x_0)$. We claim that

$$G(x, u, p) := (c + \langle q, p \rangle) F(x, u, p)$$

has all the required properties. In fact, properties (a) and (b) are obvious. To prove (c), it suffices to observe that (5.53), (5.54) and the fact that $F(x_0, u(x_0), p_0) < 0$ yield

$$\begin{aligned}
G(x_0, u(x_0), p_0) &= (c + \langle q, p_0 \rangle) F(x_0, u(x_0), p_0) \\
&\leq (c + \langle q, p \rangle) F(x_0, u(x_0), p_0) \\
&\leq (c + \langle q, p \rangle) F(x_0, u(x_0), p) = G(x_0, u(x_0), p)
\end{aligned}$$

for any $p \in D^+u(x_0)$. Finally, property (d) holds as

$$\begin{aligned}
G_p(x_0, u(x_0), p_0) &= q F(x_0, u(x_0), p_0) + (c + \langle q, p_0 \rangle) F_p(x_0, u(x_0), p_0) \\
&= q F(x_0, u(x_0), p_0) \neq 0.
\end{aligned}$$ ∎

5.5 Generalized characteristics

In this section we will show that the singularities of a solution u to (5.44) propagate according to a suitable differential inclusion. For this purpose, let us introduce the following definition.

Definition 5.5.1 *We say that a Lipschitz arc* $\mathbf{x} : [0, \rho] \to \mathbb{R}^n$ *is a generalized characteristic of equation* (5.44) *with initial point* x_0 *if* $\mathbf{x}(0) = x_0$ *and*

$$\mathbf{x}'(s) \in \mathrm{co} F_p(\mathbf{x}(s), u(\mathbf{x}(s)), D^+u(\mathbf{x}(s))) \quad a.e. \ in \ [0, \rho]. \tag{5.55}$$

Observe that the differential inclusion (5.55) is a natural generalization to the case where u is semiconcave of the usual equation of the characteristic curves (5.9). Our aim is to show that the class of generalized characteristics includes not only classical characteristics but also singular arcs.

In order to analyze the propagation of singularities along generalized characteristics we need to require more regularity on F than in the previous section. More precisely, we will assume that:

(A3)   for any $x_1, x_2 \in \Omega$, $u_1, u_2 \in \mathbb{R}$, $p \in \mathbb{R}^n$

$$|F(x_2, u_2, p) - F(x_1, u_1, p)| \leq L_0(|x_2 - x_1| + |u_2 - u_1|),$$

for some constant $L_0 > 0$;

(A4)   for any $x_1, x_2 \in \Omega$, $u_1, u_2 \in \mathbb{R}$, $p_1, p_2 \in \mathbb{R}^n$

$$|F_p(x_1, u_1, p_1) - F_p(x_2, u_2, p_2)| \leq L_1(|x_1 - x_2| + |u_1 - u_2| + |p_1 - p_2|)$$

for some constant $L_1 > 0$.

We will prove the following result.

Theorem 5.5.2 *Let $u \in \mathrm{SCL}_{loc}(\Omega)$ be a solution of equation (5.44). Let us suppose that (A1), (A2), (A3), (A4) hold and let $x_0 \in \Sigma(u)$ be such that*

$$0 \notin \mathrm{co} F_p(x_0, u(x_0), D^+u(x_0)) \,. \tag{5.56}$$

Then, there exists a generalized characteristic of equation (5.44), $\mathbf{x} : [0, \rho] \to \Omega$, with initial point x_0, such that $\mathbf{x}(s) \in \Sigma(u)$ and $\mathbf{x}(\cdot)$ is injective, for every $s \in [0, \rho]$. Moreover, for every $s \in [0, \rho]$,

$$\min\{F(\mathbf{x}(s), u(\mathbf{x}(s)), p) \,:\, p \in D^+u(\mathbf{x}(s))\} \leq \frac{1}{2}F(x_0, u(x_0), p_0), \tag{5.57}$$

where $p_0 \in \arg\min_{p \in D^+u(x_0)} F(x_0, u(x_0), p)$.

Remark 5.5.3 Assumption (5.56) is a generalization of (5.52). In the following, we will show that in some cases (5.56) is necessary and sufficient for propagation of singularities along generalized characteristics. Let us also observe that due to the upper continuity property of D^+u expressed by Proposition 3.3.4(a), assumption (5.56) implies that

$$0 \notin \mathrm{co}\, F_p(x, u(x), D^+u(x)) \qquad \forall x \in B_r(x_0) \,, \tag{5.58}$$

for some $r > 0$.

In order to prove Theorem 5.5.2 we need some technical preliminaries, the first of which is the following basic construction of singular arcs.

Lemma 5.5.4 *Let the assumptions of Theorem 5.5.2 be satisfied. Suppose also that (A4) holds with $L_1 = 1$. Let $R > 0$ be such that $\overline{B}_R(x_0) \subset \Omega$, and fix*

$$p_0 \in \arg \min_{p \in D^+u(x_0)} F(x_0, u(x_0), p) \,.$$

Let L be a Lipschitz constant for u in $\overline{B}_R(x_0)$ and let $C > 0$ be a semiconcavity constant for u in $\overline{B}_R(x_0)$. Let us set

$$C_0 := \sup\{|F_p(x, u(x), p)| \,:\, x \in B_R(x_0) \quad |p| \leq L\}; \tag{5.59}$$

$$M := 4L + 2C_0, \qquad C_1 := L_0(1 + L)M + CM^2; \tag{5.60}$$

$$\rho_0 := \min \left\{ \frac{R}{4C_0}, \frac{|F(x_0, u(x_0), p_0)|}{2C_1}, \frac{1}{2C} \right\}. \tag{5.61}$$

Then, there exist two continuous arcs $\mathbf{x} : [0, \rho_0] \to B_R(x_0)$ *and* $\mathbf{p} : [0, \rho_0] \to \mathbb{R}^n$, *with* $\mathbf{x}(0) = x_0$ *and* $\mathbf{p}(0) = p_0$, *such that*

$$\mathbf{p}(s) \in D^+ u(\mathbf{x}(s)) \tag{5.62}$$

$$\mathbf{x}(s) - x_0 = s[\mathbf{p}(s) - p_0 + F_p(x_0, u(x_0), p_0)] \tag{5.63}$$

$$F(\mathbf{x}(s), u(\mathbf{x}(s)), \mathbf{p}(s)) \le F(x_0, u(x_0), p_0) + C_1 s - \frac{1}{2}|\mathbf{p}(s) - p_0|^2 \tag{5.64}$$

$$F(\mathbf{x}(s), u(\mathbf{x}(s)), \mathbf{p}(s)) \le \frac{1}{2} F(x_0, u(x_0), p_0) \tag{5.65}$$

for any $s \in [0, \rho_0]$. *Moreover,*

$$|\mathbf{x}(t) - \mathbf{x}(s)| \le M|t - s| \qquad \forall t, s \in [0, \rho_0]. \tag{5.66}$$

Remark 5.5.5 Recalling property (5.45) it is easy to see that conditions (5.62) and (5.65) above ensure that $\mathbf{x}(s) \in \Sigma(u)$, for any $s \in [0, \rho_0]$. Moreover, we observe that $\mathbf{x}(\cdot)$ and $\mathbf{p}(\cdot)$ are related: given $\mathbf{x}(\cdot)$, arc $\mathbf{p}(\cdot)$ is uniquely defined by relation (5.63), and vice-versa.

Proof — Applying Lemma 4.2.5 to the semiconcave function u, we can construct continuous arcs $\mathbf{x} : [0, \sigma] \to B_R(x_0)$ and $\mathbf{p} : [0, \sigma] \to \mathbb{R}^n$, with $\mathbf{x}(0) = x_0$ and $\mathbf{p}(0) = p_0$, such that, for any $s \in [0, \sigma]$,

$$\mathbf{p}(s) = p_0 - F_p(x_0, u(x_0), p_0) + \frac{\mathbf{x}(s) - x_0}{s} \in D^+ u(\mathbf{x}(s)) \tag{5.67}$$

and

$$|\mathbf{x}(s) - x_0| \le 4C_0 s$$

where C_0 is given by (5.59) and $4\sigma := \min\{RC_0^{-1}, 2C^{-1}\}$. Moreover,

$$|\mathbf{x}(s) - \mathbf{x}(r)| \le M|s - r| \qquad \forall s, r \in [0, \sigma] \tag{5.68}$$

where M is given by (5.60).

In order to complete the proof, we have to verify (5.64) and (5.65). Aiming at this, we observe that, for any $x \in B_R(x_0)$ and $p \in \mathbb{R}^n$,

$$F(x, u(x), p) = F(x_0, u(x_0), p_0) + F(x, u(x), p) - F(x_0, u(x_0), p)$$
$$+ F(x_0, u(x_0), p) - F(x_0, u(x_0), p_0)$$
$$\leq F(x_0, u(x_0), p_0) + L_0(|x - x_0| + |u(x) - u(x_0)|)$$
$$+ \int_0^1 \langle F_p(x_0, u(x_0), p_0 + \lambda(p - p_0)), p - p_0 \rangle \, d\lambda$$

by assumption (A3). Moreover, by the Lipschitz continuity of u and assumption (A4) with $L_1 = 1$, we obtain

$$F(x, u(x), p)$$
$$\leq F(x_0, u(x_0), p_0) + L_0(1 + L)|x - x_0| + \langle F_p(x_0, u(x_0), p_0), p - p_0 \rangle$$
$$+ \int_0^1 \langle F_p(x_0, u(x_0), p_0 + \lambda(p - p_0)) - F_p(x_0, u(x_0), p_0), p - p_0 \rangle \, d\lambda$$

$$\leq F(x_0, u(x_0), p_0) + L_0(1 + L)|x - x_0| + \langle F_p(x_0, u(x_0), p_0), p - p_0 \rangle$$
$$+ \frac{1}{2}|p - p_0|^2.$$

Hence, taking in the above formula $(x, p) = (\mathbf{x}(s), \mathbf{p}(s))$,

$$F(\mathbf{x}(s), u(\mathbf{x}(s)), \mathbf{p}(s)) \leq F(x_0, u(x_0), p_0) + L_0(1 + L)|\mathbf{x}(s) - x_0|$$
$$+ \langle F_p(x_0, u(x_0), p_0), \mathbf{p}(s) - p_0 \rangle$$
$$+ \frac{1}{2}|\mathbf{p}(s) - p_0|^2.$$

Now, (5.67) yields

$$F(\mathbf{x}(s), u(\mathbf{x}(s)), \mathbf{p}(s)) \leq F(x_0, u(x_0), p_0) + L_0(1 + L)|\mathbf{x}(s) - x_0|$$
$$+ \left\langle \frac{\mathbf{x}(s) - x_0}{s} + p_0 - \mathbf{p}(s), \mathbf{p}(s) - p_0 \right\rangle$$
$$+ \frac{1}{2}|\mathbf{p}(s) - p_0|^2$$
$$= F(x_0, u(x_0), p_0) + L_0(1 + L)|\mathbf{x}(s) - x_0|$$
$$+ \frac{1}{s}\langle \mathbf{x}(s) - x_0, \mathbf{p}(s) - x_0 \rangle - \frac{1}{2}|\mathbf{p}(s) - p_0|^2.$$

Then, the property $\langle \mathbf{x}(s) - x_0, \mathbf{p}(s) - p_0 \rangle \leq C|\mathbf{x}(s) - x_0|^2$ which follows from semiconcavity, together with (5.68), yield (5.64). Finally, (5.65) follows from (5.64) recalling the definition of p_0. The proof is thus complete.   ∎

The next result provides a family of singular arcs satisfying suitable bounds. Such a family will be later shown to converge uniformly to a singular generalized characteristics. For brevity, let us set

$$\mathbf{p}^-(s) := \lim_{h \uparrow s} \mathbf{p}(h).$$

Lemma 5.5.6 *Let $u \in \mathrm{SCL}_{loc}(\Omega)$ be a solution of (5.44). Assume (A1), (A2), (A3), and suppose (A4) holds with $L_1 = 1$. Let $x_0 \in \Omega$ be a singular point of u satisfying*

$$0 \notin \mathrm{co}\, F_p(x, u(x), D^+ u(x)) \qquad \forall x \in B_R(x_0) \subset \Omega \qquad (5.69)$$

for some $R > 0$, and fix $p_0 \in \arg\min_{p \in D^+ u(x_0)} F(x_0, u(x_0), p)$. Then, for any $\varepsilon > 0$ there exist arcs $\mathbf{x}_\varepsilon : [0, \rho_0] \to \Omega$, $\mathbf{p}_\varepsilon : [0, \rho_0] \to \mathbb{R}^n$, with $\mathbf{x}_\varepsilon(0) = x_0$ and $\mathbf{p}_\varepsilon(0) = p_0$, and a partition $0 = s_0 < s_1 < \cdots < s_{k-1} < s_k = \rho_0$ with the following properties:

(i) $\displaystyle \max_{0 \le j \le k-1} |s_{j+1} - s_j| < \varepsilon$

(ii) $\mathbf{p}_\varepsilon(s) \in D^+ u(\mathbf{x}_\varepsilon(s)) \quad \forall s \in [0, \rho_0]$

(iii) $\mathbf{p}_\varepsilon(\cdot)$ *is continuous on* $[s_j, s_{j+1}[\quad (j = 0, \ldots, k-1)$

(iv) $|\mathbf{x}_\varepsilon(t) - \mathbf{x}_\varepsilon(s)| \le M|t - s| \quad \forall t, s \in [0, \rho_0]$

(v) $F(\mathbf{x}_\varepsilon(s), u(\mathbf{x}_\varepsilon(s)), \mathbf{p}_\varepsilon(s)) \le \dfrac{1}{2} F(x_0, u(x_0), p_0) \quad \forall s \in [0, \rho_0]$

(vi) $|\mathbf{p}_\varepsilon(s) - \mathbf{p}_\varepsilon(s')| \le 4\sqrt{\varepsilon} \quad \forall s, s' \in [s_j, s_{j+1}[\quad (j = 0, \ldots, k-1),$

where M and ρ_0 are defined in (5.60) and (5.61). Moreover, for any $s \in [0, s_1[$,

$$\mathbf{x}_\varepsilon(s) - x_0 = [\mathbf{p}_\varepsilon(s) - p_0 + F_p(\mathbf{x}_\varepsilon(0), u(\mathbf{x}_\varepsilon(0)), \mathbf{p}_\varepsilon(0))]s, \qquad (5.70)$$

while, for any $s \in [s_1, \rho_0[$,

$$\mathbf{x}_\varepsilon(s) - x_0 \qquad\qquad\qquad\qquad\qquad\qquad\qquad (5.71)$$
$$= [\mathbf{p}_\varepsilon(s) - \mathbf{p}_\varepsilon(s_{i(s)}) + F_p(\mathbf{x}_\varepsilon(s_{i(s)}), u(\mathbf{x}_\varepsilon(s_{i(s)})), \mathbf{p}_\varepsilon(s_{i(s)}))](s - s_{i(s)})$$
$$+ \sum_{j=0}^{i(s)-1} [\mathbf{p}_\varepsilon^-(s_{j+1}) - \mathbf{p}_\varepsilon(s_j) + F_p(\mathbf{x}_\varepsilon(s_j), u(\mathbf{x}_\varepsilon(s_j)), \mathbf{p}_\varepsilon(s_j))](s_{j+1} - s_j),$$

where $i(s) \in \{1, \ldots, k-1\}$ is the unique integer such that $s \in [s_{i(s)}, s_{i(s)+1}[$.

Proof — Let $\mathbf{x}(\cdot), \mathbf{p}(\cdot) : [0, \rho_0] \to \mathbb{R}^n$ be the continuous arcs given by Lemma 5.5.4 and let $\varepsilon > 0$ be fixed. Set

$$s_1 := \sup \left\{ t \in [0, \rho_0] : t + \frac{1}{2} |\mathbf{p}(s) - p_0|^2 < \varepsilon \quad \forall s \in [0, t] \right\}.$$

Then, property (i) is satisfied. If $s_1 = \rho_0$, define

$$\mathbf{x}_\varepsilon(s) := \mathbf{x}(s) \qquad \text{and} \qquad \mathbf{p}_\varepsilon(s) := \mathbf{p}(s) \qquad \forall s \in [0, s_1].$$

to conclude that properties (ii) – (v) and (5.70) can be obtained invoking Lemma 5.5.4, while (vi) follows from the trivial inequality $|\mathbf{p}_\varepsilon(s) - \mathbf{p}_\varepsilon(s')| \le |\mathbf{p}_\varepsilon(s) - p_0| + |\mathbf{p}_\varepsilon(s') - p_0|$ recalling the definition of s_1. This completes the proof in the case of $s_1 = \rho_0$.

Now, assuming that $s_1 < \rho_0$, define $\mathbf{x}_\varepsilon(\cdot)$ on $[0, s_1]$ as above, and choose

$$p_1 \in \arg \min_{p \in D^+ u(x_1)} F(x_1, u(x_1), p)$$

where $x_1 := \mathbf{x}_\varepsilon(s_1)$. Let us also set

$$\mathbf{p}_\varepsilon(s) := \begin{cases} \mathbf{p}(s) & \text{for} \quad s \in [0, s_1[\\ p_1 & \text{for} \quad s = s_1. \end{cases}$$

Notice that $s_1 + \frac{1}{2}|\mathbf{p}(s_1) - p_0|^2 = \varepsilon$ since $\mathbf{p}(\cdot)$ is continuous. Hence, (5.64) yields

$$F(x_1, u(x_1), p_1) \le (C_1 + 1)s_1 + F(x_0, u(x_0), p_0) - \varepsilon. \tag{5.72}$$

As will be clearer later, this inequality plays a crucial role in the estimate of the number k of s_j needed to cover the interval $[0, \rho]$. Moreover, in order to prove (v), it is useful to observe that

$$F(\mathbf{x}_\varepsilon(s), u(\mathbf{x}_\varepsilon(s)), \mathbf{p}_\varepsilon(s)) \le F(x_0, u(x_0), p_0) + C_1 s \qquad \forall s \in [0, s_1], \tag{5.73}$$

in view of (5.64).

Now, let us proceed to construct s_2. Let $R_1 > 0$ be such that $B_{R_1}(x_1) \subset B_R(x_0)$, and let $\mathbf{x}^1 : [0, \rho_1] \to B_{R_1}(x_1)$, $\mathbf{p}^1 : [0, \rho_1] \to \mathbb{R}^n$ be the arcs obtained applying Lemma 5.5.4 with x_0, p_0, R replaced by x_1, p_1, R_1, respectively. We note that all the properties of the conclusion of Lemma 5.5.4 hold for $\mathbf{x}^1 : [0, \rho_1] \to B_{R_1}(x_1)$ and $\mathbf{p}^1 : [0, \rho_1] \to \mathbb{R}^n$ with the same constants M, C_0, C_1 computed in (5.59) and (5.60) for the base point x_0. Set

$$r_1 = \sup \left\{ t \in [0, \rho_1] : t + \frac{1}{2}|\mathbf{p}^1(s) - p_1|^2 < \varepsilon \quad \forall s \in [0, t] \right\}$$
$$s_2 = \min\{s_1 + r_1, \rho_0\},$$

and define $\mathbf{x}_\varepsilon$ on $[0, s_2]$ as

$$\mathbf{x}_\varepsilon(s) := \mathbf{x}^1(s - s_1) \qquad \forall s \in [s_1, s_2].$$

Then, either $s_2 = \rho_0$ or $s_2 < \rho_0$. In the former case, as one can easily check, the construction can be completed defining

$$\mathbf{p}_\varepsilon(s) := \mathbf{p}^1(s - s_1) \qquad \forall s \in [s_1, s_2].$$

Next, assume $s_2 < \rho_0$, set $x_2 := \mathbf{x}_\varepsilon(s_2)$, and fix

$$p_2 \in \arg\min_{p \in D^+ u(x_2)} F(x_2, u(x_2), p).$$

Let us define

$$\mathbf{p}_\varepsilon(s) := \begin{cases} \mathbf{p}^1(s - s_1) & \text{for} \quad s \in [s_1, s_2[\\ \mathbf{p}_\varepsilon(s_2) = p_2. \end{cases}$$

We claim that $\mathbf{x}_\varepsilon(\cdot)$, $\mathbf{p}_\varepsilon(\cdot)$ and $0 = s_0 < s_1 < s_2 < \rho_0$ satisfy all the required properties. Indeed, property (i) is a consequence of the definition of s_2 while (vi) follows arguing as in the proof of the same property in the interval $[0, s_1[$. Moreover, (ii)–(iv) hold by Lemma 5.5.4. Hence, we only need to prove conclusions (v) and (5.71). From properties (5.64) and (5.73) we deduce that

$$F(\mathbf{x}_\varepsilon(s), u(\mathbf{x}_\varepsilon(s)), \mathbf{p}_\varepsilon(s)) \le F(x_1, u(x_1), p_1) + C_1(s - s_1)$$
$$\le F(x_0, u(x_0), p_0) + C_1 s$$

for any $s \in [s_1, s_2]$. Moreover, due to (5.73), the last inequality holds for every $s \in [0, s_2]$. So, (v) follows as $\rho_0 \le |F(x_0, u(x_0), p_0)|/2C_1$ by definition. In order to prove (5.71), we observe that

$$x_1 = x_0 + s_1[\mathbf{p}_\varepsilon^-(s_1) - p_0 + F_p(\mathbf{x}_\varepsilon(0), u(\mathbf{x}_\varepsilon(0)), \mathbf{p}_\varepsilon(0))]. \tag{5.74}$$

Then, by (5.63),

$$\mathbf{x}_\varepsilon(s) - x_1 = (s - s_1)[\mathbf{p}_\varepsilon(s) - p_1 + F_p(x_1, u(x_1), p_1)] \tag{5.75}$$

for all $s \in [s_1, s_2[$. So, (5.71) is a direct consequence of (5.74) and (5.75), and our claim is proved. Moreover, the same arguments used to prove (5.72), and (5.72) itself, yield

$$F(\mathbf{x}_\varepsilon(s_2), u(\mathbf{x}_\varepsilon(s_2)), \mathbf{p}^1(s_2)) \le (C_1 + 1)(s_2 - s_1) + F(x_1, u(x_1), p_1) - \varepsilon$$
$$\le (C_1 + 1)s_2 - 2\varepsilon.$$

We will now show that the desired partition of $[0, \rho_0]$, $0 = s_0 < s_1 < \cdots < s_k = \rho_0$, can be obtained by a finite iteration of the above construction. Indeed, suppose that numbers $0 = s_0 < \cdots < s_n < \rho_0$, and arcs $\mathbf{x}_\varepsilon(s)$, $\mathbf{p}_\varepsilon(s)$ ($s \in [0, s_n]$) are given so that conditions (i)–(vi), (5.70) and (5.71) are fulfilled. Assume in addition that

$$F(\mathbf{x}_\varepsilon(s_n), u(\mathbf{x}_\varepsilon(s_n)), \mathbf{p}^{n-1}(s_n)) \le (C_1 + 1)s_n - n\varepsilon. \tag{5.76}$$

Define

$$C_2 := \sup\{|F(x, u(x), p)| : x \in B_R(x_0), |p| \le L\}$$

and apply (5.76) to discover that $n\varepsilon - C_2 \le (C_1 + 1)s_n$. The upper bound

$$n \le \left[\frac{(C_1 + 1)\rho_0 + C_2}{\varepsilon}\right] + 1$$

follows, and the proof is complete.   ∎

Lemma 5.5.7 *Let* $\mathbf{x}_\varepsilon, \mathbf{p}_\varepsilon$ *be the arcs given by 5.5.6. Then there exists a constant* $C > 0$, *independent of* ε, *such that*

$$\left| \mathbf{x}_\varepsilon(r) - \mathbf{x}_\varepsilon(s) - \int_s^r F_p(\mathbf{x}_\varepsilon(t), u(\mathbf{x}_\varepsilon(t)), \mathbf{p}_\varepsilon(t)) \, dt \right| \le C\sqrt{\varepsilon} \tag{5.77}$$

for any $0 \le s < r < \rho_0$.

Proof — From formula (5.71)—applied for r and s—it follows that

$$\mathbf{x}_\varepsilon(r) - \mathbf{x}_\varepsilon(s) \tag{5.78}$$

$$= \sum_{j=i(s)}^{i(r)-1} [\mathbf{p}_\varepsilon^-(s_{j+1}) - \mathbf{p}_\varepsilon(s_j) + F_p(\mathbf{x}_\varepsilon(s_j), u(\mathbf{x}_\varepsilon(s_j)), \mathbf{p}_\varepsilon(s_j))](s_{j+1} - s_j)$$

$$+ [\mathbf{p}_\varepsilon(r) - \mathbf{p}_\varepsilon(s_{i(r)}) + F_p(\mathbf{x}_\varepsilon(s_{i(r)}), u(\mathbf{x}_\varepsilon(s_{i(r)})), \mathbf{p}_\varepsilon(s_{i(r)}))](r - s_{i(r)})$$

$$- [\mathbf{p}_\varepsilon(s) - \mathbf{p}_\varepsilon(s_{i(s)}) + F_p(\mathbf{x}_\varepsilon(s_{i(s)}), u(\mathbf{x}_\varepsilon(s_{i(s)})), \mathbf{p}_\varepsilon(s_{i(s)}))](s - s_{i(s)}).$$

Now, observe that, in view of property (i) of Lemma 5.5.6,

$$\left| \int_{s_{i(s)}}^{s_{i(r)}} F_p(\mathbf{x}_\varepsilon(t), u(\mathbf{x}_\varepsilon(t)), \mathbf{p}_\varepsilon(t)) \, dt - \int_s^r F_p(\mathbf{x}_\varepsilon(t), u(\mathbf{x}_\varepsilon(t)), \mathbf{p}_\varepsilon(t)) \, dt \right| \le C\varepsilon$$

for some constant C (hereafter C will denote any positive constant independent of ε). Also, by (5.78),

$$\left| \mathbf{x}_\varepsilon(r) - \mathbf{x}_\varepsilon(s) - \sum_{j=i(s)}^{i(r)-1} F_p(\mathbf{x}_\varepsilon(s_j), u(\mathbf{x}_\varepsilon(s_j)), \mathbf{p}_\varepsilon(s_j))(s_{j+1} - s_j) \right|$$

$$\le \sum_{j=i(s)}^{i(r)-1} |\mathbf{p}_\varepsilon^-(s_{j+1}) - \mathbf{p}_\varepsilon(s_j)|(s_{j+1} - s_j) + |\mathbf{p}_\varepsilon(s) - \mathbf{p}_\varepsilon(s_{i(s)})|(s - s_{i(s)})$$

$$+ \left[|F_p(\mathbf{x}_\varepsilon(s_{i(r)}), u(\mathbf{x}_\varepsilon(s_{i(r)})), \mathbf{p}_\varepsilon(s_{i(r)}))| + |\mathbf{p}_\varepsilon(r) - \mathbf{p}_\varepsilon(s_{i(r)})| \right](r - s_{i(r)})$$

$$+ |F_p(\mathbf{x}_\varepsilon(s_{i(s)}), u(\mathbf{x}_\varepsilon(s_{i(s)})), \mathbf{p}_\varepsilon(s_{i(s)}))|(s - s_{i(s)}).$$

Hence, using properties (i) and (vi) of Lemma 5.5.6 to estimate the right-hand side above, we obtain

$$\left| \mathbf{x}_\varepsilon(r) - \mathbf{x}_\varepsilon(s) - \sum_{j=i(s)}^{i(r)-1} F_p(\mathbf{x}_\varepsilon(s_j), u(\mathbf{x}_\varepsilon(s_j)), \mathbf{p}_\varepsilon(s_j))(s_{j+1} - s_j) \right| \le C\sqrt{\varepsilon}.$$

Therefore, to prove (5.77) it suffices to show that the quantity

$$\left| \sum_{j=i(s)}^{i(r)-1} \int_{s_j}^{s_{j+1}} [F_p(\mathbf{x}_\varepsilon(s_j), u(\mathbf{x}_\varepsilon(s_j)), \mathbf{p}_\varepsilon(s_j)) - F_p(\mathbf{x}_\varepsilon(t), u(\mathbf{x}_\varepsilon(t)), \mathbf{p}_\varepsilon(t))] \, dt \right|$$

can be bounded by $C\sqrt{\varepsilon}$ for some $C > 0$. Aiming at this, we invoke assumption (A4) and properties (i), (iv) and (vi) of Lemma 5.5.6 to deduce that

$$|F_p(\mathbf{x}_\varepsilon(s_j), u(\mathbf{x}_\varepsilon(s_j)), \mathbf{p}_\varepsilon(s_j)) - F_p(\mathbf{x}_\varepsilon(t), u(\mathbf{x}_\varepsilon(t)), \mathbf{p}_\varepsilon(t))|$$

$$\le L_1[|\mathbf{x}_\varepsilon(s_j) - \mathbf{x}_\varepsilon(t)| + |u(\mathbf{x}_\varepsilon(s_j)) - u(\mathbf{x}_\varepsilon(t))| + |\mathbf{p}_\varepsilon(s_j) - \mathbf{p}_\varepsilon(t)|$$

$$\le L_1[M(1 + L)|t - s_j| + 4\sqrt{\varepsilon}] \le C\sqrt{\varepsilon}$$

for any $t \in [s_j, s_{j+1}[$ and $i(s) \leq j \leq i(s+r) - 1$. The desired inequality follows, as well as (5.77).   ∎

For the proof of Theorem 5.5.2 we also need the following measurability result for a selection of a set-valued map (see Appendix A. 5 for the basic definitions of set-valued analysis). We set $\mathcal{P}(\mathbb{R}^n) = \{A : A \subset \mathbb{R}^n\}$.

Lemma 5.5.8 *Let $F : [0, 1] \to \mathcal{P}(\mathbb{R}^n)$ be convex with nonempty closed images. Suppose in addition that F is upper semicontinuous and let $p : [0, 1] \to \mathbb{R}^n$ be a piecewise continuous map. For $t \in [0, 1]$, let $\widehat{p}(t)$ be the projection of $p(t)$ onto $F(t)$. Then, $\widehat{p}(\cdot)$ is measurable.*

Proof — The fact that F is convex yields

$$[p(t) + d_{F(t)}(p(t))\overline{B}_1] \cap F(t) = \{\widehat{p}(t)\}   \forall t \in [0, 1]$$

where the symbol $d_{F(t)}(\cdot)$ denotes the Euclidean distance function from $F(t)$. It is easy to see that the proof of the measurability of $\widehat{p}(\cdot)$ reduces to showing that the set-valued map $d_{F(\cdot)}(p(\cdot))\overline{B}_1$ is measurable.

We claim that the function $d_{F(\cdot)}(p(\cdot))$ is lower semicontinuous from the right, i.e., for every sequence $\{t_k\} \subset [0, 1]$, with $t_k \downarrow t$,

$$\liminf_{k \to \infty} d_{F(t_k)}(p(t_k)) \geq d_{F(t)}(p(t)).$$

Indeed, take a sequence $\{t_k\} \subset [0, 1]$ such that $t_k \downarrow t$ as $k \to \infty$ and

$$\lim_{k \to \infty} d_{F(t_k)}(p(t_k)) = \liminf_{k \to \infty} d_{F(t_k)}(p(t_k)).$$

The convexity of $F(\cdot)$ implies that for any $p(t_k)$, there exists a unique $\widehat{p}(t_k) \in F(t_k)$ such that

$$d_{F(t_k)}(p(t_k)) = |\widehat{p}(t_k) - p(t_k)|.$$

Possibly extracting a subsequence, in view of the upper semicontinuity of $F(\cdot)$, we can suppose that $\widehat{p}(t_k)$ converges to a suitable $\overline{p}(t) \in F(t)$. Then, using the continuity of $p(\cdot)$, we deduce that

$$d_{F(t)}(p(t)) \leq |\overline{p}(t) - p(t)|$$
$$= \lim_{k \to \infty} d_{F(t_k)}(p(t_k)) = \liminf_{k \to \infty} d_{F(t_k)}(p(t_k)),$$

which proves our claim.

Now, to complete the proof we need only to verify that the set-valued map $d_{F(\cdot)}(p(\cdot))\overline{B}_1$ is measurable, knowing that $d_{F(\cdot)}(p(\cdot))$ is lower semicontinuous from the right. This follows by Theorem 8.1.4, p. 310, of [18] since, for every $x \in \mathbb{R}^n$, the map

$$[0, 1] \ni t \to \inf\{|x - y| : y \in d_{F(t)}(p(t))\overline{B}_1\} = (|x| - d_{F(t)}(p(t)))_+$$

where $(\cdot)_+$ denotes the positive part, is measurable.   ∎

We are now ready to prove Theorem 5.5.2.

Proof of Theorem 5.5.2 — First, let us observe that, in view of Remark 5.5.3, (5.69) is satisfied for some $R > 0$. Also, possibly replacing F by F/L_1, we can assume that assumption (A4) holds with $L_1 = 1$ without loss of generality. The general case may indeed be recovered by a change of variable.

Let ε be a positive number, and let $\mathbf{x}_\varepsilon, \mathbf{p}_\varepsilon : [0, \rho_0] \to \mathbb{R}^n$ be the arcs given by Lemma 5.5.6. Condition (iv) of the same lemma ensures that $\{\mathbf{x}_\varepsilon(\cdot)\}_\varepsilon$ is Lipschitz continuous, uniformly with respect to ε. Moreover, $\{\mathbf{x}_\varepsilon(\cdot)\}_\varepsilon$ is easily seen to be equibounded. Therefore, there exists a sequence $\varepsilon = \varepsilon_j \downarrow 0$ and a Lipschitz continuous function $\mathbf{x}(\cdot)$ such that

$$\lim_{\varepsilon \to 0} \sup_{s \in [0, \rho_0]} |\mathbf{x}_\varepsilon(s) - \mathbf{x}(s)| = 0. \qquad (5.79)$$

We claim that $\mathbf{x}(s) \in \Sigma(u)$ for any $s \in [0, \rho_0]$. Indeed, having fixed $s \in [0, \rho_0]$, let us extract a subsequence of $\mathbf{p}_\varepsilon(s)$ (still denoted by $\mathbf{p}_\varepsilon(s)$) which converges to some limit, $\mathbf{p}(s)$. By the upper-semicontinuity of the superdifferential, we have that $\mathbf{p}(s) \in D^+u(\mathbf{x}(s))$. Hence, passing to the limit as $\varepsilon \to 0$ in condition (v) of Lemma 5.5.6, we obtain

$$F(\mathbf{x}(s), u(\mathbf{x}(s)), \mathbf{p}(s)) \le \frac{1}{2} F(x_0, u(x_0), p_0) < 0$$

and the claim follows from (5.49).

Since $\mathbf{x}(\cdot)$ is Lipschitz continuous, by Rademacher's theorem $\mathbf{x}(\cdot)$ is differentiable on a set of full measure $E \subset [0, \rho_0]$. We will prove that $\mathbf{x}(\cdot)$ satisfies equation (5.55) on E.

The main step of the proof is to show that, for any $0 \le s < r < \rho_0$,

$$\mathbf{x}(r) - \mathbf{x}(s) \in \int_s^r \mathrm{co}\, F_p(\mathbf{x}(t), u(\mathbf{x}(t)), D^+u(\mathbf{x}(t)))\, dt, \qquad (5.80)$$

where the right-hand side denotes, as usual, the integrals of all measurable selections of the set-valued map $t \mapsto \mathrm{co}\, F_p(\mathbf{x}(t), u(\mathbf{x}(t)), D^+u(\mathbf{x}(t)))$. To prove (5.80), we observe first that (5.77), (5.79) and the Lipschitz continuity of u and F_p yield

$$\left| \mathbf{x}(r) - \mathbf{x}(s) - \int_s^r F_p(\mathbf{x}(t), u(\mathbf{x}(t)), \mathbf{p}_\varepsilon(t))\, dt \right| \le \omega(\varepsilon) \qquad (5.81)$$

for some positive function $\omega(\cdot)$ satisfying $\lim_{\varepsilon \to 0} \omega(\varepsilon) = 0$. Now, let us denote by $\widehat{\mathbf{p}}_\varepsilon(t)$ the projection of $\mathbf{p}_\varepsilon(t)$ onto $D^+u(\mathbf{x}(t))$ for every $t \in [s, r]$, and set

$$q_\varepsilon(t) := \mathbf{p}_\varepsilon(t) - \widehat{\mathbf{p}}_\varepsilon(t).$$

From Lemma 5.5.8 we deduce that $\widehat{\mathbf{p}}_\varepsilon(\cdot)$ is measurable, and so the same holds for $q_\varepsilon(\cdot)$. Moreover, $|q_\varepsilon(\cdot)|$ is bounded and converges to 0 almost everywhere in $[s, r]$, as $\varepsilon \to 0$, in view of the upper-semicontinuity of D^+u. Hence, by (5.81),

$$|\mathbf{x}(r) - \mathbf{x}(s) - \int_s^r F_p(\mathbf{x}(t), u(\mathbf{x}(t)), \widehat{\mathbf{p}}_\varepsilon(t))\, dt|$$

$$\le \omega(\varepsilon) + L_1 \int_s^r |q_\varepsilon(t)|\, dt =: \overline{\omega}(\varepsilon) \tag{5.82}$$

where $\lim_{\varepsilon \to 0} \overline{\omega}(\varepsilon) = 0$ as a consequence of the dominated convergence theorem. Moreover,

$$\phi_\varepsilon(t) := F_p(\mathbf{x}(t), u(\mathbf{x}(t)), \widehat{\mathbf{p}}_\varepsilon(t)) \in \text{co } F_p(\mathbf{x}(t), u(\mathbf{x}(t)), D^+u(\mathbf{x}(t)))$$

is bounded in $L^2(s, r; \mathbb{R}^n)$. Therefore, passing to a subsequence, we can assume that $\phi_\varepsilon(\cdot)$ converges to a function $\phi(\cdot)$, weakly in $L^2(s, r; \mathbb{R}^n)$. Then $\phi(\cdot)$ is also a selection of co $F_p(\mathbf{x}(\cdot), u(\mathbf{x}(\cdot)), D^+u(\mathbf{x}(\cdot)))$, and (5.80) follows taking the limit as $\varepsilon \to 0$ in (5.82).

Now, let $s \in E$ be a point of differentiability for $\mathbf{x}(\cdot)$. Then, using (5.80), the inclusion

$$\mathbf{x}'(s) \in \text{ co } F_p(\mathbf{x}(s), u(\mathbf{x}(s)), D^+u(\mathbf{x}(s))), \tag{5.83}$$

follows by the following argument, which is standard in control theory. Suppose that $\mathbf{x}'(s) \notin \text{ co } F_p(\mathbf{x}(s), u(\mathbf{x}(s)), D^+u(\mathbf{x}(s)))$. Then, there exist a positive number δ and a vector $\eta \in \mathbb{R}^n$, $|\eta| = 1$, such that

$$\langle \mathbf{x}'(s), \eta \rangle \le \langle p, \eta \rangle - \delta, \quad \forall p \in \text{ co } F_p(\mathbf{x}(s), u(\mathbf{x}(s)), D^+u(\mathbf{x}(s))).$$

By continuity, one can find $\delta_1 > 0$ such that, for any $x \in B_{\delta_1}(\mathbf{x}(s))$,

$$\langle \mathbf{x}'(s), \eta \rangle \le \langle p, \eta \rangle - \delta/2, \quad \forall p \in \text{ co } F_p(x, u(x), D^+u(x)). \tag{5.84}$$

Moreover, for any $\delta_2 > 0$ there exists $r(\delta_2) > 0$ such that

$$\left| \frac{\mathbf{x}(r) - \mathbf{x}(s)}{r - s} - \mathbf{x}'(s) \right| < \delta_2, \qquad |\mathbf{x}(r) - \mathbf{x}(s)| < \delta_1 \tag{5.85}$$

for any $r \in [s, s + r(\delta_2)]$. From (5.80) it follows that

$$\left\langle \frac{\mathbf{x}(r) - \mathbf{x}(s)}{r - s}, \eta \right\rangle \in \frac{1}{r - s} \int_s^r \langle \text{co } F_p(\mathbf{x}(t), u(\mathbf{x}(t)), D^+u(\mathbf{x}(t))), \eta \rangle\, dt.$$

Hence, by (5.84),

$$\left\langle \frac{\mathbf{x}(r) - \mathbf{x}(s)}{r - s}, \eta \right\rangle \ge \langle \mathbf{x}'(s), \eta \rangle + \frac{\delta}{2}.$$

Moreover, recalling (5.85) we conclude that

$$\delta_2 > \left| \frac{\mathbf{x}(r) - \mathbf{x}(s)}{r - s} - \mathbf{x}'(s) \right| \ge \frac{\delta}{2},$$

contrary to the fact δ_2 is any positive number. So, (5.83) follows.

It remains to prove that, possibly reducing ρ_0, the map $[0, \rho_0] \ni s \mapsto \mathbf{x}(s)$ is injective. We will argue once again by contradiction, assuming the existence of two sequences $\{s_i\}$ and $\{r_i\}$, with $s_i < r_i$, such that $s_i, r_i \to 0$ and $\mathbf{x}(s_i) = \mathbf{x}(r_i)$. By (5.80), we have that

$$0 \in \frac{1}{r_i - s_i} \int_{s_i}^{r_i} \mathrm{co} F_p(\mathbf{x}(t), u(\mathbf{x}(t)), D^+ u(\mathbf{x}(t))) \, dt \, .$$

Hence, as $i \to \infty$,

$$0 \in \mathrm{co} F_p(x_0, u(x_0), D^+ u(x_0))$$

in contradiction with assumption (5.56). This completes the proof.   ∎

5.6 Examples

In this section we apply our results for propagation of singularities to some specific examples. Let us first consider the eikonal equation

$$\begin{cases} |Du(x)|^2 = 1 & \text{a.e. in} \quad \mathbb{R}^n \setminus S \\ u(x) = 0 & \text{on} \quad \partial S, \end{cases} \tag{5.86}$$

where $S \subset \mathbb{R}^n$ is a nonempty closed set.

Remark 5.6.1 It is easily seen that d_S (the euclidean distance function from S) is a viscosity solution of the above problem. In fact, d_S is semiconcave in $\mathbb{R}^n \setminus S$ (see Proposition 2.2.2(ii)) and satisfies the equation at all points of differentiability. Using standard techniques from the theory of the viscosity solutions (see e.g., [20, Example II.4.5]) one can prove that d_S is the unique *nonnegative* viscosity solution of (5.86).

We start our analysis with a uniqueness result for a differential inclusion, which, up to a constant factor 2, is the equation of the generalized characteristics for (5.86) starting from a given point.

Lemma 5.6.2 *Let* $u \in \mathrm{SCL}_{loc}(\Omega)$. *Then for every* $x_0 \in \Omega$ *there exists* $\rho > 0$ *such that the following problem admits a unique solution*

$$\begin{cases} \mathbf{x}'(s) \in D^+ u(\mathbf{x}(s)) & \text{a.e. in} \quad [0, \rho] \\ \mathbf{x}(0) = x_0 \, . \end{cases} \tag{5.87}$$

Proof — The existence of a solution of (5.87) is a classical result on differential inclusions. A proof can be found, for example, in [17, p. 98]. In order to prove the uniqueness, let us suppose that problem (5.87) possesses two solutions, say $\mathbf{x}(\cdot)$ and $\bar{\mathbf{x}}(\cdot)$. Property (3.24) implies that, for a.e. s in $[0, \rho]$,

$$\frac{d}{ds} |\mathbf{x}(s) - \bar{\mathbf{x}}(s)|^2 = \langle \mathbf{x}(s) - \bar{\mathbf{x}}(s), \mathbf{x}'(s) - \bar{\mathbf{x}}'(s) \rangle \le C |\mathbf{x}(s) - \bar{\mathbf{x}}(s)|^2$$

for some $C > 0$. Thus, the conclusion is an immediate consequence of the Gronwall inequality.   ∎

Theorem 5.6.3 *Let $u \in$ SCL $(\mathbb{R}^n \setminus S)$ be a solution of equation (5.86) and let $x_0 \in \Sigma(u)$. Then, the following assertions are equivalent:*

(i) $0 \notin D^+ u(x_0)$;

(ii) *there exists a nonconstant singular generalized characteristic of the eikonal equation with initial point x_0, that is a solution, $\mathbf{x}(\cdot)$, of problem (5.87) satisfying $\mathbf{x}(s) \in \Sigma(u)$, for any $s \in [0, \rho]$.*

Remark 5.6.4 If in the statement of the above result we add the assumption that $u > 0$, then assertion (i) possesses a clear geometrical meaning. In fact, in light of Remark 5.6.1, we have that $u = d_S$ and, using Corollary 3.4.5(ii), one sees that $0 \notin D^+ d_S(x_0)$ if and only if $x_0 \notin$ co proj$_S(x_0)$.

Proof — Implication (i) $\Rightarrow$ (ii) is a direct consequence of Theorem 5.5.2. In order to prove (ii) $\Rightarrow$ (i) we will argue by contradiction. For this purpose, let us suppose that (ii) holds and that $0 \in D^+ u(x_0)$. Clearly, $\mathbf{x}(s) \equiv x_0$ is a singular generalized characteristic. Moreover, Lemma 5.6.2 ensures that this constant arc is the unique generalized characteristic with initial point x_0. This contradiction completes the proof of the implication that (ii) $\Rightarrow$ (i).   ∎

We now analyze the propagation of singularities for semiconcave solutions of the Hamilton–Jacobi equation

$$u_t + H(t, x, u, \nabla u) = 0 \qquad \text{a.e. in} \qquad]0, T[\times U, \tag{5.88}$$

where $U \subset \mathbb{R}^n$ is an open set. We assume that $H :]0, T[\times U \times \mathbb{R} \times \mathbb{R}^n \to \mathbb{R}$ is a continuous function such that

$$H(t, x, u, \cdot) \text{ is differentiable and strictly convex for any } (t, x, u); \tag{5.89}$$

$$H \text{ and } H_p \text{ are locally Lipschitz continuous.} \tag{5.90}$$

The following propagation result is a direct consequence of Theorem 5.4.4.

Theorem 5.6.5 *Let $u \in$ SCL$_{loc}(]0, T[\times U)$ be a solution of (5.88). Let us suppose that (5.89) holds. Then, for any $(t, x) \in \Sigma(u)$*

$$\partial D^+ u(t, x) \setminus D^* u(t, x) \neq \emptyset.$$

Consequently, all singularities propagate.

For solutions to (5.88) there is a natural direction for the propagation of singularities, described by the next result.

Theorem 5.6.6 *Let $u \in$ SC$_{loc}(]0, T[\times U)$ be a solution of (5.88). Let us suppose that (5.89), (5.90) hold and let $(t_0, x_0) \in \Sigma(u)$. Then, there exists a generalized characteristic of equation (5.88), with initial point (t_0, x_0) defined on some interval $[t_0, \rho] \subset]0, T[$, i.e., there exists a solution of*

$$\begin{cases} x'(t) \in coH_p(t, x(t), u(x(t)), \nabla^+ u(t, x(t))) & a.\,e.\ in\ (t_0, \rho) \\ x(0) = x_0 \end{cases} \tag{5.91}$$

such that $(t, x(t)) \in \Sigma(u)$, *for every* $t \in [t_0, \rho]$.

Proof — Set, for any $(t, x, u, p_t, p_x) \in\,]0, T[\, \times U \times \mathbb{R} \times \mathbb{R} \times \mathbb{R}^n$

$$F(t, x, u, p_t, p_x) := p_t + H(t, x, u, p_x).$$

It is immediate to the see that the above assumptions ensure that (A1), (A3) and (A4) of Theorem 5.5.2 are fulfilled. Now, to prove that (A2) holds, suppose that

$$[(p_t^1, p_x^1), (p_t^2, p_x^2)] \subset \{(p_t, p_x) \in \mathbb{R}^{n+1} : F(t, x, u, p_t, p_x) = 0\}. \tag{5.92}$$

We want to show that $(p_t^1, p_x^1) = (p_t^2, p_x^2)$. In fact, (5.92) yields

$$\lambda_1 p_t^1 + \lambda_2 p_t^2 + H(t, x, u, \lambda_1 p_x^1 + \lambda_2 p_x^2) = 0$$

for every $\lambda_1, \lambda_2 > 0$ such that $\lambda_1 + \lambda_2 = 1$. Hence, recalling that

$$p_t^i + H(t, x, u, p_x^i) = 0 \qquad (i = 1, 2) \tag{5.93}$$

we obtain

$$H(t, x, u, \lambda_1 p_x^1 + \lambda_2 p_x^2) = \lambda_1 H(t, x, u, p_x^1) + \lambda_2 H(t, x, u, p_x^2).$$

So, the above equality and assumption (5.89) yield $p_x^1 = p_x^2$ which in turn implies that $p_t^1 = p_t^2$ on account of (5.93), i.e., (A2) holds. Finally, since

$$F_p((t, x), u, (p_t, p_x)) = (1, H_p(t, x, u, p_x)), \tag{5.94}$$

assumption (5.56) is trivially fulfilled.

Now, from Theorem 5.5.2 we deduce that there exists a singular generalized characteristic of equation (5.88) with initial point (t_0, x_0). To complete the proof, it remains to show that the x-projection of such a generalized characteristic solves (5.91). This follows from the identity

$$F_p((t, x), u, D^+ u(t, x)) = (1, H_p(t, x, u, \nabla^+ u(t, x)))$$

which can be easily derived from (5.94) applying Lemma 3.3.16. ∎

We point out that one cannot expect that $\rho = +\infty$ in the hypotheses of the previous theorem. In other words, in general, singularities may disappear, as shown in the next example.

Example 5.6.7 Let us consider the function

$$u(t, x) = -a(t)^2 |x|, \qquad (t, x) \in \mathbb{R}_+ \times \mathbb{R},$$

where $a(t) = \min\{0, t - 1\}$. Then, u is semiconcave (it is the minimum of smooth functions) and is a solution of the equation

$$u_t(t, x) + |u_x(t, x)|^2 + 2a(t)|x| - a(t)^4 = 0 \qquad \text{a.e. in } \mathbb{R}_+ \times \mathbb{R}.$$

Observe that $\Sigma(u) = \{(t, 0) \ : \ 0 < t < 1\}$. Therefore, the singularities do not propagate beyond time $t = 1$.

Remark 5.6.8 As proved in [6], one can show that $\rho = +\infty$ if u is a concave solution of the Hamilton–Jacobi equation $u_t + H(\nabla u) = 0$. Also, if $n = 1$ and H does not depend on u, then equation (5.88) can be transformed into a conservation law; then the results of Dafermos [68] imply that $\rho = +\infty$ provided H is of class C^2 (a regularity assumption which is not satisfied in the above example).

Bibliographical notes

The method of characteristics goes back to Cauchy and is already described in the text [49] by Carathéodory. For more details the reader can consult [86, 71] and the references therein.

Viscosity solutions have been introduced by Crandall and Lions [67] at the beginning of the 1980s. Some ideas of the theory were already present in Kruzhkov's work on hyperbolic conservation laws [101]; for instance, the idea of replacing the equation by a family of inequalities, or the doubling of the variables in the uniqueness proof. Apart from these formal similarities, however, the two theories are substantially independent. Shortly afterwards, Lions' book [110] unified and generalized the existing results on first order equations and put in a rigorous setting the connections with control theory.

A simplified approach to the theory of viscosity solutions was given by Crandall, Evans and Lions in [65]. Our presentation of Theorems 5.2.3 and 5.2.6 follows that paper. In the following years, the theory has been considerably expanded; the most significant achievement has been the treatment of second order degenerate elliptic equations at the end of the 1980s. A comprehensive exposition of these developments can be found in the textbooks and survey papers listed at the beginning of Section 5.2.

The close relation between semiconcavity and viscosity solutions was clearly indicated in the first works on viscosity solutions [67, 110]. Afterwards it was recognized that the semiconcavity property could provide improved convergence rate for approximating schemes [48] and a fine description of the singularities [45]. The semiconcavity property was also used in [29] in the study of Hamilton–Jacobi equations by the semigroup approach. In [46] it was shown that the mild regularity assumptions of Theorem 5.3.8 suffice to obtain the generalized one-sided estimate of Section 2.5, while in [127] it was proved that such an estimate implies semiconcavity. Theorem 5.3.9 is due to Ishii [93]; following [20], we have proved the theorem under slightly stronger hypotheses than the original ones in order to avoid some technicalities in the proof.

The results of the remaining sections are taken from [7]. Previous results on propagation of singularities for solutions of Hamilton–Jacobi equations have been obtained in [45, 46, 13]. Related results when the modulus of semiconcavity of the solutions is not necessarily linear can be found in [1]. Some results on the propagation of singularity along surfaces for the value function of optimal control problems have been obtained in [43] and [51].

We recall that the notion of generalized characteristic is well known in the context of hyperbolic conservation laws from the work of Dafermos [68] (see also Wu [142]). Dafermos' analysis of the singular set (which will be summarized at the beginning of Section 6.5) is very detailed, but is restricted to evolution problems in the one-dimensional space.

6

Calculus of Variations

We begin now the analysis of some optimization problems where the results about semiconcave functions and their singularities can be applied. In this chapter we consider what Fleming and Rishel [80] call "the simplest problem in the calculus of variations," a case where the dynamic programming approach is particularly powerful, and which will serve as a guideline for the analysis of optimal control problems in the following.

The problem in the calculus of variations we study here has been introduced in Chapter 1, where some results have been given in the case where the integrand is independent of t, x. Here, we consider a general integrand, so that the minimizers no longer admit an explicit description as in the Hopf formula. However, the structure of minimizers is described by classical results: we give a result about existence and regularity of minimizers, and we show that they satisfy the well-known Euler–Lagrange equations. We then apply the dynamic programming approach and introduce the value function of the problem, which is semiconcave and is a viscosity solution of an associated Hamilton–Jacobi equation. The main purpose of our analysis is to study the singularities of the value function and their interpretation in the calculus of variations. For instance, we derive a correspondence between generalized gradients of the value function and the minimizing trajectories of the problem. This shows, in particular, that the singularities of the value function are exactly the endpoints for which the minimizer of the variational problem is not unique. In addition, we can bound the size of $\overline{\Sigma}$, the closure of the singular set, proving that it enjoys the same rectifiability properties of Σ itself. This result is interesting because on the complement of $\overline{\Sigma}$ the value function has the same regularity as the data and can be computed by the method of characteristics.

The chapter is organized as follows. In Section 6.1 we give the statement of the problem and the existence result for minimizers. In Section 6.2 we show that minimizers are regular and derive the Euler–Lagrange equations. Starting from Section 6.3 we focus our attention on problems with one free endpoint; we introduce the notions of irregular and conjugate point and we prove Jacobi's necessary optimality condition. Then, in Section 6.4, we apply the dynamic programming approach to this problem. We show that the value function u solves the associated Hamilton–Jacobi equation in the viscosity sense and that it is semiconcave. In addition, the minimizers for the variational problem with a given final endpoint (t, x) are in one-to-one correspondence with the elements of $D^*u(t, x)$; in particular, the differentiability of u is equivalent to the uniqueness of the minimizer. We also show that u is as regular as the data of the problem in the complement of the closure of its singular set Σ.

The rest of the chapter is devoted to study the structure of $\overline{\Sigma}$. Since the properties of Σ are well known from the general analysis of Chapter 4, we focus our attention on the set $\overline{\Sigma} \setminus \Sigma$. The starting point is given in Section 6.5, where we show that $\overline{\Sigma} = \Sigma \cup \Gamma$, where Γ is the set of conjugate points. In addition, we prove some results about conjugate points showing, roughly speaking, that these are the points at which the singularities of u are generated. Then, in Section 6.6, we prove that $\overline{\Sigma}$ has the same rectifiability property as Σ, i.e., it is a countably $\mathcal{H}^n$-rectifiable subset of $\mathbb{R} \times \mathbb{R}^n$. By the previous remarks, this is equivalent to proving the rectifiability of $\Gamma \setminus \Sigma$. Combining a careful analysis of the hamiltonian system satisfied by the minimizing arcs with some tools from geometric measure theory, we obtain the finer estimate

$$\mathcal{H}^{n-1+\frac{2}{k-1}}(\Gamma \setminus \Sigma) = 0,$$

where $k \geq 3$ is the differentiability class of the data. This yields in particular the desired Hausdorff estimate on $\overline{\Sigma}$, which shows that u is as smooth as the data in the complement of a closed rectifiable set of codimension one.

6.1 Existence of minimizers

Let us consider the problem in the calculus of variations of Chapter 1 in a more general setting. We fix $T > 0$, a connected open set $\Omega \subset \mathbb{R}^n$ and two closed subsets $S_0, S_T \subset \overline{\Omega}$. We denote by $AC([0, T], \mathbb{R}^n)$ the class of all absolutely continuous arcs $\xi : [0, T] \to \mathbb{R}^n$ and define the set of *admissible arcs* by

$$\mathcal{A} = \{\xi \in AC([0, T], \mathbb{R}^n) : \xi(t) \in \overline{\Omega} \text{ for all } t \in [0, T], \xi(0) \in S_0, \xi(T) \in S_T\}.$$

Moreover, we define the functionals Λ, J on $\mathcal{A}$ by

$$\Lambda(\xi) = \int_0^T L(s, \xi(s), \dot{\xi}(s))ds \tag{6.1}$$

and

$$J(\xi) = \Lambda(\xi) + u_0(\xi(0)) + u_T(\xi(T)). \tag{6.2}$$

Here $L : [0, T] \times \overline{\Omega} \times \mathbb{R}^n \to \mathbb{R}$ and $u_0, u_T : \overline{\Omega} \to \mathbb{R}$ are given continuous functions called *running cost*, *initial cost* and *final cost*, respectively, and Λ is the *action* functional. We then consider the following minimization problem:

(CV) Find $\xi_* \in \mathcal{A}$ such that $J(\xi_*) = \min\{J(\xi) : \xi \in \mathcal{A}\}$.

Later in this chapter we will restrict our attention to the case with no constraints, i.e., $\Omega = \mathbb{R}^n$. However, we present the basic results on the existence and regularity of minimizing arcs in this more general setting. Let us first recall some basic properties of weak derivatives.

Lemma 6.1.1

(a) *Let $f, g : [a, b] \to \mathbb{R}$ be integrable functions and suppose that*

$$\int_a^b [f(t) \cdot v(t) + g(t) \cdot v'(t)]dt = 0$$

for all $v \in \mathrm{Lip}([a, b])$ such that $v(a) = v(b) = 0$. Then there exists a continuous representative $\tilde{g}$ of g such that $\tilde{g} \in AC([a, b])$ and $\tilde{g}' = f$ almost everywhere.
(b) *Let $f, g : [a, b] \to \mathbb{R}$ be continuous functions and suppose that*

$$\int_a^b [f(t) \cdot v(t) + g(t) \cdot v'(t)]dt = 0$$

for all $v \in C_0^1([a, b])$. Then $g \in C^1([a, b])$ and $g' = f$.

Proof — (a) Suppose first that $f \equiv 0$. For any $n \geq 1$, set

$$g_n(t) = \begin{cases} g(t) & \text{if } |g(t)| \leq n, \\ 0 & \text{otherwise,} \end{cases}$$

$$c_n = \frac{1}{b-a} \int_a^b g_n(s)ds, \qquad c_0 = \frac{1}{b-a} \int_a^b g(s)ds.$$

Then $g_n(t) \to g(t)$ and $c_n \to c_0$ as $n \to \infty$; in addition

$$(g(t) - c_0) \cdot (g_n(t) - c_n) = g_n(t)^2 + c_0 \cdot c_n - c_0 \cdot g_n(t) - c_n \cdot g(t)$$
$$\geq -C(|g(t)| + 1)$$

for some positive constant C. If we set

$$v_n(t) = \int_a^t (g_n(s) - c_n)ds,$$

we have that $v_n \in \mathrm{Lip}([a, b])$, $v_n(a) = v_n(b) = 0$ and $v_n'(t) = g_n(t) - c_n$ almost everywhere. Hence, by Fatou's Lemma and our assumption on g,

$$\int_a^b (g(t) - c_0)^2 dt = \int_a^b \lim_{n \to \infty} (g(t) - c_0) \cdot (g_n(t) - c_n)dt$$

$$\leq \lim_{n \to \infty} \int_a^b (g(t) - c_0) \cdot v_n'(t)dt = 0$$

and so $g(t) = c_0$ almost everywhere. Thus, we can define $\tilde{g}(t) = c_0$ for all $t \in [a, b]$.

In the general case, let $F(t) = \int_a^t f(s)ds$. Then $F'(t) = f(t)$ a.e. and so, using our assumption, we obtain that, for all $v \in \mathrm{Lip}([a, b])$ such that $v(a) = v(b) = 0$,

$$0 = \int_a^b [f(t) \cdot v(t) + g(t) \cdot v'(t)]dt$$

$$= \int_a^b \frac{d}{dt}[F(t) \cdot v(t)]dt + \int_a^b [g(t) - F(t)] \cdot v'(t)dt$$

$$= \int_a^b [g(t) - F(t)] \cdot v'(t)dt.$$

Hence, by the previous step, $g - F$ is equal to some constant c_0 almost everywhere. We can then set $\tilde{g}(t) = F(t) + c_0$.

(b) Suppose first that $f \equiv 0$. Define

$$c_0 = \frac{1}{b - a} \int_a^b g(\tilde{s})d\tilde{s}, \qquad \bar{v}(t) = \int_a^t (g(s) - c_0)ds.$$

Then $\bar{v} \in C^1([a, b])$ and $\bar{v}(a) = \bar{v}(b) = 0$. By our assumption,

$$0 = \int_a^b g(t) \cdot \bar{v}'(t)dt = \int_a^b (g(t) - c_0) \cdot \bar{v}'(t)dt = \int_a^b (g(t) - c_0)^2 dt.$$

Hence $g \equiv c_0$ and $g' = 0 = f$. The general case can then be obtained as in part (a).
∎

We are now ready to prove an existence result for the minimizers of problem (CV).

Theorem 6.1.2 *Suppose that the data L, u_0, u_T of problem (CV) satisfy the following assumptions:*

(i) L is a convex function with respect to the third variable;
(ii) There exists $c_0 \geq 0$ and $\theta : \mathbb{R}_+ \to \mathbb{R}_+$ such that $\theta(q)/q \to +\infty$ as $q \to +\infty$ and

$$L(t, x, v) \geq \theta(|v|) - c_0, \qquad u_0(x), u_T(x) \geq -c_0; \qquad (6.3)$$

(iii) for all $r > 0$ there exists $C(r) > 0$ such that

$$|L(t, x, v) - L(t, y, v)| < C(r)\omega(|x - y|)\theta(|v|)$$

for all $t \in [0, T]$, $x, y \in B_r \cap \bar{\Omega}$, $v \in \mathbb{R}^n$, where $\omega : \mathbb{R}_+ \to \mathbb{R}_+$ is a nondecreasing function such that $\omega(r) \to 0$ as $r \to 0$.

If either S_0 or S_T is a compact set, then problem (CV) admits a solution.

A function θ satisfying property (ii) above is often called a *Nagumo function* for the functional Λ.

Proof — Suppose, for instance, that S_0 is compact and consider any minimizing sequence $\{\xi_k\}$ for J, that is, a sequence such that $J(\xi_k) \to m := \inf\{J(\xi) : \xi \in \mathcal{A}\}$ as $k \to \infty$. We want to show that this sequence admits a cluster point which is the required minimizer.

As a first step, let us show that the sequence of derivatives $\{\dot{\xi}_k\}$ is *equiabsolutely integrable*, i.e., for every $\varepsilon > 0$ there exists $\delta > 0$ such that, if $E \subset [0, T]$, we have

$$|E| < \delta \quad \Longrightarrow \quad \int_E |\dot{\xi}_k| dt < \varepsilon \text{ for all } k.$$

By assumption (ii) we have that, for all $\mu > 0$, there exists $C_\mu > 0$ such that $r \le \theta(r)/\mu$ for all $r \ge C_\mu$. Then, for any measurable set $E \subset [0, T]$, we obtain, again by (ii),

$$\int_E |\dot{\xi}_k(t)| dt \le \frac{1}{\mu} \int_{E \cap \{|\dot{\xi}_k| \ge C_\mu\}} \theta(|\dot{\xi}_k(t)|) dt + \int_{E \cap \{|\dot{\xi}_k| < C_\mu\}} |\dot{\xi}_k(t)| dt$$

$$\le \frac{1}{\mu}(J(\xi_k) + c_0 T + 2c_0) + |E| C_\mu \le \frac{\tilde{C}}{\mu} + |E| C_\mu, \qquad (6.4)$$

for some $\tilde{C} > 0$. The right-hand side can be made arbitrarily small by choosing μ large and $|E|$ small, and this proves our claim.

Since the sequence $\{\dot{\xi}_k\}$ is equiabsolutely integrable, by the Dunford–Pettis Theorem (see e.g., [71, Theorem 4.21.2]) there exists a subsequence, which we still denote by $\{\xi_k\}$, and a function $\eta^* \in L^1([0, T], \mathbb{R}^n)$ such that $\dot{\xi}_k \rightharpoonup \eta^*$ in the weak-L^1 topology. It is easy to see that the equiabsolute integrability of $\dot{\xi}_k$ and the compactness of S_0 imply that the sequence ξ_k is equicontinuous and uniformly bounded. Hence, by the Ascoli–Arzelà Theorem, we can also assume that the sequence $\{\xi_k\}$ converges uniformly to some absolutely continuous function ξ_∞. Since ξ_∞ is a uniform limit, we have that $\xi_\infty(t) \in \overline{\Omega}$ for all $t \in [0, T]$, $\xi_\infty(0) \in S_0$, $\xi_\infty(T) \in S_T$, which means that ξ_∞ is an admissible arc. Observe that, for any $f \in C_0^1([0, T], \mathbb{R}^n)$,

$$\int_0^T f \cdot \eta^* dt = \lim_{k \to \infty} \int_0^T f \cdot \dot{\xi}_k dt = -\lim_{k \to \infty} \int_0^T \dot{f} \cdot \xi_k dt = -\int_0^T \dot{f} \cdot \xi_\infty dt.$$

By Lemma 6.1.1 we can conclude that $\dot{\xi}_\infty \equiv \eta^*$ almost everywhere.

We show now that $J(\xi_\infty) \le \liminf_{k \to \infty} J(\xi_k)$, and this will conclude the proof. By the continuity of u_0 and u_T, it suffices to prove that $\Lambda(\xi_\infty) \le \liminf_{k \to \infty} \Lambda(\xi_k)$. Now,

$$\Lambda(\xi_\infty) - \Lambda(\xi_k) = \int_0^T [L(t, \xi_\infty(t), \dot{\xi}_\infty(t)) - L(t, \xi_k(t), \dot{\xi}_k(t))] dt$$

$$= \int_0^T [L(t, \xi_\infty(t), \dot{\xi}_\infty(t)) - L(t, \xi_\infty(t), \dot{\xi}_k(t))] dt \qquad (6.5)$$

$$+ \int_0^T [L(t, \xi_\infty(t), \dot{\xi}_k(t)) - L(t, \xi_k(t), \dot{\xi}_k(t))] dt,$$

where, by assumptions (ii) and (iii),

$$\int_0^T [L(t, \xi_\infty(t), \dot{\xi}_k(t)) - L(t, \xi_k(t), \dot{\xi}_k(t))]dt$$

$$\leq C(r)\omega(\|\xi_\infty - \xi_k\|_\infty) \int_0^T \theta(|\dot{\xi}_k(t)|)dt$$

$$\leq C(r)\tilde{C}\omega(\|\xi_\infty - \xi_k\|_\infty) \to 0 \quad \text{as } k \to \infty.$$

where r is an upper bound on $\|\xi_k\|_\infty$ and $\tilde{C}$ is the same constant as in estimate (6.4). To prove that the other term in (6.5) also tends to zero, it suffices to show that the functional $\Lambda^* : L^1([0, T], \mathbb{R}^n) \to \mathbb{R}$ defined by

$$\Lambda^*(u) = \int_0^T L(t, \xi_\infty(t), u(t))dt,$$

is weakly lower semicontinuous with respect to the L^1 topology. To this end, let us set, for $\lambda \in \mathbb{R}$,

$$X_\lambda = \{u \in L^1([0, T], \mathbb{R}^n) : \Lambda^*(u) \leq \lambda\}.$$

The sets X_λ are convex, by assumption (i), and are closed in the strong L^1 topology. Indeed, if $\{u_k\} \subset X_\lambda$ is such that $u_k \to u_\infty$ in $L^1([0, T], \mathbb{R}^n)$, then $u_k \to u_\infty$ almost everywhere (up to subsequences). Since L is continuous, $L(t, \xi_\infty(t), u_k(t)) \to L(t, \xi_\infty(t), u_\infty(t))$ almost everywhere. By assumption (ii), L is bounded from below and so, by Fatou's Lemma, $\Lambda^*(u_\infty) \leq \liminf_{k \to \infty} \Lambda^*(u_k) \leq \lambda$, that is, $u_\infty \in X_\Lambda$.

A well-known property of weak topologies implies that the sets X_λ, being convex and strongly closed, are closed also in the weak-L^1 topology. From this we easily obtain the weak lower semicontinuity of Λ^*. ∎

Remark 6.1.3 Hypothesis (iii) in the previous theorem can be replaced by much weaker assumptions, e.g., that $L(x, u, p)$ is lower semicontinuous with respect to (u, p) (see [30, Sect. 3.1] and the references therein). However, the proof becomes considerably harder. On the other hand, the following two examples show that conditions (i) and (ii) cannot be removed.

Example 6.1.4 (*L is not convex*) Consider the map $L : [0, 1] \times \mathbb{R} \times \mathbb{R} \to \mathbb{R}$, defined by $L(t, x, v) = |x|^2 + (|v|^2 - 1)^2$, $S_0 = S_1 = \{0\}$, and $u_0, u_1 \equiv 0$. We have $J(\xi) \geq 0$ for all $\xi \in \mathcal{A}$. Moreover, if we consider the arcs ξ_k defined by

$$\xi_k(t) = \begin{cases} t - \frac{j}{k} & \text{for } t \in \left[\frac{j}{k}, \frac{2j+1}{2k}\right], \ j = 0, \ldots, k-1, \\ -t + \frac{j+1}{k} & \text{for } t \in \left]\frac{2j+1}{2k}, \frac{j+1}{k}\right], \ j = 0, \ldots, k-1, \end{cases}$$

we have that $\xi_k \in \mathcal{A}$ and

$$\int_0^1 L(s, \xi_k(s), \dot{\xi}_k(s))ds = \int_0^1 \xi_k(s)^2 ds \leq \frac{1}{4k^2} \to 0 \text{ as } k \to \infty.$$

Hence $\inf\{J(\xi) : \xi \in \mathcal{A}\} = 0$. On the other hand the minimum cannot be attained, since a minimizing arc should satisfy $\xi = 0$, $|\dot{\xi}| = 1$ almost everywhere, which is impossible. ∎

Example 6.1.5 (L is not superlinear) Consider the map $L : [0, 1] \times \mathbb{R} \times \mathbb{R} \to \mathbb{R}$, defined by $L(t, x, v) = \sqrt{x^2 + v^2}$, $S_0 = \{0\}$, $S_1 = \{1\}$ and $u_0, u_1 \equiv 0$. We observe that, for all $\xi \in \mathcal{A}$

$$J(\xi) = \int_0^1 \sqrt{\xi(s)^2 + \dot{\xi}(s)^2}\,ds \geq \int_0^1 \dot{\xi}(s)ds = 1.$$

However, we cannot have $J(\xi) = 1$ on any admissible arc. In fact, because of the constraint $\xi(1) = 1$, the inequality $\sqrt{\xi(s)^2 + \dot{\xi}(s)^2} \geq \dot{\xi}(s)$ is strict at least in a neighborhood of $s = 1$.

On the other hand, if we consider the maps $\xi_k \in \mathcal{A}$, given by

$$\xi_k(t) = \begin{cases} 0 & \text{for } t \in [0, 1 - 1/k], \\ k(t - 1 + 1/k) & \text{for } t \in]1 - 1/k, 1], \end{cases}$$

we find that

$$\int_0^1 L(s, \xi_k(s), \dot{\xi}_k(s))ds = \int_{1-1/k}^1 \sqrt{k^2 + \xi_k(s)^2}\,ds \leq \frac{\sqrt{k^2 + 1}}{k} \to 1,$$

as $k \to \infty$. Thus $\inf J(\xi) = 1$, and no minimizer exists. ∎

6.2 Necessary conditions and regularity

We show in this section that the minimizing arcs for problem (CV) are regular and solve a system of equations called the *Euler–Lagrange* equations. Our analysis is restricted to those minimizers which are contained in Ω; minimizers touching the boundary of Ω would require a longer analysis. We need some further assumptions on the data. Namely, we assume that the lagrangian L is of class C^1 and that for all $r > 0$ there exists $\tilde{C}(r) > 0$ such that

$$|L_x(t, x, v)| + |L_v(t, x, v)| \leq \tilde{C}(r)\theta(|v|)$$
$$\forall t \in [0, T], \ x \in \overline{\Omega} \cap B_r, \ v \in \mathbb{R}^n, \tag{6.6}$$

where θ is the Nagumo function appearing in hypothesis (ii) of Theorem 6.1.2. Observe that property (iii) of the same theorem is implied by (6.6). In addition, we assume that θ satisfies

$$\theta(q + m) \leq K_M[1 + \theta(q)] \qquad \forall m \in [0, M], \ q \geq 0. \tag{6.7}$$

It is easily checked that assumption (6.7) is satisfied for many classes of superlinear functions θ, such as powers or exponentials. It is violated in cases where θ grows "very fast", e.g., $\theta(q) = e^{e^q}$, $q \in \mathbb{R}$.

Definition 6.2.1 *Let $\xi, \eta \in AC([0, T], \mathbb{R}^n)$ be such that $\xi + s\eta \in \mathcal{A}$ and $J(\xi + s\eta) < \infty$ for s in a neighborhood of 0. Then we set*

$$\frac{\partial J}{\partial \eta}(\xi) := \frac{d}{ds} J(\xi + s\eta)|_{s=0}$$

provided the derivative at the right-hand side exists. The above quantity is called the first variation *of J at ξ with respect to η.*

Lemma 6.2.2 *Assume that the lagrangian L is of class C^1 and that there exists a function θ satisfying assumption (ii) of Theorem 6.1.2 and properties (6.6), (6.7). If $\xi \in \mathcal{A}$ is such that $\xi(t) \in \Omega$ for all $t \in [0, T]$ and $J(\xi) < \infty$, then for all $\eta \in Lip([0, T], \mathbb{R}^n)$ such that $\eta(0) = \eta(T) = 0$ the first variation of J at ξ with respect to η exists and is equal to*

$$\frac{\partial J}{\partial \eta}(\xi) = \int_0^T \left(L_x(t, \xi(t), \dot{\xi}(t)) \cdot \eta(t) + L_v(t, \xi(t), \dot{\xi}(t)) \cdot \dot{\eta}(t) \right) dt.$$

Proof — From the assumptions that $\xi(t) \in \Omega$ and that $\eta(0) = \eta(T) = 0$ it easily follows that the arc $\xi + s\eta$ belongs to the class $\mathcal{A}$ of admissible arcs if s is sufficiently small. Let us prove that if $J(\xi) < \infty$, then $J(\xi + s\eta) < \infty$. We have

$$L(t, \xi + s\eta, \dot{\xi} + s\dot{\eta}) = L(t, \xi, \dot{\xi}) + \int_0^s \frac{d}{d\lambda}[L(t, \xi + \lambda\eta, \dot{\xi} + \lambda\dot{\eta})]d\lambda$$

and we can estimate, using (6.6), (6.7) and (6.3),

$$\left| \int_0^s \frac{d}{d\lambda}[L(t, \xi + \lambda\eta, \dot{\xi} + \lambda\dot{\eta})]d\lambda \right|$$

$$= \left| \int_0^s [L_x(t, \xi + \lambda\eta, \dot{\xi} + \lambda\dot{\eta}) \cdot \eta + L_v(t, \xi + \lambda\eta, \dot{\xi} + \lambda\dot{\eta}) \cdot \dot{\eta}]d\lambda \right|$$

$$\leq \int_0^s C_1(|\eta| + |\dot{\eta}|)\theta(|\dot{\xi} + \lambda\dot{\eta}|)d\lambda \leq C_2 \int_0^s \theta(|\dot{\xi} + \lambda\dot{\eta}|)d\lambda$$

$$\leq C_3 \int_0^s (1 + \theta(|\dot{\xi}|))d\lambda \leq C_3 s(1 + L(t, \xi, \dot{\xi}) + c_0),$$

for suitable constants $C_1, C_2, \ldots$ depending only on ξ and on the Lipschitz constant of η. Thus, the integral of L along the arc $\xi + s\eta$ can be estimated by the integral of L along ξ, and this shows that $J(\xi + s\eta)$ is finite.

Now, the first variation of J can be computed by differentiating under the integral sign, since by the previous estimates we also obtain that

$$\left| \frac{d}{ds} L(t, \xi + s\eta, \dot{\xi} + s\dot{\eta}) \right|$$

$$\leq \left| L_x(t, \xi + \lambda\eta, \dot{\xi} + \lambda\dot{\eta}) \cdot \eta \right| + \left| L_v(t, \xi + \lambda\eta, \dot{\xi} + \lambda\dot{\eta}) \cdot \dot{\eta} \right|$$

$$\leq C_3(1 + \theta(|\dot{\xi}|)),$$

where $\theta(|\dot{\xi}|)$ is integrable on $[0, T]$ by assumption (6.3) and the finiteness of $J(\xi)$. ∎

We can consider variations where the endpoints are also changed, when it is allowed by the constraints. Arguing as in the previous proof we obtain for instance the following result, dealing with variations where the first endpoint is moved.

Lemma 6.2.3 *Under the hypotheses of the previous lemma, assume in addition that $\xi(0)$ belongs to the interior of S_0 and that $u_0 \in C^1(\Omega)$. Then the first variation of J exists for all $\eta \in \text{Lip}([0, T], \mathbb{R}^n)$ such that $\eta(T) = 0$ and is equal to*

$$\frac{\partial J}{\partial \eta}(\xi) = \int_0^T \left(L_x(t, \xi(t), \dot{\xi}(t)) \cdot \eta(t) + L_v(t, \xi(t), \dot{\xi}(t)) \cdot \dot{\eta}(t)\right) dt$$
$$+ Du_0(\xi(0)) \cdot \eta(0).$$

Theorem 6.2.4 *Suppose that L is of class C^1 and that there exists θ satisfying assumption (ii) of Theorem 6.1.2 and properties (6.6), (6.7). If $\xi \in \mathcal{A}$ is a solution of (CV) and $\xi(t) \in \Omega$ for all $t \in [0, T]$, then*

(i) *the map $t \mapsto L_v(t, \xi(t), \dot{\xi}(t))$ is absolutely continuous;*
(ii) *ξ is a solution of the* Euler–Lagrange *equations*

$$\frac{d}{dt} L_v(t, \xi(t), \dot{\xi}(t)) = L_x(t, \xi(t), \dot{\xi}(t)) \tag{6.8}$$

almost everywhere in $[0, T]$;
(iii) *if $\xi(0)$ lies in the interior of S_0 and $u_0 \in C^1(\Omega)$, then ξ also satisfies the transversality condition*

$$L_v(0, \xi(0), \dot{\xi}(0)) = Du_0(\xi(0)). \tag{6.9}$$

Proof — Let us take any $\eta \in \text{Lip}([0, T], \mathbb{R}^n)$ such that $\eta(0) = \eta(T) = 0$. Then, since ξ is a solution of (CV), the function $s \mapsto J(\xi + s\eta)$ attains the minimum at $s = 0$. Thus we obtain, by Lemma 6.2.2,

$$\frac{\partial J}{\partial \eta}(\xi) = \int_0^T \left(L_x(t, \xi(t), \dot{\xi}(t)) \cdot \eta(t) + L_v(t, \xi(t), \dot{\xi}(t)) \cdot \dot{\eta}(t)\right) dt = 0.$$

Hence statements (i) and (ii) are consequences of Lemma 6.1.1. To prove the transversality condition, let us again take any $\eta \in \text{Lip}([0, T], \mathbb{R}^n)$, assuming this time only that $\eta(T) = 0$. Then Lemma 6.2.3 implies

$$\int_0^T \left(L_x(t, \xi, \dot{\xi}) \cdot \eta + L_v(t, \xi, \dot{\xi}) \cdot \dot{\eta}\right) dt + Du_0(\xi(0)) \cdot \eta(0) = 0.$$

Integrating by parts and using the Euler–Lagrange equations proved above, we obtain

$$0 = \int_0^T \left(L_x(t, \xi(t), \dot{\xi}(t)) - \frac{d}{dt} L_v(t, \xi(t), \dot{\xi}(t)) \right) \cdot \eta(t) dt$$
$$+ \left[L_v(t, \xi(t), \dot{\xi}(t)) \cdot \eta(t) \right]\Big|_0^T + Du_0(\xi(0)) \cdot \eta(0)$$
$$= \left[-L_v(0, \xi(0), \dot{\xi}(0)) + Du_0(\xi(0)) \right] \cdot \eta(0).$$

Since $\eta(0)$ is arbitrary, (6.9) is proved. ∎

Theorem 6.2.5 *Let the hypotheses of Theorems 6.1.2 and 6.2.4 be satisfied. If $\xi \in \mathcal{A}$ is a solution of* (CV) *and $\xi(t) \in \Omega$ for all $t \in [0, T]$, then ξ is Lipschitz continuous. Moreover, if $L(t, x, v)$ is strictly convex with respect to v, then $\xi \in C^1([0, T], \mathbb{R}^n)$.*

Proof — Let us prove the first statement. First we observe that, by Euler–Lagrange equations (6.8), there exists $p_0 \in \mathbb{R}^n$ such that

$$L_v(t, \xi(t), \dot{\xi}(t)) = \int_0^t L_x(r, \xi(r), \dot{\xi}(r)) dr + p_0, \qquad \forall t \in [0, T]. \qquad (6.10)$$

Using properties (6.3), (6.6) and the convexity of L we obtain

$$\theta(|\dot{\xi}(t)|) - c_0 \le L(t, \xi(t), \dot{\xi}(t)) \le L(t, \xi(t), 0) + L_v(t, \xi(t), \dot{\xi}(t)) \cdot \dot{\xi}(t)$$
$$\le L(t, \xi(t), 0) + \left(\int_0^t L_x(r, \xi(r), \dot{\xi}(r)) dr + p_0 \right) \cdot \dot{\xi}(t)$$
$$\le k_1 + k_2 |\dot{\xi}(t)|,$$

where k_1, k_2 are independent of t. Thus, $|\dot{\xi}|$ must be bounded on $[0, T]$, otherwise we get a contradiction with the superlinearity of θ.

To prove the second statement, let N be the set of zero Lebesgue measure where $\dot{\xi}$ does not exist. For $\bar{t} \in [0, T]$, choose a sequence $\{t_k\} \in [0, T] \setminus N$ such that $t_k \to \bar{t}$. Then $\dot{\xi}(t_k) \to \bar{v}$ for some $\bar{v} \in \mathbb{R}^n$ (up to subsequences) and

$$L_v(\bar{t}, \xi(\bar{t}), \bar{v}) = \lim_{k \to \infty} L_v(t_k, \xi(t_k), \dot{\xi}(t_k))$$

$$= \lim_{k \to \infty} \int_0^{t_k} L_x(r, \xi(r), \dot{\xi}(r)) dr + p_0 = \int_0^{\bar{t}} L_x(r, \xi(r), \dot{\xi}(r)) dr + p_0.$$

From the strict convexity of L it follows that the map $v \mapsto L_v(t, \xi(t), v)$ is injective (see Theorem A. 2.4). Hence, the above relations show that the value of $\bar{v}$ is uniquely determined and this implies that

$$\lim_{[0,T] \setminus N \ni t \to \bar{t}} \dot{\xi}(t) = \bar{v}.$$

The assertion is then a consequence of the following lemma. ∎

Lemma 6.2.6 *If $f : [a, b] \to \mathbb{R}$ is a Lipschitz function and $N \subset [a, b]$ is the set of zero Lebesgue measure where the derivative $\dot{f}$ does not exist, then for all $\bar{t} \in [a, b]$*

$$\liminf_{[a,b]\backslash N \ni t \to \bar{t}} \dot{f}(t) \leq \liminf_{t \to \bar{t}} \frac{f(t) - f(\bar{t})}{t - \bar{t}}$$

and

$$\limsup_{[a,b]\backslash N \ni t \to \bar{t}} \dot{f}(t) \geq \limsup_{t \to \bar{t}} \frac{f(t) - f(\bar{t})}{t - \bar{t}}.$$

Therefore, if $\lim_{[a,b]\backslash N \ni t \to \bar{t}} \dot{f}(t)$ exists, then $\dot{f}(\bar{t})$ exists and coincides with $\lim_{[a,b]\backslash N \ni t \to \bar{t}} \dot{f}(t)$.

Proof — Both inequalities can be easily obtained by writing

$$\frac{f(t) - f(\bar{t})}{t - \bar{t}} = \frac{1}{t - \bar{t}} \int_{\bar{t}}^{t} \dot{f}(s) ds$$

and observing that the set N is negligible when estimating the integral. ∎

Remark 6.2.7 For a later application, let us observe that the previous results hold under slightly more general assumptions than the ones we have given. In fact, in the proof of Lemma and 6.2.2 and Theorem 6.2.5, the lagrangian L is only estimated for values of x belonging to a neighborhood of the trajectory ξ. Therefore, these results still hold if hypothesis (ii) of Theorem 6.1.2 is replaced by the following: there exists $\theta : \mathbb{R}_+ \to \mathbb{R}_+$ such that $\theta(q)/q \to +\infty$ as $q \to +\infty$ and such that for every $r > 0$ the inequality

$$L(t, x, v) \geq \theta(|v|) - c_r, \qquad t \in [0, T], \ x \in \Omega \cap B_r, \ v \in \mathbb{R}^n$$

holds for a suitable constant $c_r \geq 0$. Also, let us observe that we are never using the differentiability of L with respect to t in the above mentioned proofs, and so the C^1 assumption on L can be weakened by requiring that $L_x(t, x, v), L_v(t, x, v)$ exist for all t, x, v and are continuous with respect to t, x, v.

Theorem 6.2.8 *In addition to the hypotheses of the previous theorem, assume that the lagrangian L is of class C^k, for some $k \geq 2$, and that $L_{vv}(t, x, v)$ (the hessian of L with respect to v) is positive definite for all t, x, v. Then any minimizer ξ^* of J such that $\xi^*(t) \in \Omega$ for all $t \in [0, T]$ is of class $C^k([0, T], \mathbb{R}^n)$ and is a classical solution of the Euler–Lagrange equations.*

Proof — By Theorem 6.2.5 we already know that $\xi^* \in C^1([0, T], \mathbb{R}^n)$. Thus, it suffices to show that if $\xi^* \in C^{h-1}([0, T], \mathbb{R}^n), 2 \leq h \leq k$, then $\xi^* \in C^h([0, T], \mathbb{R}^n)$. Set

$$P(t) = \int_0^t L_x(r, \xi^*(r), \dot{\xi}^*(r)) dr$$

and

$$\Phi(t, v) = L_v(t, \xi^*(t), v) - P(t), \qquad (t, v) \in [0, T] \times \mathbb{R}^n.$$

Then $\Phi \in C^{h-1}([0, T], \mathbb{R}^n)$ and Euler–Lagrange equations give

$$\Phi(t, \dot{\xi}^*(t)) = p_0 \qquad \forall t \in [0, T],$$

for some constant p_0. Applying the implicit function theorem and using the assumption on the hessian of L with respect to v we easily obtain that $\dot{\xi}^*$ is a C^{h-1} function. ∎

Remark 6.2.9 Consider the case $L = L(v)$, $\Omega = \mathbb{R}^n$. Then Euler–Lagrange equations (6.8) are equivalent to $\ddot{\xi} = 0$, and so the above results imply that any minimizing arc for the problem (CV) is linear. Using this property it is easy to the recover the Hopf formula derived in Chapter 1. Let us remark that in this case the minimization problem (CV) is reduced to a minimization problem on a finite dimensional space.

Proposition 6.2.10 *Under the hypotheses of the previous theorem, any minimizer ξ^* of J such that $\xi^*(t) \in \Omega$ for all $t \in [0, T]$ satisfies the* du Bois–Reymond *equations (or* Erdmann *conditions)*

$$\frac{d}{dt}\left[L(t, \xi^*(t), \dot{\xi}^*(t)) - \dot{\xi}^*(t) \cdot L_v(t, \xi^*(t), \dot{\xi}^*(t))\right] = \frac{\partial}{\partial t}L(t, \xi^*(t), \dot{\xi}^*(t)),$$

for all $t \in [0, T]$.

Proof — By the previous theorem it is possible to take the first derivative of $t \mapsto L(t, \xi^*(t), \dot{\xi}^*(t)) - \dot{\xi}^*(t) \cdot L_v(t, \xi^*(t), \dot{\xi}^*(t))$. We obtain

$$\frac{d}{dt}\left[L(t, \xi^*, \dot{\xi}^*) - \dot{\xi}^* \cdot L_v(t, \xi^*, \dot{\xi}^*)\right]$$

$$= \frac{\partial}{\partial t}L(t, \xi^*, \dot{\xi}^*) + L_x(t, \xi^*, \dot{\xi}^*) \cdot \dot{\xi}^* + L_v(t, \xi^*, \dot{\xi}^*) \cdot \ddot{\xi}^*$$

$$- \ddot{\xi}^* \cdot L_v(t, \xi^*, \dot{\xi}^*) - \dot{\xi}^* \cdot \frac{d}{dt}L_v(t, \xi^*, \dot{\xi}^*)$$

$$= \frac{\partial}{\partial t}L(t, \xi^*, \dot{\xi}^*) + \dot{\xi}^* \cdot \left(L_x(t, \xi^*, \dot{\xi}^*) - \frac{d}{dt}L_v(t, \xi^*, \dot{\xi}^*)\right)$$

$$= \frac{\partial}{\partial t}L(t, \xi^*, \dot{\xi}^*),$$

since ξ^* satisfies the Euler–Lagrange equations. ∎

6.3 The problem with one free endpoint

Starting with this section we focus our attention on the case where the admissible trajectories of our problem have the final endpoint fixed and the initial one free; in addition, we no longer consider space constraints and our trajectories will be defined in the whole space. We assume that the data of the problem satisfy the assumption needed for the results of the previous section; for the reader's convenience, we list them together here.

For a given $T > 0$, we set $Q_T =]0, T[\times \mathbb{R}^n$. We assume that the running cost $L : \overline{Q}_T \times \mathbb{R}^n \to \mathbb{R}$ and the initial cost $u_0 : \mathbb{R}^n \to \mathbb{R}$ satisfy the following assumptions.

(L1) $L \in C^{R+1}(\overline{Q}_T \times \mathbb{R}^n)$, $u_0 \in C^{R+1}(\mathbb{R}^n)$ for some given $R \geq 1$.
(L2) There exist $c_0 \geq 0$ and $\theta : \mathbb{R}_+ \to \mathbb{R}_+$ with $\lim_{r \to +\infty} \theta(r)/r = +\infty$ such that

$$L(t, x, v) \geq \theta(|v|) - c_0, \quad u_0(x) \geq -c_0 \qquad \forall t, x, v.$$

In addition, θ is such that for any $M > 0$ there exists $K_M > 0$ with

$$\theta(r + m) \leq K_M[1 + \theta(r)]$$

for all $m \in [0, M]$ and all $r \geq 0$.
(L3) $L_{vv}(t, x, v)$ is positive definite for all t, x, v.
(L4) For all $r > 0$ there exists $C(r) > 0$ such that

$$|L_x(t, x, v)| + |L_v(t, x, v)| < C(r)\theta(|v|) \quad \forall t \in [0, T], \ x \in B_r, \ v \in \mathbb{R}^n.$$

It will be convenient to consider problem (CV) on a general interval of the form $[0, t] \subset [0, T]$. We consider arcs whose initial endpoint is free and whose final one is prescribed. More precisely, for any fixed $(t, x) \in \overline{Q}_T$, we consider the following class of admissible arcs

$$\mathcal{A}_{t,x} = \{\xi \in AC([0, t], \mathbb{R}^n) : \xi(t) = x\}.$$

We then introduce the functional

$$J_t(\xi) = \int_0^t L(s, \xi(s), \dot{\xi}(s))ds + u_0(\xi(0)), \qquad \xi \in \mathcal{A}_{t,x}, \tag{6.11}$$

and consider the problem that the $(CV)_{t,x}$ minimize J_t over all arcs $\xi \in \mathcal{A}_{t,x}$.

Thus, we are considering problem (CV) of the previous sections (with the final time denoted by t instead of T) with $S_0 = \Omega = \mathbb{R}^n$, $S_t = \{x\}$, $u_t \equiv 0$. Applying the results of the previous sections we obtain that there exists a minimizer, that any minimizer is regular and satisfies the Euler–Lagrange equations together with the transversality condition. A useful property of minimizers, which can be proved with arguments similar to Theorem 6.2.5, is the following.

Theorem 6.3.1 *For any $r > 0$, there exists $M(r) > 0$ such that, if $(t, x) \in [0, T] \times B_r$ and ξ^* is a minimizer for (t, x), then*

$$\sup_{s \in [0,t]} |\dot{\xi}^*(s)| \leq M(r).$$

Definition 6.3.2 Any arc $\xi \in \mathcal{A}_{t,x}$ which solves the Euler–Lagrange equations (6.8) together with the transversality condition (6.9) is called an *extremal* for problem $(CV)_{t,x}$.

Euler's equations (6.8) can be equivalently restated as a first order system of $2n$ equations, as it is shown below. Such a formulation, called *hamiltonian*, is more convenient for our later computations. Let us set

$$H(t, x, p) = \max_{v \in \mathbb{R}^n}[p \cdot v - L(t, x, v)]. \tag{6.12}$$

The function H is the Legendre–Fenchel transform of L with respect to the third argument (see Appendix A. 2) and is called the *hamiltonian* associated with L.

Theorem 6.3.3 *Let* $\xi^* \in C^2([0, t])$ *be an extremal for problem* $(CV)_{t,x}$ *and let us set*

$$\eta^*(s) = L_v(s, \xi^*(s), \dot{\xi}^*(s)), \qquad s \in [0, t].$$

Then $\eta^*(0) = Du_0(\xi^*(0))$, *and the pair* (ξ^*, η^*) *satisfies*

$$\begin{cases} \dot{\xi}^*(s) = H_p(s, \xi^*(s), \eta^*(s)) \\ \dot{\eta}^*(s) = -H_x(s, \xi^*(s), \eta^*(s)). \end{cases} \tag{6.13}$$

Conversely, suppose that $\xi^*, \eta^* \in C^2([0, t])$ *solve system* (6.13) *together with the conditions* $\xi^*(t) = x$, $\eta^*(0) = Du_0(\xi^*(0))$. *Then* ξ^* *is an extremal for problem* $(CV)_{t,x}$.

Proof — Since $L_v(t, x, \cdot)$ and $H_p(t, x, \cdot)$ are reciprocal inverses, see (A. 28), the definition of η^* implies that

$$\dot{\xi}^*(t) = H_p(t, \xi^*(t), \eta^*(t)) \qquad t \in [0, T].$$

On the other hand, since ξ^* is an extremal, we have

$$\dot{\eta}^*(t) = L_x(t, \xi^*(t), \dot{\xi}^*(t)) \qquad \eta^*(0) = Du_0(\xi^*(0)). \tag{6.14}$$

Recalling that $-H_x(t, x, L_v(t, x, v)) = L_x(t, x, v)$ (see (A. 26)) we obtain the first part of the assertion. The converse part is obtained by similar arguments. ∎

Following the terminology of control theory, we call η^* the *dual arc* or *co-state* associated with ξ^*. For any $z \in \mathbb{R}^n$, we denote by $(\xi(\cdot, z), \eta(\cdot, z))$ the solution of

$$\begin{cases} \dot{\xi}(s) = H_p(s, \xi(s), \eta(s)) \\ \dot{\eta}(s) = -H_x(s, \xi(s), \eta(s)) \qquad s \geq 0, \end{cases} \tag{6.15}$$

with initial data

$$\begin{cases} \xi(0) = z \\ \eta(0) = Du_0(z). \end{cases} \tag{6.16}$$

Our regularity assumptions imply that ξ, η and their time derivatives are of class C^R in both arguments. Observe that, by Theorem 6.3.3, an arc is an extremal for problem $(\text{CV})_{t,x}$ if and only if it is of the form $\xi = \xi(\cdot, z)$ for some $z \in \mathbb{R}^n$.

By differentiating (6.15)–(6.16) with respect to z we obtain that ξ_z and η_z satisfy the linearized system

$$\begin{cases} \dot{\xi}_z = H_{px}\xi_z + H_{pp}\eta_z \\ \dot{\eta}_z = -H_{xx}\xi_z - H_{xp}\eta_z, \end{cases} \tag{6.17}$$

with initial conditions

$$\begin{cases} \xi_z(0) = I \\ \eta_z(0) = D^2u_0(z). \end{cases} \tag{6.18}$$

Let us observe a consequence of the above relations which will be important in our later analysis. For fixed $z \in \mathbb{R}^n$, $\theta \in (\mathbb{R}^n \setminus \{0\})$, let us denote by $w(t)$ the $2n$-vector with components $(\xi_z(t, z)\theta, \eta_z(t, z)\theta)$. Then $w(t)$ satisfies a linear differential system with nonzero initial conditions, since the first n components of $w(0)$ are given by θ. By well-known properties of linear systems, we have $w(t) \neq 0$ for all $t > 0$. We conclude that for any $t > 0$, $z \in \mathbb{R}^n$, $\theta \in (\mathbb{R}^n \setminus \{0\})$,

$$\xi_z(t, z)\theta = 0 \implies \eta_z(t, z)\theta \neq 0. \tag{6.19}$$

We now recall further classical notions from calculus of variations.

Definition 6.3.4 (Conjugate and irregular points)

(i) *A point $(t, x) \in Q_T$ is called* regular *if there exists a unique minimizer of* $(\text{CV})_{t,x}$. *All other points are called* irregular.

(ii) *A point $(t, x) \in Q_T$ is called* conjugate *if $z \in \mathbb{R}^n$ exists such that $\xi(t, z) = x$, $\xi(\cdot, z)$ is a minimizer of* $(\text{CV})_{t,x}$, *and*

$$\det \xi_z(t, z) = 0.$$

We denote by Σ the set of irregular points and by Γ the set of conjugate points.

It is easily checked that for any $R > 0$, one can find $T_R > 0$ such that for all $t \leq T_R$, $\xi(t, z_1) \neq \xi(t, z_2)$ for all $z_1, z_2 \in B_R$ and $\det \xi_z(t, z) \neq 0$ for all $z \in B_R$. Thus, in an initial time interval (whose size may depend on R) all points are regular and not conjugate.

Before continuing our analysis, let us consider a specific example to illustrate the meaning of the notions introduced above.

Example 6.3.5 Let us suppose that $n = 1$ and $L(v) = v^2/2$. We consider an initial cost $u_0 \in C^2(\mathbb{R})$ satisfying the following assumption.

(A) The second derivative u_0'' attains a minimum value $c^* < 0$ at some point z^*, is strictly decreasing in $]-\infty, z^*]$, strictly increasing in $[z^*, +\infty[$, and tends to 0 as $z \to \pm\infty$.

The solution to system (6.15)–(6.16) is given by

$$\xi(t, z) = z + u_0'(z)t, \qquad \eta(t, z) = u_0'(z), \qquad t \geq 0, z \in \mathbb{R}.$$

For a fixed $(\bar{t}, \bar{x}) \in \mathbb{R}_+ \times \mathbb{R}$, we want to find the minimizers of problem $(CV)_{\bar{t}, \bar{x}}$. We know that we can restrict our attention to the arcs of the form $\xi(\cdot, z)$ for some z. Therefore we will first find the values of z such that $\xi(\bar{t}, z) = \bar{x}$ and then find among them the minimizing ones. Let us set $T^* = -1/c^*$. Since $\xi_z(t, z) = 1 + u_0''(z)t$, we see that the analysis is different depending on whether $\bar{t}$ is smaller or greater than T^*.

Case (i): $\bar{t} \leq T^*$. In this case the function $z \to \xi(\bar{t}, z)$ is strictly increasing in $\mathbb{R}$ and so there exists a unique $\bar{z}$ such that $\xi(\bar{t}, \bar{z}) = \bar{x}$. We deduce that $\xi(\cdot, \bar{z})$ is the unique minimizer for $(\bar{t}, \bar{x})$. Thus, all points $(\bar{t}, \bar{x})$ with $\bar{t} \leq T^*$ are regular. In addition, $\xi_z(t, z) > 0$ for all $z \in \mathbb{R}$ and $t \leq T^*$ except for the case $z = z^*, t = T^*$. Therefore, setting $x^* = z^* + u_0'(z^*)T^* = \xi(z^*, T^*)$, we have that the point (T^*, x^*) is conjugate, while all other points $(\bar{t}, \bar{x})$ with $\bar{t} \leq T^*$ are not conjugate.

Case (ii): $\bar{t} > T^*$. This case requires a longer analysis. By assumption (A), there are z_1 and z_2, with $z_1 < z^* < z_2$, such that $u_0''(z_1) = u_0''(z_2) = -1/\bar{t}$. In addition, the function $z \to \xi(\bar{t}, z)$ is increasing for $z \leq z_1$, decreasing for $z \in [z_1, z_2]$, and increasing for $z \geq z_2$. Let us set $x_- = \xi(\bar{t}, z_2)$, $x_+ = \xi(\bar{t}, z_1)$. If $\bar{x} \notin [x_-, x_+]$, there exists a unique value $\bar{z} \in \mathbb{R}$ such that $\xi(\bar{t}, \bar{z}) = \bar{x}$; thus $\xi(\cdot, \bar{z})$ is the unique minimizer for $(\bar{t}, \bar{x})$ and $(\bar{t}, \bar{x})$ is regular. In addition, $\xi_z(\bar{t}, \bar{z}) \neq 0$ and so $(\bar{t}, \bar{x})$ is not conjugate in this case either. If $\bar{x} \in [x_-, x_+]$ there is instead more than one value of z such that $\xi(\bar{t}, z) = \bar{x}$ and a further analysis is required to determine which value yields a minimizer.

We call z_0 the value to the left of z_1 such that $\xi(\bar{t}, z_0) = x_-$ and z_3 the value to the right of z_2 such that $\xi(\bar{t}, z_3) = x_+$. Let us first consider the case $x = x_-$. Then there are two values of z such that $\xi(\bar{t}, z) = x_-$, namely $z = z_0$ and $z = z_2$. To determine whether $\xi(\cdot, z_0)$ or $\xi(\cdot, z_2)$ is minimizing for $(\bar{t}, x_-)$, let us compare directly the two values of the functional. Setting $\xi_i := \xi(\bar{t}, z_i)$ we obtain

$$J_{\bar{t}}(\xi_2) - J_{\bar{t}}(\xi_0) = u_0(z_2) - u_0(z_0) + \frac{(x_- - z_2)^2}{2\bar{t}} - \frac{(x_- - z_0)^2}{2\bar{t}}$$

$$= \int_{z_0}^{z_2} \left(u_0'(z) + \frac{z - x_-}{\bar{t}} \right) dz > 0 \qquad (6.20)$$

because, by our definitions, $u_0'(z) + (z - x_-)/\bar{t}$ vanishes for $z = z_0$, $z = z_2$ and is positive if $z_0 < z < z_2$. It follows that $\xi(\cdot, z_0)$ is the unique minimizing trajectory for x_-. By a similar computation, one obtains that $\xi(\cdot, z_3)$ is the unique minimizing trajectory for x_+.

Let us consider now the case when $\bar{x} \in]x_-, x_+[$. Then there are three values w_1, w_2, w_3 such that $\xi(\bar{t}, w_i) = \bar{x}$; we label them in such a way that $w_i \in]z_{i-1}, z_i[$

for $i = 1, 2, 3$. We can observe that the trajectory starting from w_2 is not a local minimizer. In fact

$$\frac{d^2}{dz^2}\Bigg|_{z=w_2} \left(u_0(z) + \frac{(\bar{x} - z)^2}{2\bar{t}} \right) = u_0''(w_2) + \frac{1}{\bar{t}} < 0$$

since $z_1 < w_2 < z_2$. Therefore only the trajectories starting from w_1 and w_3 are candidates to be a minimizer.

Let us first observe that the values w_1 and w_3 are smooth functions of $x \in \,]x_-, x_+[$, both strictly increasing. They satisfy

$$w_1(x) \rightarrow z_0, \, w_3(x) \rightarrow z_2 \text{ as } x \rightarrow x_-$$

$$w_1(x) \rightarrow z_1, \, w_3(x) \rightarrow z_3 \text{ as } x \rightarrow x_+.$$

Let us set

$$I(x) = u_0(w_3(x)) - u_0(w_1(x)) + \frac{(x - w_3(x))^2}{2\bar{t}} - \frac{(x - w_1(x))^2}{2\bar{t}}.$$

Then we have, taking into account (6.20),

$$\lim_{x \to x_-} I(x) > 0, \qquad \lim_{x \to x_+} I(x) < 0.$$

We claim that I is strictly decreasing. Indeed,

$$I'(x) = \frac{w_1(x) - w_3(x)}{\bar{t}} + \left(u_0'(w_3(x)) + \frac{w_3(x) - x}{\bar{t}} \right) w_3'(x)$$
$$- \left(u_0'(w_1(x)) + \frac{w_1(x) - x}{\bar{t}} \right) w_1'(x).$$

Since by definition $u_0'(w_i(x)) + (w_i(x) - x)/\bar{t} = 0$ and $w_1(x) < w_3(x)$ our claim follows. We deduce that $I(x)$ changes sign only once at some value $\hat{x} \in \,]x_-, x_+[$. By the definition of $I(x)$, we see that the sign of $I(x)$ tells us whether the functional is greater when evaluated at $\xi(\cdot, w_1)$ or at $\xi(\cdot, w_3)$. Therefore, if $\bar{x} \in \,]x_-, \hat{x}[$, the unique minimizer is the arc $\xi(\cdot, w_1(\bar{x}))$, while if $\bar{x} \in \,]\hat{x}, x_+[$, the unique minimizer is the arc $\xi(\cdot, w_3(\bar{x}))$. If $x = \hat{x}$, then both arcs are minimizers and so $(\bar{t}, \hat{x})$ is an irregular point.

Observe that x_-, x_+ and $\hat{x}$ depend smoothly on t for $t > T^*$. In addition $x_-(t)$ and $x_+(t)$ both tend to x^* as $t \rightarrow T^{*+}$ and so does $\hat{x}(t)$ by comparison. In conclusion, there is one conjugate point, i.e., (T^*, x^*), a smooth curve $x = \hat{x}(t)$ of irregular points starting from the conjugate point (T^*, x^*) and all other points are regular. If we introduce the value function of the problem as in Chapter 1

$$u(t, x) = \min\{J_t(\xi) \, : \, \xi \in \mathcal{A}_{t,x}\},$$

then it can be checked that u is smooth on the set of regular points, is not differentiable at the irregular points, and is differentiable once but not twice at the conjugate point. Thus the nature of a point (t, x) (regular, irregular, conjugate) appears to be related to the differentiability of u.

One could analyze in a similar way the cases when $\mathbb{R}$ can be partitioned into a finite number of intervals such that u_0'' is strictly monotone on every interval. One would find that the value function is smooth outside a finite number of curves of irregular points, each one starting from a conjugate point. Hence the solution is piecewise smooth, and so it is far from being as irregular as a general Lipschitz continuous or semiconcave function can be. In the remainder of this chapter we investigate how much of the behavior of this special example can be found in the general case.   ∎

We now show how the notions of irregular and conjugate points play a role in optimality conditions.

Theorem 6.3.6 *Assume* (L1)–(L4). *If* ξ^* *is a minimizer for problem* $(CV)_{t,x}$, *then* $(s, \xi^*(s))$ *is regular and not conjugate for any* $0 \leq s < t$.

Remark 6.3.7 The above result is classical in the calculus of variations (see [88, 53]); however, the proof is usually given for problems where both endpoints of the arc are fixed. The statement that along a minimizer there are no conjugate points is known as *Jacobi's necessary optimality condition*.

Before giving the proof of Theorem 6.3.6, we need to introduce another functional. Given an extremal $\xi^* \in C^2([0, t])$, we set

$$\Lambda^*(s, x, v) = \langle L_{xx}(s, \xi^*(s), \dot{\xi}^*(s))x, x \rangle$$
$$+ 2\langle L_{xv}(s, \xi^*(s), \dot{\xi}^*(s))x, v \rangle + \langle L_{vv}(s, \xi^*(s), \dot{\xi}^*(s))v, v \rangle, \qquad (6.21)$$

$$\Phi^*(v) = \frac{\partial^2 u_0}{\partial v^2}(\xi^*(0)) = \langle D^2 u_0(\xi^*(0))v, v \rangle, \qquad (6.22)$$

for all $s \in [0, t]$ and $x, v \in \mathbb{R}^n$. Then we define, for $\alpha \in AC([0, t], \mathbb{R}^n)$,

$$J^*(\alpha) = \int_0^t \Lambda^*(s, \alpha(s), \dot{\alpha}(s)) \, ds + \Phi^*(\alpha(0)). \qquad (6.23)$$

Observe that this is a functional of the form (6.2). The integrand Λ^* is continuous with respect to all variables, quadratic with respect to x, v, and uniformly convex with respect to v, by the positiveness assumption on L_{vv}.

Proposition 6.3.8 *Let* ξ^* *be a minimizer for problem* $(CV)_{t,x}$ *and let* J^* *be the functional associated with* ξ^* *defined in* (6.23). *Then, for all* $\alpha \in Lip([0, t], \mathbb{R}^n)$ *such that* $\alpha(t) = 0$ *we have* $J^*(\alpha) \geq 0$.

Proof — We observe that J^* gives the *second variation* of J_t at ξ^*, that is

$$\frac{d^2}{d\varepsilon^2}\bigg|_{\varepsilon=0} J_t(\xi^* + \varepsilon\alpha) = J^*(\alpha),$$

for any $\alpha \in \mathrm{Lip}\,([0, t], \mathbb{R}^n)$ such that $\alpha(t) = 0$. This follows from a direct computation, where we can differentiate J_t under the integral sign using the C^2 regularity of L and ξ^*. If ξ^* is a minimizer, then the second variation of J is nonnegative, and this proves the assertion. ∎

Lemma 6.3.9 *Any arc* $\alpha \in AC([0, t], \mathbb{R}^n)$ *minimizing the functional* J^* *is of class* C^1.

Proof — In general, we cannot immediately apply Theorem 6.2.5 to obtain the assertion. In fact, Λ^* does not satisfy a lower bound of the form (6.3) valid for all $x \in \mathbb{R}^n$, since the quadratic part with respect to x can be unbounded below. Also, Λ may not be differentiable with respect to t if L is no more regular than C^2.

However, as observed in Remark 6.2.7, the differentiability with respect to t is not needed, and the bound in (6.3) can be replaced by a local one. For this purpose, let us take $\lambda_0 > 0$ such that

$$\langle L_{vv}(s, \xi^*(s), \dot{\xi}^*(s))v, v \rangle \geq \lambda_0 |v|^2, \qquad s \in [0, t], \; v \in \mathbb{R}^n$$

and let us set $\theta(q) = 1 + \frac{\lambda_0}{2} q^2$. It is easy to see that θ satisfies properties (6.6)–(6.7). In addition, if we fix $r > 0$ and take any $x \in B_r$, $s \in [0, t]$, $v \in \mathbb{R}^n$, we obtain

$$\Lambda^*(s, x, v) \geq -C(|x|^2 + |x||v|) + \lambda_0|v|^2 \geq -Cr^2 - Cr|v| + \lambda_0|v|^2$$

$$\geq -Cr^2 - \frac{C^2 r^2}{2\lambda_0} + \frac{\lambda_0}{2}|v|^2 = -Cr^2 - \frac{C^2 r^2}{2\lambda_0} - 1 + \theta(|v|),$$

for a suitable constant C. Thus, taking into account Remark 6.2.7, we can apply Theorem 6.2.5 and conclude that any minimizer of the functional J^* is an arc of class C^1. ∎

We are now ready to give the proof of Theorem 6.3.6.

Proof of Theorem 6.3.6 — Given a minimizer ξ^* for the point (t, x), let us first show that $(t_0, \xi^*(t_0))$ is regular for all $t_0 \in \,]0, t[$. Given $t_0 \in \,]0, t[$, let ξ_0 be a minimizer for the point $(t_0, \xi^*(t_0))$. Then it is easily seen that the arc

$$\gamma(s) = \begin{cases} \xi_0(s) & 0 \leq s \leq t_0 \\ \xi^*(s) & t_0 < s \leq t \end{cases}$$

is a minimizer for (t, x). Since minimizers are of class C^2, we obtain that $\dot{\xi}^*(t_0) = \dot{\xi}_0(t_0)$. But then ξ^* and ξ_0 coincide on $[0, t_0]$, since they solve Euler's equation with the same terminal conditions. This shows that there are no minimizers for the point $(t_0, \xi^*(t_0))$ different from ξ^*, and so the point is regular.

To prove the second part of the theorem, let us argue by contradiction and suppose that there exists $t_0 \in \,]0, t[$ such that the point $(t_0, \xi^*(t_0))$ is conjugate. Let us set for simplicity $x_0 = \xi^*(t_0)$ and $z^* = \xi^*(0)$. By Theorem 6.3.3, ξ^* coincides with

$\xi(\cdot, z^*)$. By the first part of the proof, ξ^* is the unique minimizer for (t_0, x_0). Then, by the definition of conjugate point, there exists a nonzero vector θ such that

$$\xi_z(t_0, z^*)\theta = 0.$$

Let us define $\alpha : [0, t] \to \mathbb{R}^n$ as follows:

$$\alpha(s) = \begin{cases} \xi_z(s, z^*)\theta & 0 \le s \le t_0 \\ 0 & t_0 < s \le t. \end{cases}$$

Then $\alpha \in \text{Lip}([0, t], \mathbb{R}^n)$, since it is piecewise smooth and continuous at t_0. However, α it is not differentiable at t_0. In fact, its left derivative at t_0 is nonzero, by (6.17) and (6.19).

Let J^* be the functional associated with ξ^* defined in (6.21)–(6.23). We claim that $J^*(\alpha) = 0$. If this is true, then Proposition 6.3.8 implies that α minimizes J^*. On the other hand, by Lemma 6.3.9, any minimizer of J^* is of class C^1. The contradiction proves that (t_0, x_0) cannot be conjugate.

Therefore, to conclude the proof of the theorem we only need to prove our claim that $J^*(\alpha) = 0$. This is done by a direct computation, as we show below. In the following formulas, to simplify notation, we sometimes write L_{xx} instead of $L_{xx}(s, \xi(s, z^*), \dot{\xi}(s, z^*))$, ξ instead of $\xi(s, z^*)$, etc. We obtain

$$J^*(\alpha) = \int_0^t \Lambda^*(s, \alpha(s), \dot{\alpha}(s))\, ds + \Phi^*(\alpha(0))$$

$$= \int_0^{t_0} \left(\langle L_{xx}\xi_z\theta, \xi_z\theta \rangle + 2\langle L_{xv}\xi_z\theta, \dot{\xi}_z\theta \rangle + \langle L_{vv}\dot{\xi}_z\theta, \dot{\xi}_z\theta \rangle \right) ds$$
$$+ \langle D^2 u_0(z^*)\theta, \theta \rangle$$

$$= \int_0^{t_0} \langle \partial_z |_{z=z^*} L_x(s, \xi(s, z), \dot{\xi}(s, z))\, \theta, \xi_z\theta \rangle ds$$
$$+ \int_0^{t_0} \langle \partial_z |_{z=z^*} L_v(s, \xi(s, z), \dot{\xi}(s, z))\, \theta, \dot{\xi}_z\theta \rangle ds + \langle D^2 u_0(z^*)\theta, \theta \rangle$$

$$= \int_0^{t_0} \langle \partial_z |_{z=z^*} (L_x - \dot{L}_v)\theta, \xi_z\theta \rangle ds + \langle \partial_z |_{z=z^*} L_v\theta, \xi_z\theta \rangle |_{s=0}^{s=t_0}$$
$$+ \langle D^2 u_0(z^*)\theta, \theta \rangle$$

$$= -\langle (L_{vx}\xi_z + L_{vv}\dot{\xi}_z)\theta, \xi_z\theta \rangle |_{s=0} + \langle D^2 u_0(z^*)\theta, \theta \rangle$$

$$= \langle (-(L_{vx} + L_{vv}\dot{\xi}_z)|_{s=0} + D^2 u_0(z^*))\theta, \theta \rangle.$$

Here we have used the fact that $L_x - \dot{L}_v \equiv 0$ along ξ since ξ satisfies Euler's equations, the property that $\xi_z = I$ for $s = 0$, and the assumption $\xi_z(s, z^*)\theta = 0$ for $s = t_0$. Now, if we differentiate with respect to z the transversality condition $L_v(0, z, \dot{\xi}(0, z)) = Du_0(z)$ we obtain

$$(L_{vx} + L_{vv}\dot{\xi}_z)|_{s=0} = D^2 u_0(z),$$

and so the right-hand side in the previous formula vanishes, proving our claim. ∎

6.4 The value function

As in Chapter 1, we further develop the study of problem (CV) looking at the *value function* associated to the problem. We define

$$u(t, x) = \min_{\xi \in \mathcal{A}_{t,x}} J_t(\xi). \tag{6.24}$$

We recall that u satisfies the dynamic programming principle (Theorem 1.2.2). We begin our analysis by studying the semiconcavity of u. We first give a result where the initial value is required to be semiconcave; in the following we will show that such an assumption can be omitted.

Theorem 6.4.1 *Assume properties* (L1)–(L4) *and suppose in addition that* $u_0 \in$ SCL$_{loc}(\mathbb{R}^n)$ *of the functions which are locally semiconcave with linear modulus. Then the value function u defined in* (6.24) *belongs to* SCL$_{loc}(\overline{Q}_T)$.

It is convenient to prove first the semiconcavity inequality in a special case, as in the next lemma.

Lemma 6.4.2 *Under the hypotheses of the previous theorem, for any* $\rho > 0$ *there exists C such that*

$$u(2h, x + z) + u(0, x - z) - 2u(h, x) \leq C(h^2 + |z|^2)$$

for all $x, z \in B_\rho$ *and* $h \in [0, T/2]$.

Proof — Let x, z, h be as in the statement of the lemma and let $\xi : [0, t] \to \mathbb{R}^n$ be a minimizing arc for the point (h, x). We define an admissible trajectory ξ_0 for $(2h, x + z)$ in the following way:

$$\xi_0(s) = z - x + 2\xi\left(\frac{s}{2}\right), \qquad s \in [0, 2h].$$

It is immediately checked that $\xi_0(2h) = x + z$. It follows that

$$u(2h, x + z) + u(0, x - z) - 2u(h, x)$$
$$\leq \int_0^{2h} L(s, \xi_0(s), \dot{\xi}_0(s))\, ds - 2\int_0^h L(s, \xi(s), \dot{\xi}(s))\, ds$$
$$+ u_0(x - z) + u_0(\xi_0(0)) - 2u_0(\xi(0))$$
$$= 2\int_0^h \{L(2s, z - x + 2\xi(s), \dot{\xi}(s)) - L(s, \xi(s), \dot{\xi}(s))\}\, ds$$
$$+ u_0(x - z) + u_0(2\xi(0) - x + z) - 2u_0(\xi(0)).$$

By Theorem 6.3.1, $|\xi|$ and $|\dot{\xi}|$ are bounded by some constant M depending only on ρ. Thus we obtain

$$|\xi(s) - x + z| = |\xi(s) - \xi(h) + z| \leq Mh + |z|, \qquad s \in [0, h].$$

Hence, using the Lipschitz continuity of L with respect to the first two arguments and the semiconcavity of u_0 we obtain

$$u(2h, x + z) + u(0, x - z) - 2u(h, x)$$
$$\leq 2C_1 \int_0^h (s + |\xi(s) - x + z|)\, ds + C_2 |\xi(0) - x + z|^2$$
$$\leq 2C_1 h(h + Mh + |z|) + C_2 (Mh + |z|)^2. \qquad \blacksquare$$

Proof of Theorem 6.4.1 — For fixed $r > 0$, we take any $x, z \in \mathbb{R}^n$, $t, h \geq 0$, such that $x \pm z \in B_r$ and $t \pm h \in [0, T]$. Let $\xi : [0, t] \to \mathbb{R}^n$ a minimizing arc for the point (t, x). Let us define

$$\xi_-(s) = -z + \xi(s + h), \qquad s \in [0, t - h],$$

$$\xi_+(s) = z + \xi(s - h), \qquad s \in [2h, t + h].$$

Since $\xi_\pm(t \pm h) = x \pm z$, we obtain from the dynamic programming principle

$$u(t + h, x + z) + u(t - h, x - z) - 2u(t, x)$$
$$\leq \int_{2h}^{t+h} L(s, \xi_+(s), \dot\xi_+(s))\, ds + u(2h, \xi_+(2h))$$
$$+ \int_0^{t-h} L(s, \xi_-(s), \dot\xi_-(s))\, ds + u(0, \xi_-(0))$$
$$- 2\int_h^t L(s, \xi(s), \dot\xi(s))\, ds - 2u(h, \xi(h))$$
$$= u(2h, \xi(h) + z) + u(0, \xi(h) - z) - 2u(h, \xi(h))$$
$$+ \int_h^t \{L(s - h, \xi(s) - z, \dot\xi(s)) + L(s + h, \xi(s) + z, \dot\xi(s))$$
$$- 2L(s, \xi(s), \dot\xi(s))\}\, ds.$$

By Theorem 6.3.1, $|\xi(s)|$ and $|\dot\xi(s)|$ are uniformly bounded by some constant depending only on r. We can therefore use the semiconcavity of L in the first two arguments and the estimate given by Lemma 6.4.2 to obtain, for suitable constants C_1, C_2,

$$u(t + h, x + z) + u(t - h, x - z) - 2u(t, x)$$
$$\leq C_1(h^2 + |z|^2) + \int_h^t C_2(h^2 + |z|^2)\, ds$$
$$\leq C_1(h^2 + |z|^2) + TC_2(h^2 + |z|^2). \qquad \blacksquare$$

We now show that the above semiconcavity results hold assuming no other regularity property of u_0 than continuity.

Theorem 6.4.3 *Assume that $L \in C^2(\overline{Q}_T \times \mathbb{R}^n)$, that $u_0 \in C(\mathbb{R}^n)$ and that properties (L2)–(L4) hold. Then, for any $t \in {]0, T]}$, $u(\cdot, t)$ belongs to $SCL_{loc}(\mathbb{R}^n)$.*

Proof — For fixed $t, r > 0$, let us consider $x, z \in \mathbb{R}^n$ such that $x \pm z \in B_r$ and let ξ be a minimizing arc for the point (t, x). Define

$$\xi_\pm(s) = \xi(s) \pm \frac{s}{t} z.$$

Since L is of class C^2, it is locally semiconcave, and so we can estimate

$$u(t, x + z) + u(t, x - z) - 2u(t, x)$$

$$\leq \int_0^t L(s, \xi_+(s), \dot{\xi}_+(s)) \, ds + u_0(\xi_+(0))$$

$$+ \int_0^t L(s, \xi_-(s), \dot{\xi}_-(s)) \, ds + u_0(\xi_-(0))$$

$$-2 \int_0^t L(s, \xi(s), \dot{\xi}(s)) \, ds - 2u_0(\xi(0))$$

$$= \int_0^t \left\{ L\left(s, \xi(s) + \tfrac{s}{t}z, \dot{\xi}(s) + \tfrac{z}{t}\right) + L\left(s, \xi(s) - \tfrac{s}{t}z, \dot{\xi}(s) - \tfrac{z}{t}\right) \right.$$

$$\left. \int_0^t -2L(s, \xi(s), \dot{\xi}(s)) \right\} ds$$

$$\leq \int_0^t C \tfrac{s^2+1}{t^2} |z|^2 \, ds = C \left(\tfrac{t}{3} + \tfrac{1}{t}\right) |z|^2. \qquad \blacksquare$$

Corollary 6.4.4 *Under the hypotheses of the previous theorem, u belongs to* $SCL_{loc}(]0, T] \times \mathbb{R}^n)$.

Proof — Let us fix $t' \in]0, T[$. Using Theorem 6.4.3 and formula (1.8), we see that the restriction of u to $[t', T] \times \mathbb{R}^n$ can be represented as the value function of a problem with a semiconcave initial cost. Therefore it is semiconcave by Theorem 6.4.1. By the arbitrariness of t' we obtain the assertion. $\qquad \blacksquare$

The above corollary shows that the value function u not only inherits the semiconcavity of the initial cost u_0, but is semiconcave even if u_0 is not. In the latter case, of course, we have semiconcavity only away from the initial time $t = 0$. In the case $L = L(v)$ this property was already proved in Corollary 1.6.2. Such a gain of regularity is related with the analogous result for solutions of the Hamilton–Jacobi equation (see Theorem 5.3.5); as we show below, our value function u is also solution to a Hamilton–Jacobi equation where H is the hamiltonian function.

In the following we always assume that properties (L1)–(L4) hold. Let us remark that the C^{R+1} regularity of L, u_0 is needed only in Theorem 6.4.11, while the other results hold also under milder assumptions.

Theorem 6.4.5 *The value function u defined in (6.24) is a viscosity solution of the dynamic programming equation*

$$\begin{cases} \partial_t u(t, x) + H(t, x, \nabla u(t, x)) = 0 & \text{in } Q_T \\ u(0, x) = u_0(x) & \text{in } \mathbb{R}^n. \end{cases} \qquad (6.25)$$

Proof — Let us take $(t, x) \in Q_T$ and $(p_t, p_x) \in D^+u(t, x)$. Then we have, for any $v \in \mathbb{R}^n$,

$$\limsup_{h \to 0^+} \frac{u(t - h, x - hv) - u(t, x) + h(p_t + v \cdot p_x)}{h\sqrt{1 + |v|^2}} \leq 0$$

which is equivalent to

$$\limsup_{h \to 0^+} \frac{u(t - h, x - hv) - u(t, x)}{h} \leq -p_t - v \cdot p_x.$$

Now let us set $\zeta(\sigma) = x + (\sigma - t)v$ for $\sigma \leq t$. We obtain from the dynamic programming principle

$$u(t, x) \leq u(t - h, \zeta(t - h)) + \int_{t-h}^{t} L(\sigma, \zeta(\sigma), \dot{\zeta}(\sigma)) d\sigma$$

$$= u(t - h, x - hv) + \int_{t-h}^{t} L(\sigma, x + (\sigma - t)v, v) d\sigma,$$

which yields

$$\limsup_{h \to 0^+} \frac{u(t - h, x - hv) - u(t, x)}{h}$$

$$\geq \lim_{h \to 0^+} -\frac{1}{h} \int_{t-h}^{t} L(\sigma, x + (\sigma - t)v, v) d\sigma = -L(t, x, v).$$

We conclude that

$$p_t + v \cdot p_x - L(t, x, v) \leq 0, \qquad \forall v \in \mathbb{R}^n,$$

and so $p_t + H(t, x, p_x) \leq 0$, which proves that u is a viscosity subsolution of the equation.

To prove that u is a supersolution, let us take $(p_t, p_x) \in D^-u(t, x)$. Let $\xi \in C^2([0, t])$ be a minimizing arc for (t, x) and let us set $w = \dot{\xi}(t)$. Then we have, using the Lipschitz continuity of u,

$$\liminf_{h \to 0^+} \frac{u(t - h, \xi(t - h)) - u(t, x)}{h}$$

$$= \liminf_{h \to 0^+} \frac{u(t - h, x - hw) - u(t, x)}{h} \geq -p_t - w \cdot p_x.$$

On the other hand, by the optimality of ξ,

$$u(t, x) = u(t - h, \xi(t - h)) + \int_{t-h}^{t} L(s, \xi(s), \dot{\xi}(s)) ds,$$

for all $0 \leq h \leq t$, which implies that

$$\lim_{h \to 0^+} \frac{u(t - h, \xi(t - h)) - u(t, x)}{h} = -L(t, x, w).$$

We conclude that $p_t + w \cdot p_x - L(t, x, w) \geq 0$, which implies that $p_t + H(t, x, p_x) \geq 0$. This shows that u is a viscosity supersolution as well. Since u satisfies the initial condition by definition, this proves our theorem.   ∎

Theorem 6.4.6 *The value function u is the minimal selection of the multifunction obtained applying the method of characteristics to equation (6.25), as described in (5.6).*

Proof — Comparing (5.3) and (6.15)–(6.16), we see that the extremals of problem (CV) coincide with the characteristics of equation (6.25). In addition, since H is the Legendre transform of L, it satisfies the identity

$$H(t, x, p) = p H_p(t, x, p) - L(t, x, H_p(t, x, p)),$$

(see (A. 21), (A. 19)). Hence, we obtain from (5.4)

$$
\begin{aligned}
U(t, z) &= U(0, z) + \int_0^t [-H(s, X, P) + P \cdot H_p(s, X, P)] \, ds \\
&= u_0(z) + \int_0^t L(s, X, H_p(s, X, P)) \, ds \\
&= u_0(z) + \int_0^t L(s, X, \dot{X}) \, ds = J_t(X).
\end{aligned}
$$

Hence, along an extremal, the functional J_t coincides with U. Since the minimum in $(CV)_{t,x}$ is necessarily attained by an extremal, we deduce that the value function of (CV) coincides with the minimal selection defined in (5.6).   ∎

An interesting property of the minimizing trajectories is that they are contained, except possibly for the endpoints, in the set where the value function is differentiable.

Theorem 6.4.7 *Let $(t, x) \in Q_T$ be given, and let ξ be a minimizing arc for $(CV)_{t,x}$. Then u is differentiable at $(s, \xi(s))$ for all $s \in {]}0, t[$.*

Proof — Let $s \in {]}0, t]$ be given, and let us set for simplicity $y = \xi(s)$. We know from Theorem 6.4.5 that the elements of $D^+ u(s, y)$ satisfy the Hamilton–Jacobi equation as an inequality. We claim that if $s < t$, then the converse inequality holds as well. In fact we have, by the dynamic programming principle,

$$u(s + h, \xi(s + h)) = u(s, y) + \int_s^{s+h} L(\sigma, \xi(\sigma), \dot{\xi}(\sigma)) d\sigma$$

for all $0 \leq h \leq t - s$. Therefore, setting $w = \dot{\xi}(s)$, we obtain

$$\lim_{h \to 0^+} \frac{u(s + h, \xi(s + h)) - u(s, y)}{h} = L(s, y, w). \tag{6.26}$$

On the other hand, for any $(p_t, p_x) \in D^+ u(s, y)$ we have

$$\lim_{h \to 0^+} \frac{u(s + h, \xi(s + h)) - u(s, y)}{h}$$

$$= \lim_{h \to 0^+} \frac{u(s + h, y + hw) - u(s, y)}{h} \le p_t + w \cdot p_x,$$

which implies, together with (6.26), that $p_t + H(s, y, p_x) \ge 0$. We already know that the converse inequality holds, and so we conclude that

$$p_t + H(s, \xi(s), p_x) = 0, \qquad s \in \,]0, t[\,, \ (p_t, p_x) \in D^+u(s, \xi(s)). \tag{6.27}$$

Since H is strictly convex in the third argument and $D^+u(s, \xi(s))$ is a convex set, we deduce that $D^+u(s, \xi(s))$ consists of a single element. Recalling that u is semiconcave, we obtain that u is differentiable at $(s, \xi(s))$ (see Proposition 3.3.4-(d)). ∎

Theorem 6.4.8 *Let $(t, x) \in Q_T$ be given, and let ξ be a minimizing trajectory for* $(CV)_{t,x}$. *Let η be the dual arc associated with ξ as in Theorem 6.3.3. Then we have*

$$\eta(t) \in \nabla^+u(t, x); \tag{6.28}$$

$$\eta(s) = \nabla u(s, \xi(s)), \qquad s \in \,]0, t[\,. \tag{6.29}$$

Proof — To prove (6.28), it suffices to show that for any nonzero vector $v \in \mathbb{R}^n$, we have

$$\limsup_{h \to 0} \frac{u(t, x + hv) - u(t, x) - hp(t) \cdot v}{h} \le 0. \tag{6.30}$$

Let us set $x_0 = \xi(0)$. For fixed h, the arc $s \to \xi(s) + hv$ is admissible for our functional with endpoint $(t, x + hv)$ and therefore

$$u(t, x + hv) \le u_0(x_0 + hv) + \int_0^t L(s, \xi(s) + hv, \dot\xi(s))\, ds.$$

Recalling also (6.14), it follows that

$$\limsup_{h \to 0} \frac{u(t, x + hv) - u(t, x)}{h}$$

$$\le \lim_{h \to 0} \frac{u_0(x_0 + hv) - u_0(x_0)}{h}$$

$$+ \lim_{h \to 0} \int_0^t \frac{L(s, \xi(s) + hv, \dot\xi(s)) - L(s, \xi(s), \dot\xi(s))}{h}\, ds$$

$$= v \cdot \left(Du_0(x_0) + \int_0^t L_x(s, \xi(s), \dot\xi(s))\, ds \right) = v \cdot \eta(t),$$

which proves (6.30).

If we now consider $s \in \,]0, t[$ we have that the trajectory ξ is optimal for the point $(s, \xi(s))$. By the first part of the proof we deduce that $\eta(s) \in \nabla^+ u(s, \xi(s))$. But u is differentiable at $(s, \xi(s))$ by Theorem 6.4.7 and so (6.29) follows.   ∎

The previous property is not surprising if one observes that $\eta(t)$ evolves according to the same equation satisfied by $P(t) = \nabla u(t, X(t))$ in the method of characteristics. In the terminology of control theory, a result like (6.28) is called a *co-state inclusion* in the superdifferential of the value function.

Theorem 6.4.9 *For any $(t, x) \in Q_T$ the map that associates with any $(p_t, p_x) \in D^* u(t, x)$ the arc ξ obtained by solving problem (6.15) with the terminal conditions*

$$\begin{cases} \xi(t) = x \\ \eta(t) = p_x \end{cases} \tag{6.31}$$

provides a one-to-one correspondence between $D^* u(t, x)$ and the set of minimizers of $(\mathrm{CV})_{t,x}$.

Proof — Let us first prove the conclusion when (t, x) is a point of differentiability for u. In this case the set $D^+ u(t, x)$ contains only the gradient $(u_t(t, x), \nabla u(t, x))$. Since u is semiconcave, the same is true for $D^* u(t, x)$. We must therefore show that problem $(\mathrm{CV})_{t,x}$ has a unique minimizer, obtained by solving problem (6.15) with final conditions

$$\xi(t) = x, \qquad \eta(t) = \nabla u(t, x). \tag{6.32}$$

To see this, let us suppose that ξ^* is a minimizer, and let η^* be the associated dual arc. Then the pair (ξ^*, η^*) solves system (6.15). In addition, by Theorem 6.4.8 and the differentiability of u at (t, x), the pair (ξ^*, η^*) also satisfies conditions (6.32). This proves the theorem in this case.

Let us now consider the case of a general (t, x). Let (p_t, p_x) be any point in $D^* u(t, x)$. Then there exists a sequence $\{(t_k, x_k)\}$ of points where u is differentiable which converges to (t, x) and is such that $(p_t, p_x) = \lim_{k \to \infty} Du(t_k, x_k)$. By Theorem 6.4.5, we have that

$$\partial_t u(t_k, x_k) + H(t_k, x_k, \nabla u(t_k, x_k)) = 0,$$

which implies, letting $k \to \infty$, that $p_t + H(t, x, p_x) = 0$. For fixed k, by the first part of the proof, there exists a unique minimizer ξ_k for the point (t_k, x_k), obtained by solving system (6.15) with the conditions $\xi_k(t_k) = x_k$, $\eta_k(t_k) = \nabla u(t_k, x_k)$. By continuous dependence, $\xi_k \to \xi$ as $k \to \infty$, where ξ is the first component of the solution of problems (6.15)–(6.31). It follows that by the continuity of u and the optimality of ξ_k

$$u(t, x) = \lim_{k \to \infty} u(t_k, x_k) = \lim_{k \to \infty} \left(\int_0^{t_k} L(s, \xi_k(s), \dot{\xi}_k(s))\, ds + u_0(\xi_k(0)) \right)$$

$$= \int_0^t L(s, \xi(s), \dot{\xi}(s))\, ds + u_0(\xi(0)),$$

and so the arc ξ is a minimizer for the point (t, x). This proves that the correspondence $(p_t, p_x) \rightarrow \xi$ is indeed a map from $D^*u(t, x)$ to the set of minimizers for (t, x).

Next, we prove the injectivity of our map. Let (p_t^1, p_x^1) and (p_t^2, p_x^2) be two distinct elements of $D^*u(t, x)$. As observed before, $p_t^i = -H(t, x, p_x^i)$, and so p_x^1 and p_x^2 must be distinct. Then the corresponding solutions (ξ_1, η_1) and (ξ_2, η_2) of problem (6.15)–(6.31) are also distinct. This implies that $\xi_1 \neq \xi_2$, since otherwise η_1 and η_2 would also coincide, as a consequence of the definition of η (see Theorem 6.3.3).

Finally, we prove surjectivity. Let ξ^* be any minimizer for (t, x) and let η^* be the associated dual arc. Then the pair (ξ^*, η^*) solves system (6.15). In addition, by Theorems 6.4.5 and 6.4.8, u is differentiable at the points of the form $(s, \xi^*(s))$ with $0 < s < t$ and satisfies

$$\nabla u(s, \xi^*(s)) = \eta^*(s),$$

$$\partial_t u(s, \xi^*(s)) = -H(s, \xi^*(s), \nabla u(s, \xi^*(s))) = -H(s, \xi^*(s), \eta^*(s)).$$

Therefore, setting

$$p_x = \eta^*(t) = \lim_{s \to t} \nabla u(s, \xi^*(s)),$$

$$p_t = -H(t, x, p_x) = \lim_{s \to t} \partial_t u(s, \xi^*(s)),$$

we have that $(p_t, p_x) \in D^*u(t, x)$. This shows that ξ^* belongs to the range of the map defined in the statement of the theorem. $\blacksquare$

In particular, the above theorem shows the equivalence between regularity (in the sense of Definition 6.3.4) and differentiability of u.

Corollary 6.4.10 *A point (t, x) is regular for problem* (CV) *if and only if u is differentiable at x. Thus, the set Σ of irregular points coincides with the singular set of u.*

Let us denote by Σ the set of irregular points (equivalently, the set of points where u is not differentiable) and by Γ the set of conjugate points.

Theorem 6.4.11 *Assume* (L1)–(L4). *Then $\Sigma \cup \Gamma$ is a closed set and $u \in C^{R+1}(Q_T \setminus (\Sigma \cup \Gamma))$.*

Proof — Let us consider a point $(\bar{t}, \bar{x})$ in the complement of $\Sigma \cup \Gamma$. By Corollary 6.4.10 there exists a unique minimizer for problem (CV)$_{\bar{t}, \bar{x}}$, which has the form $\xi(\cdot, \bar{z})$ for some $\bar{z} \in \mathbb{R}^n$; in addition $\det \xi_z(\bar{t}, \bar{z}) \neq 0$. Then we can find a neighborhood W of $(\bar{t}, \bar{z})$ such that the map $(t, z) \rightarrow (t, \xi(t, z))$ is a C^R diffeomorphism.

We claim now that there exists a neighborhood N of $(\bar{t}, \bar{x})$ with the following property: if $(t, x) \in N$ and ξ^* is a minimizer for (t, x), then $\xi^* = \xi(\cdot, z)$ for some z such that $(t, z) \in W$.

We prove the claim by contradiction. If no such neighborhood exists, we can find a sequence $\{(t_k, x_k)\}$ converging to $(\bar{t}, \bar{x})$ with the following property: each (t_k, x_k)

has a minimizer ξ_k of the form $\xi_k = \xi(\cdot, z_k)$ with $(t_k, z_k) \notin W$. Now let η_k be the dual arc associated with ξ_k; since u is differentiable at $(\bar{t}, \bar{x})$ we deduce from Theorem 6.4.7 and the upper semicontinuity of D^+u (Proposition 3.3.4(a)) that

$$\eta_k\left(t_k - \frac{1}{k}\right) = \nabla u\left(t_k - \frac{1}{k}, \xi_k\left(t_k - \frac{1}{k}\right)\right) \to \nabla u(\bar{t}, \bar{x}).$$

By continuous dependence, $\xi_k \to \xi$, where ξ is obtained by solving system (6.15) with terminal conditions $\xi(\bar{t}) = \bar{x}, \eta(\bar{t}) = \nabla u(\bar{t}, \bar{x})$. Then ξ coincides with $\xi(\cdot, \bar{z})$, which is the unique minimizer for $(\bar{t}, \bar{x})$ by Theorem 6.4.9. In particular, this implies that $z_k = \xi_k(0) \to \xi(0) = \bar{z}$. But this contradicts the assumption that $(z_k, t_k) \notin W$ for all k.

From the claim we deduce that for $(t, x) \in N$, we have

$$u(t, x) = u_0(z) + \int_0^t L(s, \xi(s, z), \dot{\xi}(s, z))\, ds,$$

where z is uniquely determined by the conditions $(t, z) \in W$, $x = \xi(t, z)$. This shows that u is of class C^R in N and that all points of N are regular and not conjugate. In addition, by Theorem 6.4.8 we have $\nabla u(t, x) = \eta(t, z)$ and $\partial_t u(t, x) = -H(t, x, \nabla u(t, x))$; therefore the derivatives of u are also of class C^R and $u \in C^{R+1}(N)$. ∎

We conclude the section giving a characterization of the extreme points of D^+u (see Definitions 3.3.12 and 3.3.14).

Theorem 6.4.12 *For any* $(t, x) \in Q_T$ *we have*

$$Eu(t, x) = \text{Ext}(D^+u(t, x)) = D^*u(t, x)$$
$$= \{(q, p) \in D^+u(t, x) \mid p_t + H(t, x, p_x) = 0\}.$$

Proof — The inclusion $Eu(t, x) \subset \text{Ext}(D^+u(t, x))$ follows from the definition, while the inclusions $Eu(t, x) \subset D^*u(t, x)$ and $D^*u(t, x) \subset D^+u(t, x)$ hold for a general semiconcave function (see (3.26) and Proposition 3.3.4(b)). The fact that any $(p_t, p_x) \in D^*u(t, x)$ satisfies $p_t + H(t, x, p_x) = 0$ has been already observed during the proof of Theorem 6.4.9. To complete the proof we only need to show that every $(p_t, p_x) \in D^+u(t, x)$ which satisfies $p_t + H(t, x, p_x) = 0$ is an exposed vector for $D^+u(t, x)$. This is a consequence of the strict convexity of H. If we set $h = H_p(t, x, p_x)$ we have in fact

$$H(t, x, q_x) > H(t, x, p_x) + h \cdot (q_x - p_x), \qquad q_x \in \mathbb{R}^n, q_x \neq p_x.$$

On the other hand, since u is a viscosity solution, we have

$$q_t + H(t, x, q_x) \leq 0, \qquad (q_t, q_x) \in D^+u(t, x).$$

Therefore

$$-p_t - h \cdot p_x < -q_t - h \cdot q_x, \qquad (q_t, q_x) \in D^+u(t, x), q_x \neq p_x,$$

and therefore (p_t, p_x) is an exposed vector with respect to the direction $(-1, -h)$. ∎

6.5 The singular set of u

In this section we begin the study of the singular set of the value function u of problem (CV). We assume that conditions (L1)–(L4) are satisfied throughout the section. For brevity, we omit recalling these hypotheses in the statements below.

Before starting our analysis, let us briefly review some of the results of Dafermos [68]. He considers the Cauchy problem for the one-dimensional scalar conservation law

$$
\begin{cases}
\partial_t v(t, x) + \partial_x f(t, x, v(t, x)) = 0 & (t, x) \in \mathbb{R}_+ \times \mathbb{R} \\
v(0, x) = v_0(x) & x \in \mathbb{R},
\end{cases}
\tag{6.33}
$$

where f and v_0 satisfy suitable regularity assumptions and f is strictly convex in the third variable. We have considered this equation in Section 1.7 in the case where $f = f(v)$. We have seen that if $f = H$, then $u(t, x)$ is a semiconcave solution of (6.25) if and only if $v(t, x) = \partial_x u(t, x)$ is an entropy solution of (6.33) with initial data $v_0(x) = u'_0(x)$. It can be proved that the same equivalence holds also when f depends on t, x, and so in the one-dimensional case our equation belongs to the class considered by Dafermos.

The results of [68], rephrased according to our terminology, are the following. The set $\Sigma \cup \Gamma$ is the union of an at most countable family of arcs of the form $\{(t, x) \mid t \geq t_i, x = \gamma_i(t)\}$, with γ_i Lipschitz and $t_i > 0$. All points of the form $(t, \gamma_i(t))$ with $t > t_i$ are irregular. The initial point $(t_i, \gamma(t_i))$ is called a *shock generation point*. It is a conjugate point, and may be either regular or irregular. In the latter case it is called the center of a *compression wave*, because there are infinitely many optimal trajectories ending at that point and they have the form $\xi(\cdot, z)$ with $z \in [a, b]$ for some $[a, b] \subset \mathbb{R}$. The function γ_i may be characterized as the unique solution of the differential inclusion

$$
\begin{cases}
\gamma'(t) \in [H_p(t, \gamma(t), u_x(t, \gamma(t)+)), H_p(t, \gamma(t), u_x(t, \gamma(t)-))], & t \geq t_0 \\
\gamma(t_0) = \gamma_i(t_0).
\end{cases}
$$

Here we have denoted by $u_x(t, x\pm)$ the right and left limit of u_x in the x variable. It follows from the semiconcavity of u and from the convexity of H that these limits exist and satisfy $H_p(t, x, u_x(t, x+)) \leq H_p(t, x, u_x(t, x-))$ for any $(t, x) \in Q$. Observe that the arcs γ_i solutions of the above differential inclusion are generalized characteristics according to our definition (5.55).

The above results require the special structure of the one-dimensional case. We will see, however, that some of them possess a multidimensional analogue, although proofs must be based on a different method.

To begin with, we study the structure of the set $\Sigma \cup \Gamma$. The following lemma states that any point (t_0, x_0), which is singular or conjugate, is the cluster point of a family of singular points, propagating forward in time. The interesting case is when (t_0, x_0) is conjugate, since the result is then a consequence of Theorem 5.6.6.

Lemma 6.5.1 *Let* $(t_0, x_0) \in \Sigma \cup \Gamma$ *be given. Then there exist* $M > 0$, $h > 0$ *with the following property: given any* $\varepsilon \in (0, h]$ *there exists* $x_\varepsilon \in \overline{B_{\varepsilon M}(x_0)}$ *such that* $(t_0 + \varepsilon, x_\varepsilon) \in \Sigma$.

Proof — By Theorem 6.3.1 there exists a constant M such that if $(t, x) \in [t_0, t_0 + 1] \times B_1(x_0)$ and ξ^* is a minimizer for problem $(CV)_{t,x}$, then $|\dot{\xi}^*(s)| < M$ for any s. We then define $h = \min\{1, M^{-1}\}$.

We claim that M and h have the desired property. Let us suppose, on the contrary, that there exists $\varepsilon \in (0, h]$ such that all points of the form $(t_0 + \varepsilon, x)$ with $x \in \overline{B_{\varepsilon M}(x_0)}$ are regular. Then, by Corollary 6.4.10 and Proposition 3.3.4-(e), $Du(t_0 + \varepsilon, x)$ exists for any $x \in \overline{B_{\varepsilon M}(x_0)}$ and depends continuously on x.

Let us consider the hamiltonian system (6.15) with final conditions (6.31) and denote by $\xi(s; t, x, p_x)$ the first component of the solution. By Theorem 6.4.9, for any $x \in B_{\varepsilon M}(x_0)$ the arc $\xi(\cdot; t_0 + \varepsilon, x, \nabla u(t_0 + \varepsilon, x))$ is the unique minimizer for problem (CV) with endpoint $(t_0 + \varepsilon, x)$. We now define a map $\Lambda : \overline{B_{\varepsilon M}(x_0)} \to \mathbb{R}^n$ in the following way:

$$\Lambda(x) = x_0 + x - \xi(t_0; t_0 + \varepsilon, x, \nabla u(t_0 + \varepsilon, x)).$$

The previous remarks show that Λ is continuous. In addition,

$$|x - \xi(t_0; t_0 + \varepsilon, x, \nabla u(t_0 + \varepsilon, x))|$$
$$= |\xi(t_0 + \varepsilon; t_0 + \varepsilon, x, \nabla u(t_0 + \varepsilon, x)) - \xi(t_0; t_0 + \varepsilon, x, \nabla u(t_0 + \varepsilon, x))|$$
$$\leq \varepsilon \sup |\dot{\xi}| \leq \varepsilon M$$

and so Λ maps $\overline{B_{\varepsilon M}(x_0)}$ into itself. Hence, by Brouwer's theorem, $x_\varepsilon \in \overline{B_{\varepsilon M}(x_0)}$ exists such that $\Lambda(x_\varepsilon) = x_\varepsilon$. Recalling the definition of Λ, we conclude that the minimizer for $(t_0 + \varepsilon, x_\varepsilon)$ passes through the point (t_0, x_0). This fact contradicts Theorem 6.3.6 since we are assuming that (t_0, x_0) is irregular or conjugate. ∎

From the above lemma we recover the converse of the inclusion $\overline{\Sigma} \subset \Sigma \cup \Gamma$ (see Theorem 6.4.11) and obtain the following.

Corollary 6.5.2 $\overline{\Sigma} = \Sigma \cup \Gamma$.

We now give a sufficient condition in terms of reachable gradients to detect conjugate points.

Theorem 6.5.3 *If* $D^* u(t, x)$ *is an infinite set, then* $(t, x) \in \Gamma$.

Proof — Let us assume that (t, x) is not conjugate. Then, if we set

$$Z = \{z \in \mathbb{R}^n \mid \xi(\cdot, z) \text{ is a minimizer for } (t, x)\},$$

we have that the points of Z are isolated. In fact, if $z \in Z$, then $\xi_z(t, z)$ is nonsingular, and so $\xi(t, z') \neq \xi(t, z) = x$ for z' in a neighborhood of z. Moreover, by Theorem 6.3.1, Z is bounded. We deduce that Z is finite.

On the other hand, there are infinitely many minimizers of (CV) with endpoint (t, x), by Theorem 6.4.9. In addition, Theorem 6.3.3 implies that two different minimizers have different initial points. Thus Z is infinite. We have reached a contradiction.   ∎

The next result says, roughly speaking, that the second derivative of u is infinite at a conjugate point.

Theorem 6.5.4 *Let* $(t_0, x_0) \in \Gamma$ *be given, and let* $\xi(\cdot, z_0)$ *be a minimizer for* (t_0, x_0) *such that* $\det \xi_z(t_0, z_0) = 0$. *Then*

$$\lim_{t \to t_0^-} \|\nabla^2 u(t, \xi(t, z_0))\| = +\infty. \tag{6.34}$$

If in addition (t_0, x_0) *is regular and* $\{x_k\}$, $\{p_k\}$ *satisfy*

$$
\begin{cases}
\lim_{k \to \infty} x_k = x_0, \\[2ex]
\lim_{k \to \infty} \dfrac{x_k - x_0}{|x_k - x_0|} = v \text{ for some } v \notin Im\,\xi_z(t_0, z_0), \\[2ex]
(-H(t_0, x_k, p_k), p_k) \in D^* u(t_0, x_k),
\end{cases}
\tag{6.35}
$$

then

$$\lim_{k \to \infty} \frac{|p_k - \nabla u(t_0, x_0)|}{|x_k - x_0|} = +\infty.$$

Proof — Let $\theta \in (\mathbb{R}^n \setminus \{0\})$ be given such that $\xi_z(t_0, z_0)\theta = 0$. Then $\eta_z(t_0, z_0)\theta \neq 0$ by (6.19). Given $t < t_0$, the point $(t, \xi(t, z_0))$ is regular and not conjugate by Theorem 6.3.6, and so are the points in a neighborhood, by Theorem 6.4.11. Then we have, by Theorem 6.4.9

$$\nabla u(t, \xi(t, z)) = \eta(t, z)$$

for z in a neighborhood of z_0. It follows that

$$\nabla^2 u(t, \xi(t, z_0))\xi_z(t, z_0) = \eta_z(t, z_0), \qquad t < t_0.$$

Since $\eta_z(t, z_0)\theta \neq 0$ we conclude that

$$\|\nabla^2 u(t, \xi(t, z_0))\| \geq \frac{|\eta_z(t, z_0)\theta|}{|\xi_z(t, z_0)\theta|} \to \infty \text{ as } t \to t_0^-.$$

Now suppose (t_0, x_0) is regular. Then $\xi(\cdot, z_0)$ is the unique minimizer for (t_0, x_0) and we have $\nabla u(t_0, x_0) = \eta(z_0, t_0)$ by Theorem 6.4.9. Let $\{x_k\}$, $\{p_k\}$ satisfy assumptions (6.35). Again by Theorem 6.4.9 there exists a sequence $\{z_k\}$ such that $x_k = \xi(t_0, z_k)$, $p_k = \eta(t_0, z_k)$. The sequence $\{z_k\}$ is bounded, by Theorem 6.3.1; moreover, any cluster point $\bar{z}$ of $\{z_k\}$ is such that the trajectory $\xi(\cdot, \bar{z})$ is a minimizer for (t_0, x_0). Since such a minimizer is unique, we deduce that $\lim z_k = z_0$.

By passing to a subsequence we may assume

$$\lim_{k\to\infty} \frac{z_k - z_0}{|z_k - z_0|} = v$$

for some unit vector v. Since

$$x_k = x_0 + \xi_z(t_0, z_0)(z_k - z_0) + o(|z_k - z_0|)$$

we deduce

$$\lim_{k\to\infty} \frac{x_k - x_0}{|z_k - z_0|} = \xi_z(t_0, z_0)v.$$

If $\xi_z(t_0, z_0)v \neq 0$, then the above relation implies that $v \in \operatorname{Im}\xi_z(t_0, z_0)$, in contradiction with our assumptions. Therefore we have $\xi_z(t_0, z_0)v = 0$, which implies $\eta_z(t_0, z_0)v \neq 0$ by (6.19). We conclude that

$$\lim_{k\to\infty} \frac{|p_k - \nabla u(t_0, x_0)|}{|x_k - x_0|}$$

$$= \lim_{k\to\infty} \frac{|\eta(t_0, z_k) - \eta(t_0, z_0)|}{|z_k - z_0|} \frac{|z_k - z_0|}{|\xi(t_0, z_k) - \xi(t_0, z_0)|}$$

$$= \lim_{k\to\infty} \frac{|\eta_z(t_0, z_0)v|z_k - z_0| + o(|z_k - z_0|)|}{|z_k - z_0|} \frac{|z_k - z_0|}{o(|z_k - z_0|)} = +\infty. \qquad \blacksquare$$

Finally, we describe the structure of u about a singular point which is not conjugate.

Theorem 6.5.5 *Let $(\bar{t}, \bar{x}) \in \Sigma \setminus \Gamma$ be given. Then there exists a neighborhood N of $(\bar{t}, \bar{x})$ and a finite number of C^{R+1} functions $v_1, \dots, v_k : N \to \mathbb{R}$ such that $Dv_i \neq Dv_j$ if $i \neq j$ and $u = \min\{v_1, \dots, v_k\}$ in N.*

Proof — Let $(\bar{t}, \bar{x}) \in \Sigma \setminus \Gamma$. Then, by Theorem 6.5.3, $D^*u(\bar{t}, \bar{x})$ contains a finite number of distinct elements $(q_1, p_1), \dots, (q_k, p_k) \in \mathbb{R} \times \mathbb{R}^n$. By Theorem 6.4.12 we have $H(\bar{t}, \bar{x}, p_i) + q_i = 0$ for all i and therefore $p_1, \dots, p_k$ are all different. By Theorem 6.4.9 the minimizers of (CV) with endpoint $(\bar{t}, \bar{x})$ are obtained solving (6.15) with the terminal conditions $\xi(\bar{t}) = \bar{x}$, $\eta(\bar{t}) = p_i$ for $i = 1, \dots, k$. Let us denote with $z_1, \dots, z_k$ the initial point of these trajectories; then the minimizers can be written as $\xi(\cdot, z_1), \dots, \xi(\cdot, z_k)$. Moreover $\eta(t, z_i) = p_i$ for $i = 1, \dots, k$.

We now argue as in the proof of Theorem 6.4.11, with the difference that we have here k minimizers instead of one. Using the assumption that the point $(\bar{t}, \bar{x})$ is not conjugate we can find $W_1, \dots, W_k$ neighborhoods of $(\bar{t}, z_i)$ and a V neighborhood of $(\bar{t}, \bar{x})$ such that the map $(t, z) \to (t, \xi(t, z))$ is a diffeomorphism between W_i and V for all $i = 1, \dots, n$. We choose V small enough to have that $W_i \cap W_j = \emptyset$ if $i \neq j$ and that $\eta(t, z) \neq \eta(t', z')$ for any $(t, z) \in W_i$, $(t', z') \in W_j$ with $i \neq j$. For every $i = 1, \dots, n$, we define $\zeta_i : V \to \mathbb{R}^n$ in the following way: $\zeta_i(t, x) = z$ where z is the unique value such that $(t, z) \in W_i$ and that $\xi(t, z) = x$. Let us set

$$v_i(t, x) = u_0(z) + \int_0^t L(s, \xi(s, z), \dot\xi(s, z))ds, \quad \text{with } z = \zeta_i(t, x).$$

Then, in a neighborhood $N \subset V$ of $(\bar{t}, \bar{x})$ we have that $u = \min\{v_1, \ldots, v_k\}$. Otherwise we could find points (t, x) arbitrarily close to $(\bar{t}, \bar{x})$ which have a minimizer of the form $\xi(\cdot, z)$ with (t, z) not belonging to any W_i and we would obtain a contradiction as in the proof of Theorem 6.4.11. In addition, the functions v_i are of class C^{R+1} and satisfy $\nabla v_i(t, x) = \eta(t, \zeta_i(t, x))$ for all i, and so $Dv_i(t, x) \neq Dv_j(t, x)$ if $i \neq j$. $\blacksquare$

From the above result we can deduce the following representation of the set $\Sigma \setminus \Gamma$.

Corollary 6.5.6 *Given* $(t, x) \in \Sigma \setminus \Gamma$, *there exists a neighborhood* N *of* (t, x) *such that* $N \cap \Sigma$ *is contained in a finite union of n-dimensional hypersurfaces of class* C^{R+1}.

Proof — Let $v_1, \ldots, v_k$ be the functions given by Theorem 6.5.5. Since $u = \min_{i=1,\ldots,k} v_i$, u is not differentiable at a point $(s, y) \in N$ if and only if there exist $h, j \in \{1, \ldots, k\}$ such that $v_h(s, y) = v_j(s, y) = \min_{i=1,\ldots,k} v_i(s, y)$. But since $Dv_h \neq Dv_j$ in N if $h \neq j$, the set $\{(s, y) \mid v_h(s, y) = v_j(s, y)\}$ is a C^{R+1} hypersurface for any pair h, j. $\blacksquare$

6.6 Rectifiability of $\overline{\Sigma}$

We continue in this section the analysis of the regularity of the value function u of problem $(CV)_{t,x}$. We have seen that u is semiconcave with linear modulus (see Theorem 6.4.1); therefore, its gradient Du belongs to the class BV (see Theorem 2.3.1–(ii)). In addition, its singular set Σ is countably n-rectifiable (Corollary 4.1.13). It is natural to investigate whether u enjoys further regularity properties. We recall that in Example 6.3.5 we analyzed a class of data for which the value function is piecewise smooth. Such a type of regularity has been investigated by Schaeffer [125] in the one-dimensional case: he showed that the value function is piecewise smooth for a generic class of data, but that there exist C^∞ data for which this property fails.

Therefore, we cannot expect the value function to be piecewise smooth in general. However, we will prove a regularity property which is in some sense close to piecewise smoothness, showing that u is regular on the complement of a closed set of codimension one. Our starting point is the result, stated in Theorem 6.4.11, that u has the same smoothness of the initial data in the complement of the set $\overline{\Sigma} = \Sigma \cup \Gamma$. Our aim will be to prove that this set is countably $\mathcal{H}^n$-rectifiable. Since we already know that Σ is countably n-rectifiable, we only need to prove the rectifiability of Γ, or of $\Gamma \setminus \Sigma$. This will be done in this section; actually, we also give a Hausdorff estimate on $\Gamma \setminus \Sigma$ which is much sharper than the one needed for the rectifiability of Γ. In fact we prove that $\mathcal{H}^{n-1+2/R}(\Gamma \setminus \Sigma) = 0$, showing that the set of conjugate points (at least the regular ones) is very "small" compared with the set of irregular points.

Let us first outline the main ideas of our approach. We introduce the set

$$G = \{(t, z) \mid \det \xi_z(t, z) = 0\}$$

and the function

$$\overline{\xi}(t, z) = (t, \xi(t, z)).$$

Then, by definition, $\overline{\xi} \in C^R$ and $\Gamma \subset \overline{\xi}(G)$. We can use this property to obtain upper bounds on the size of Γ. A possible way, pursued in [78], is to apply the following Sard-type result ([76, Theorem 3.4.3]).

Theorem 6.6.1 *Let $F : \mathbb{R}^N \to \mathbb{R}^M$ be a map of class C^R for some $R \geq 1$. For any $k \in \{0, 1, \ldots, N - 1\}$ set*

$$A_k = \{x \in \mathbb{R}^N \mid \operatorname{rk} DF(x) \leq k\},$$

where $\operatorname{rk} DF$ denotes the rank of DF. Then $\mathcal{H}^{k+(N-k)/R}(F(A_k)) = 0$.

This result, applied to $\overline{\xi}$, yields

$$\mathcal{H}^{n+1/R}(\Gamma) = 0.$$

However, the above estimate does not imply the rectifiability of Γ even if the data of our problem are C^∞.

Another approach is to try and prove the rectifiability of G using the implicit function theorem. In the case of one-dimensional space this idea can be implemented in a very simple way.

Theorem 6.6.2 *Suppose $n = 1$. Then Γ is a countably 1-rectifiable set.*

Proof — Since $\Gamma \subset \overline{\xi}(G)$, it suffices to prove the rectifiability of G. We recall that $H_{pp} > 0$ by the convexity of H. Thus, if $(t_0, z_0) \in G$ we have, by (6.17) and (6.19),

$$\xi_{zt}(t_0, z_0) = H_{pp}(\xi(t_0, z_0), \eta(t_0, z_0))\eta_z(t_0, z_0) \neq 0.$$

By the implicit function theorem, we can cover G by a countable union of arcs of class C^1. ∎

When we try to extend this argument to the case of $n > 1$ we find two difficulties. The first is technical, and consists of the fact that we have to deal with $\det \xi_z(t, z)$ instead of $\xi_z(t, z)$. A more substantial problem is the fact that $\det \xi_z(t, z)$ may well vanish simultaneously with its time derivative. This fact is explained by the following result.

Theorem 6.6.3 *Let $n > 1$. Then, for any $(t, z) \in G$,*

$$\partial_t (\det \xi_z(t, z)) \neq 0 \quad \Longleftrightarrow \quad \operatorname{rk} \xi_z(t, z) = n - 1.$$

Before giving the proof of Theorem 6.6.3, let us show how this result can be used to obtain the rectifiability of Γ.

Theorem 6.6.4 *Suppose $n > 1$. Then Γ is a countably $\mathcal{H}^n$-rectifiable set.*

Proof — Let us define

$$G' = \{(t, z) \mid \mathrm{rk}(\xi_z(t, z)) = n - 1\}, \tag{6.36}$$
$$G'' = \{(t, z) \mid \mathrm{rk}(\xi_z(t, z)) < n - 1\}. \tag{6.37}$$

Then, by Theorem 6.6.3 and the implicit function theorem, we deduce that G' is a countably n-rectifiable set. On the other hand, by Theorem 6.6.1,

$$\mathcal{H}^n(\overline{\xi}(G'')) \leq \mathcal{H}^{n-1+2/R}(\overline{\xi}(G'')) = 0.$$

Since $\Gamma \subset \overline{\xi}(G' \cup G'')$, our assertion follows.   ∎

The proof of Theorem 6.6.3 requires many intermediate steps as we show below. The first consists of giving a suitable expression for the derivative of a determinant. For this purpose, we recall below some elementary facts from linear algebra.

Let $A = (a_{ij})$ be an $n \times n$ matrix (we assume $n > 1$). We denote by A^+ the transpose of the matrix of the cofactors of A. Equivalently, if we denote by a_{ij}^+ the generic element of A^+ and by $\hat{A}_{ij}$ the matrix obtained removing from A the i-th row and the j-th column, then $a_{ij}^+ = (-1)^{i+j} \det \hat{A}_{ji}$. Then, Laplace's identity states that

$$\sum_{j=1}^{n} a_{ij} a_{jk}^+ = \delta_{ik} \det A,$$

or

$$AA^+ = A^+ A = (\det A)\, I. \tag{6.38}$$

Therefore

$$\frac{\partial \det A}{\partial a_{ij}} = \frac{\partial}{\partial a_{ij}} \left(\sum_{k=1}^{n} a_{ik} a_{ki}^+ \right) = a_{ji}^+. \tag{6.39}$$

Moreover, (6.38) implies that

$$\mathrm{rk}\, A = n \quad \Rightarrow \quad \mathrm{rk}\, A^+ = n. \tag{6.40}$$

Similarly, it is easy to check that

$$\mathrm{rk}\, A = n - 1 \quad \Rightarrow \quad \mathrm{rk}\, A^+ = 1 \tag{6.41}$$
$$\mathrm{rk}\, A < n - 1 \quad \Rightarrow \quad \mathrm{rk}\, A^+ = 0 \tag{6.42}$$

Indeed, recalling the estimate

$$\mathrm{rk}\, BC \geq \mathrm{rk}\, B + \mathrm{rk}\, C - n,$$

which holds for any two $n \times n$ matrices B and C, from (6.38) we conclude that $\mathrm{rk}\, A^+ \leq 1$ if $\mathrm{rk}\, A = n - 1$. On the other hand, by definition, $A^+ = 0$ if and only if $\mathrm{rk}\, A < n - 1$.

Finally, if $A(\cdot)$ is time-dependent and of class C^1, then (6.39) yields

$$\frac{d}{dt}\det A(t) = \sum_{i,j=1}^{n} \frac{\partial \det A(t)}{\partial a_{ij}} a'_{ij}(t)$$

$$= \sum_{i,j=1}^{n} a_{ji}^{+}(t) a'_{ij}(t) = \operatorname{tr}(A'(t) A^{+}(t)). \tag{6.43}$$

Let us now consider the hamiltonian system (6.15) again. In the following we denote by A^* the transpose of a given matrix A.

Lemma 6.6.5 *We have $\xi_z^*(t, z)\eta_z(t, z) = \eta_z^*(t, z)\xi_z(t, z)$ for any (t, z).*

Proof — Let us set $R(t) = \xi_z^*(t, z)\eta_z(t, z)$. Since

$$R(0) = D^2 u_0(z) = R^*(0),$$

$$R'(t) = \eta_z^*(t, z)H_{pp}\eta_z(t, z) - \xi_z^*(t, z)H_{xx}\xi_z(t, z) = (R^*)'(t),$$

we have $R(t) = R^*(t)$ for any $t \geq 0$, which proves our assertion. ∎

Lemma 6.6.6 *Suppose $n > 1$. Let $\operatorname{rk}\xi_z(t_0, z_0) = n - 1$ for some (t_0, z_0) and let θ be a generator of $\operatorname{Ker}\xi_z(t_0, z_0)$. Then $\eta_z(t_0, z_0)\theta$ generates the orthogonal complement of $\operatorname{Im}\xi_z(t_0, z_0)$ and*

$$\xi_z^{+}(t_0, z_0) = c\theta \otimes \eta_z(t_0, z_0)\theta \tag{6.44}$$

for some $c \in \mathbb{R} \setminus \{0\}$.

Proof — We know from (6.19) that $\eta_z(t_0, z_0)\theta \neq 0$. Given any $w \in \mathbb{R}^n$, we have

$$\langle \eta_z(t_0, z_0)\theta, \xi_z(t_0, z_0)w \rangle = \langle \xi_z^*(t_0, z_0)\eta_z(t_0, z_0)\theta, w \rangle$$

$$= \langle \eta_z^*(t_0, z_0)\xi_z(t_0, z_0)\theta, w \rangle = 0.$$

Since w is arbitrary, we obtain the first assertion of the lemma.

To prove (6.44) we first observe that by (6.41), $\xi_z^{+}(t_0, z_0)$ has rank one, and so $\xi_z^{+}(t_0, z_0) = v_1 \otimes v_2$ for some nonzero vectors v_1, v_2. Moreover,

$$v_1\mathbb{R} = \operatorname{Im}\xi_z^{+}(t_0, z_0) = \operatorname{Ker}\xi_z(t_0, z_0).$$

Indeed, equality (6.38) yields $\operatorname{Im}\xi_z^{+}(t_0, z_0) \subset \operatorname{Ker}\xi_z(t_0, z_0)$; then, the second equality above follows from the fact that both $\operatorname{Im}\xi_z^{+}(t_0, z_0)$ and $\operatorname{Ker}\xi_z(t_0, z_0)$ are one-dimensional spaces. Similarly,

$$\{v_2\}^{\perp} = \operatorname{Ker}\xi_z^{+}(t_0, z_0) = \operatorname{Im}\xi_z(t_0, z_0).$$

Therefore v_1 and v_2 must be parallel to θ and $\eta_z(t_0, z_0)\theta$, respectively. ∎

Proof of Theorem 6.6.3 — By (6.43) we have, for any (t, z), (we omit for simplicity the dependence of H on (t, ξ, η))

$$\frac{\partial}{\partial t} \det \xi_z(t, z) = \operatorname{tr}\left(\xi_{zt}(t, z)\xi_z^+(t, z)\right)$$
$$= \operatorname{tr}\left(H_{px}\xi_z(t, z)\xi_z^+(t, z)\right) + \operatorname{tr}\left(H_{pp}\eta_z(t, z)\xi_z^+(t, z)\right)$$
$$= \det \xi_z(t, z)\operatorname{tr}(H_{px}) + \operatorname{tr}(H_{pp}\eta_z(t, z)\xi_z^+(t, z)).$$

Now let us suppose that $\operatorname{rk} \xi_z(t, z) = n - 1$. Then, if θ is such that $\operatorname{Ker} \xi_z(t_0, z_0) = \theta \mathbb{R}$ and $q = \eta_z(t_0, z_0)\theta$, we have by Lemma 6.6.6,

$$\frac{\partial}{\partial t} \det \xi_z(t_0, z_0) = \operatorname{tr}\left(cH_{pp}\eta_z(t, z)\theta \otimes q\right)$$
$$= cH_{pp}q \cdot q \neq 0$$

since H_{pp} is positive definite and $q \neq 0$. On the other hand, if $\operatorname{rk} \xi_z(t, z) < n - 1$, then $\xi_z^+(t, z) = 0$ by (6.42) and thus the derivative of $\det \xi_z$ is zero. ∎

Remark 6.6.7 As a consequence of the implicit function theorem and Theorem 6.6.3, we conclude that the set G' defined in (6.36) is locally a graph, that is, for any $(t_0, z_0) \in G'$, there exist $r, \varepsilon > 0$ and a map $\phi : B_r(z_0) \to \,]t_0 - \varepsilon, t_0 + \varepsilon[$ of class C^{R-1}, such that

$$\det \xi_z(t, z) = 0 \text{ if and only if } t = \phi(z) \tag{6.45}$$

for any $(t, z) \in \,]t_0 - \varepsilon, t_0 + \varepsilon[\times B_r(z_0)$. If we choose r small enough, we may also assume that $\xi_z(\phi(z), z)$ has rank $n - 1$ for any $z \in B_r(z_0)$. Then, there also exists a vector field $\theta : B_r(z_0) \to (\mathbb{R}^n \setminus \{0\})$ of class C^{R-1} such that for any $z \in B_r(z_0)$, $\theta(z)$ is a generator of $\operatorname{Ker} \xi_z(\phi(z), z)$. Indeed, for θ one can take the exterior product of $n - 1$ linearly independent rows of $\xi_z(\phi(z), z)$.

The geometric relationship between vector fields $D\phi$ and θ, about a regular conjugate point, is made clear by the following result.

Proposition 6.6.8 *Let G' be as in (6.36) and, given $(t_0, z_0) \in G'$, let r, ε and ϕ, θ be defined as in Remark 6.6.7. If $\bar{z} \in B_r(z_0)$ is such that $(\phi(\bar{z}), \xi(\phi(\bar{z}), \bar{z})) \notin \Sigma$, then*

$$D\phi(\bar{z}) \cdot \theta(\bar{z}) = 0.$$

Before proving this result, let us show how it can be used to improve the information about $\Gamma \setminus \Sigma$.

Theorem 6.6.9 *Assume conditions (L1)–(L4). Then*

$$\mathcal{H}^{n-1+2/R}(\Gamma \setminus \Sigma) = 0.$$

In particular, if

$$L \in C^\infty(\overline{Q_T} \times \mathbb{R}^n), \quad u_0 \in C^\infty(\mathbb{R}^n),$$

then

$$\mathcal{H} - \dim(\Gamma \setminus \Sigma) \leq n - 1.$$

Proof — Let G' and G'' be defined as in (6.36) and (6.37), respectively, and let $(z_0, t_0) \in G'$ be fixed. Then, by Remark 6.6.7, there exist $r, \varepsilon > 0$ such that $(\,]t_0 - \varepsilon, t_0 + \varepsilon[\, \times B_r(z_0)) \cap G$ coincides with the graph of some function $\phi : B_r(z_0) \to \,]t_0 - \varepsilon, t_0 + \varepsilon[$. In other words, $\tilde{\xi}((\,]t_0 - \varepsilon, t_0 + \varepsilon[\, \times B_r(z_0)) \cap G) = \tilde{\xi}(B_r(z_0))$, where

$$\tilde{\xi}(z) = \overline{\xi}(\phi(z), z), \qquad z \in B_r(z_0).$$

Moreover, Proposition 6.6.8 implies that for any $z \in B_r(z_0)$ satisfying $\tilde{\xi}(z) \notin \Sigma$, there exists $\theta \in (\mathbb{R}^n \setminus \{0\})$ such that $\xi_z(\phi(z), z)\theta = D\phi(z) \cdot \theta = 0$. A direct computation shows that for such a z,

$$D\tilde{\xi}(z)\theta = D\left(\begin{matrix} \phi(z) \\ \xi(\phi(z), z) \end{matrix} \right)\theta = \left(\begin{matrix} D\phi \\ \xi_t \otimes D\phi + \xi_z \end{matrix} \right)\theta = 0,$$

and so $\operatorname{rk} D\tilde{\xi}(z) \leq n - 1$. Recall that $\tilde{\xi}$ is defined in an n-dimensional space and is of class C^{R-1}. Hence, we can apply Theorem 6.6.1 to conclude that $\mathcal{H}^{n-1+\frac{1}{R-1}}(\tilde{\xi}(G') \setminus \Sigma) = 0$. On the other hand, from the proof of Theorem 6.6.4 we know that $\mathcal{H}^{n-1+2/R}(\tilde{\xi}(G'')) = 0$. We are assuming $R \geq 2$, and so $\frac{1}{R-1} \leq \frac{2}{R}$. Thus, observing that

$$\Gamma \setminus \Sigma \subset \overline{\xi}(G'') \cup (\tilde{\xi}(G') \setminus \Sigma),$$

the conclusion follows.   ∎

For the proof of Proposition 6.6.8 we need the following result.

Lemma 6.6.10 *Let* $(t_0, z_0) \in G'$, *and let* r, ε *and* ϕ, θ *be defined as in Remark 6.6.7. Then, for any* $z \in B_r(z_0)$,

$$D\phi(z) \cdot \theta(z) = 0 \iff \frac{\partial^2 \xi}{\partial \theta(z)^2}(\phi(z), z) \cdot (\eta_z(\phi(z), z)\theta(z)) = 0. \quad (6.46)$$

Proof — By definition we have, for any $z \in B_r(z_0)$ and $i = 1, \ldots, n$, that

$$\sum_{j=1}^{n} \frac{\partial \xi_i}{\partial z_j}(\phi(z), z)\theta_j(z) = 0. \quad (6.47)$$

Let us set for simplicity $q(z) = \eta_z(\phi(z), z)\theta(z)$. Differentiating (6.47) with respect to z_k and using (6.17) we obtain

$$\sum_{h=1}^{n} \frac{\partial^2 H}{\partial p_i \partial p_h} q_h \frac{\partial \phi}{\partial z_k} + \sum_{j=1}^{n} \left\{ \frac{\partial^2 \xi_i}{\partial z_j \partial z_k}\theta_j + \frac{\partial \xi_i}{\partial z_j}\frac{\partial \theta_j}{\partial z_k} \right\} = 0 \quad (6.48)$$

for any $i, k = 1, \ldots, n$. We then multiply equality (6.48) by $q_i \theta_k$ and sum over $i, k = 1, \ldots, n$, to obtain

$$\sum_{i,h=1}^{n} \frac{\partial^2 H}{\partial p_i \partial p_h} q_h q_i \sum_{k=1}^{n} \frac{\partial \phi}{\partial z_k}\theta_k + \sum_{i,j,k=1}^{n} \frac{\partial^2 \xi_i}{\partial z_j \partial z_k}\theta_j \theta_k q_i = 0 \quad (6.49)$$

recalling that in light of Lemma 6.6.6, $q \perp \mathrm{Im}\xi_z$. This last identity may be rewritten as

$$(H_{pp}q \cdot q)(D\phi \cdot \theta) + \frac{\partial^2 \xi}{\partial \theta^2} \cdot q = 0.$$

Since H_{pp} is positive definite and $q \neq 0$, we have that $H_{pp}q \cdot q \neq 0$. Therefore, $D\phi \cdot \theta = 0$ if and only if $\frac{\partial^2 \xi}{\partial \theta^2} \cdot q = 0$.   ∎

In the proof of Proposition 6.6.8 we will apply a known lemma on local invertibility of mappings, which we recall below. For its proof see e.g., [11, Lemma 3.2.5].

Lemma 6.6.11 *Let* $F : \mathbb{R}^n \to \mathbb{R}^n$ *be a map of class* C^2. *Let* DF *have rank* $n - 1$ *at some* $\bar{z} \in \mathbb{R}^n$ *and set* $\bar{x} = F(\bar{z})$. *Let* θ *be a generator of* $\mathrm{Ker}DF(\bar{z})$, *and let* w *be a nonzero vector orthogonal to* $\mathrm{Im}DF(\bar{z})$. *Suppose that*

$$\frac{\partial^2 F}{\partial \theta^2}(\bar{z}) \cdot w > 0.$$

Then there exist $\rho, \sigma > 0$ *such that the equation*

$$F(z) = \bar{x} + sw, \quad z \in B_\rho(\bar{z})$$

has two solutions if $0 < s < \sigma$, *and no solutions if* $-\sigma < s < 0$.

Proof of Proposition 6.6.8. Let us take $\bar{z} \in B_r(z_0)$ and set

$$\bar{t} = \phi(\bar{z}), \qquad \bar{x} = \xi(\bar{t}, \bar{z}).$$

Suppose that $D\phi(\bar{z}) \cdot \theta(\bar{z}) \neq 0$. Then, in light of Lemmas 6.6.6 and 6.6.10 the map $F = \xi(\bar{t}, \cdot)$ satisfies all assumptions of Lemma 6.6.11 with

$$\theta = \theta(\bar{z}), \quad w = \pm\eta_z(\bar{t}, \bar{z})\theta(\bar{z}).$$

For $x_k = \bar{x} - w/k$, let ξ_k be a minimizer of functional $J_{\bar{t}}$ in (6.2) with terminal condition $\xi_k(\bar{t}) = x_k$. Let us consider the sequence $\{z_k\}$ of all initial points of such minimizers, i.e., $z_k = \xi_k(0)$. Then ξ_k coincides with the characteristic starting at z_k and in particular $x_k = \xi(\bar{t}, z_k)$. So, Lemma 6.6.11 implies that $z_k \notin B_\rho(\bar{z})$ if k is large enough.

On the other hand, $\{z_k\}$ is bounded. Hence, we conclude that there exists a subsequence of $\{z_k\}$ converging to some $\hat{z} \neq \bar{z}$. Then $\xi(\cdot, \hat{z})$ is a minimizing trajectory for $J_{\bar{t}}$ at $(\bar{t}, \bar{x})$, different from $\xi(\cdot, \bar{z})$. This fact contradicts our assumption that $(\bar{t}, \bar{x})$ be a regular point.   ∎

Let us remark that the rectifiability results of this section imply that the gradient Du of the value function belongs to the class SBV of special functions of bounded variation (see Appendix A. 6).

Corollary 6.6.12 *The gradient* Du *of the value function of problem* (CV) *belongs to* $\mathrm{SBV}_{loc}(Q_T)$.

Proof — The semiconcavity of u with a linear modulus implies that $Du \in BV_{loc}(Q_T)$. To see that $Du \in SBV$, observe that Du is of class C^1 on the complement of $\overline{\Sigma}$, by Theorem 6.4.11, and that $\overline{\Sigma}$ is countably $\mathcal{H}^n$-rectifiable, by the results of this section. The assertion is then a direct consequence of Proposition A. 6.6. ∎

To conclude, we give an example showing that if the smoothness of the data of $(CV)_{t,x}$ is below a certain threshold, then $\overline{\Sigma}$ can fail to be countably $\mathcal{H}^n$-rectifiable.

Example 6.6.13 *There exists $u_0 \in W^{2,\infty}(\mathbb{R}) \setminus C^2(\mathbb{R})$ with the following property: the value function of problem* (CV) *with initial cost u_0 and lagrangian $L(t, x, v) = v^2/2$ has a singular set $\Sigma \subset \mathbb{R}_+ \times \mathbb{R}$ such that $\mathcal{H}$–$\dim(\overline{\Sigma}) \geq 1 + \log 2/\log 3$.*

Proof — We first introduce some notation. Let i, j be integers with $i \geq 0$ and $0 \leq j \leq 2^i - 1$. Then j can be written in the form $j = \sum_{h=0}^{i-1} e_h 2^h$ with $e_h \in \{0, 1\}$. We define

$$c_{i,j} = \sum_{h=1}^{i}(2e_{i-h})3^{-h} + 1/3^i 2, \quad I_{i,j} =]c_{i,j} - 1/(3^i 6), c_{i,j} + 1/(3^i 6)[.$$

We then consider the *Cantor set C*

$$C = [0, 1] \setminus \bigcup_{i=0,\dots,\infty,\ j=0,\dots,2^i-1} I_{i,j}.$$

It is well known that C consists of all numbers $x \in [0, 1]$ that can be written in the form $x = \sum_{h=1}^{\infty} d_h 3^{-h}$ with $d_h \in \{0, 2\}$. Thus C is uncountable; moreover $\mathcal{H}$–$\dim(C) = \log 2/\log 3 > 0$ (see e.g., [74, example 2.7]).

A property we will need in the sequel is the fact that $C \subset \overline{\{c_{i,j}\}}$. Indeed, take any $x = \sum_{h=1}^{\infty} d_h 3^{-h}$ with $d_h \in \{0, 2\}$. Given a positive integer i, define $j(i) = \sum_{h=0}^{i-1}(d_{i-h}/2)2^h$. Then

$$|c_{i,j(i)} - x| = \left| \sum_{h=1}^{\infty} d_h 3^{-h} - \sum_{h=1}^{i-1} d_h 3^{-h} \right| \leq 3^{-i+1}.$$

Since i is arbitrary, we have proved our claim.

We now define, for $\lambda \in]0, 1[$,

$$b(\lambda, z) = \begin{cases} \left(\left(\dfrac{z^2}{\lambda} - \lambda\right)\right)^2 & -\lambda < z < \lambda \\ 0 & z \geq \lambda \text{ or } z \leq -\lambda. \end{cases}$$

Note that $b(\lambda, z) \in W^{2,\infty}(\mathbb{R})$. The second derivative $b_{zz}(\lambda, z)$ exists for $z \neq \pm\lambda$ and satisfies $b_{zz}(\lambda, z) = b_{zz}(1, z/\lambda)$. It follows that $|b_{zz}(\lambda, z)| \leq K$ for some constant K independent of λ. This implies that $b_z(\lambda, \cdot)$ has Lipschitz constant K and that

$$\sup_{z \in [-\lambda, -\lambda + \varepsilon[} b(\lambda, z) \leq \int_{-\lambda}^{-\lambda + \varepsilon} \left(\int_{-\lambda}^{z} |b_{zz}(\lambda, w)| dw \right) dz \leq K \varepsilon^2 / 2 \qquad (6.50)$$

for every $\lambda > 0$, $\varepsilon > 0$.

For a fixed λ, let us first consider problem (CV) with lagrangian $L(v) = v^2/2$ and initial cost given by $b(\lambda, x)$. We denote by u^λ the value function, given by

$$u^\lambda(t, x) = \min_{\xi \in \mathcal{A}_{t,x}} \left[\int_0^t \frac{1}{2} (\dot{\xi}(s))^2 dt + b(\lambda, \xi(0)) \right]. \qquad (6.51)$$

We know that all minimizers are of the form $\xi(\cdot, z)$ for some z. In our case the hamiltonian is $H(p) = p^2/2$ and the solution of system (6.15) is

$$\xi(t, z) = z + t b_z(\lambda, z) = z + 4t \left(\left(\frac{z}{\lambda} \right)^2 - 1 \right) z \chi_{[-\lambda, \lambda]}(z), \qquad (6.52)$$

where $\chi_{[-\lambda, \lambda]}$ is the characteristic function of the interval $[-\lambda, \lambda]$. Let us compute the minimizers in the case $x = 0$. If $t \leq 1/4$, then (6.52) shows that $\xi(t, z) = 0$ only if $z = 0$, while if $t > 1/4$ we have $\xi(t, z) = 0$ for $z = 0$ and for $z = \pm z(t, \lambda)$ where

$$z(t, \lambda) = \lambda \sqrt{1 - \frac{1}{4t}}.$$

A direct computation shows that if $t > 1/4$, the minimum in (6.51) is attained for $z = \pm z(t, \lambda)$, while the trajectory $\xi(\cdot, 0)$ is not minimal. Thus, if $t > 1/4$ the minimizer for the point $(t, 0)$ is not unique, and so the half line $]1/4, \infty[\times \{0\}$ is contained in the set Σ of irregular points.

The initial value for our counterexample is obtained repeating this construction on each interval $I_{i,j}$. We define

$$u_0(z) = \begin{cases} 0 & \text{if } z \in C \text{ or } z > 1 \text{ or } z < -1 \\ b(1/3^i 6, z - c_{i,j}) & \text{if } z \in I_{i,j}. \end{cases} \qquad (6.53)$$

Let us show that $u_0 \in W^{2,\infty}(\mathbb{R})$. By construction u_0 is differentiable in the complement of C; on the other hand we claim that if $z \in C$, then $|u_0(y) - u_0(z)| \leq K|y - z|^2/2$, which implies that $u_0'(z) = 0$. In fact, let us take for instance $y > z$. If $y \in C$, then $u(y) = u(z) = 0$. Otherwise let y' be the left endpoint of the interval $I_{i,j}$ to which y belongs; we have, by estimate (6.50),

$$u(y) - u(z) = u(y) - u(y') \leq K|y - y'|^2/2 \leq K|y - z|^2/2,$$

which proves our claim. In a similar way, we can prove that u_0' is Lipschitz continuous by using the uniform Lipschitz continuity of $b_z(\lambda, \cdot)$.

However, u_0 is not of class C^2. In fact, u_0'' attains at each interval $I_{i,j}$ a maximum and a minimum value which are independent of i, j. Since the length of $I_{i,j}$ tends to zero as $i \to \infty$, this shows that u_0'' cannot be uniformly continuous in $[-1, 1]$.

Let us now consider problem (CV) with lagrangian $L(v) = v^2/2$ and with initial value u_0 given by (6.53). It is easy to see by the definition of u_0 that the solution is given by

$$u(t, x) = \begin{cases} 0 & x \in C \text{ or } x > 1 \text{ or } x < -1 \\ u^{1/3^i 6}(t, x - c_{i,j}) & \text{if } x \in I_{i,j}, \end{cases}$$

with $u^\lambda(t, x)$ given by (6.51). Our previous analysis implies that

$$\Sigma \supset]1/4, \infty[\times \{c_{i,j} \mid i \geq 0, 0 \leq j \leq 2^i - 1\}.$$

But since $C \subset \overline{\{c_{i,j}\}}$ we obtain that $\overline{\Sigma} \supset]1/4, \infty[\times C$. Thus Σ has Hausdorff dimension of at least $1 + \log 2/\log 3$. ∎

Bibliographical notes

The results about the existence and regularity of minimizers of Section 6.1, 6.2 and 6.3 are classical; among the many textbooks in the field the reader may consult [49, 143, 88, 54, 53, 30]. General references about the dynamic programming approach considered in Section 6.4 can be found among the bibliographical notes to Chapter 1; our presentation partly follows [80, 81]. The result of Corollary 6.4.10 was first obtained by Kuznetsov and Šiškin (see [104]), while Theorem 6.4.9 was given in [46]. Fleming derived Theorem 6.4.11 in [78] and pointed out that it could provide a basis for studying the fine regularity properties of the value function.

The results of Sections 6.5 are taken from [38], although some of them were more or less explicitly present in the previous literature. For instance, the property that conjugate points are related to a blowup of the second derivatives was well known. Also, Corollary 6.5.6 was already given in [78]. The rectifiability results of Section 6.6 were first proved in [38]; our presentation follows that paper. The impulse for studying that kind of regularity was given by some previous works about image segmentation (see e.g., [14]) where the class SBV was introduced.

7

Optimal Control Problems

In this chapter and in the following one we describe the dynamic programming approach to control problems, focusing our attention on the role played by the semi-concavity property and the structure of the singularities of the value function. The theory of optimal control is very broad and has a large variety of applications; we do not aim to give an exhaustive treatment of the subject, but we choose some model problems and develop the theory in these cases.

The problem in the calculus of variations of the previous chapter can be regarded as a special case of optimal control problems. It is natural to check how much of the previous analysis can be extended to this more general setting. It will turn out that most of the results of the previous chapter have some analogue in optimal control, at least for certain classes of systems. The extension often requires new tools; an example is the so-called Pontryagin maximum principle, a key result in control theory which replaces, roughly speaking, the Euler–Lagrange conditions for extremal arcs in the calculus of variations.

This chapter is devoted to optimal control problems with finite time horizon and unrestricted state space, namely the so-called Mayer and Bolza problems. In Section 7.1 we introduce our control system, and state the Mayer problem, where one wants to minimize a functional depending only on the final endpoint of the trajectory. We then give an existence result for optimal controls. In Section 7.2 we introduce the value function V for the Mayer problem, prove that V is Lipschitz continuous, semi-concave, and is a viscosity solution of an associated Hamilton–Jacobi equation. We note that many results hold in a weaker form compared with the case of the calculus of variations; the main reason is that the hamiltonian function associated with a Mayer problem is not strictly convex. Thus, we no longer have the regularizing effect induced by the Hamilton–Jacobi equation which ensures that V be semiconcave even if the initial value is only Lipschitz. Also, we do not have regularity of the value function along optimal arcs but only a bound on the dimension of the superdifferential D^+V.

In Section 7.3 we prove the Pontryagin maximum principle and other optimality conditions for our control problem. We analyze whether optimal trajectories can be obtained by solving a hamiltonian system and whether they are in correspondence with the elements of D^*V, as for calculus of variations. Such properties do not hold in full generality for the Mayer problem, because of the lack of smoothness of the hamiltonian. However, if the control system satisfies suitable assumptions, it is possible to obtain similar results.

Finally, in Section 7.4 we study the Bolza problem, where the functional to minimize includes an integral term with a running cost. For this problem we can prove results which are mostly similar to the ones of the previous sections. We also consider cases where the presence of the integral term yields better regularity properties than in the Mayer problem, and the analysis of optimality conditions can be simplified.

7.1 The Mayer problem

We begin by giving the definition of a control system.

Definition 7.1.1 *A control system* consists of a pair (f, U), *where* $U \subset \mathbb{R}^m$ *is a closed set and* $f : \mathbb{R}^n \times U \to \mathbb{R}^n$ *is a continuous function. The set* U *is called the* control set, *while* f *is called the* dynamics *of the system. The* state equation *associated with the system is*

$$\begin{cases} y'(t) = f(y(t), u(t)), & t \in [t_0, +\infty[\, a.e. \\ y(t_0) = x, \end{cases} \tag{7.1}$$

where $t_0 \in \mathbb{R}$, $x \in \mathbb{R}^n$ *and* $u \in L^1_{loc}([t_0, \infty[\, , U)$. *The function* u *is called a* control strategy *or simply a* control. *We denote the solution of* (7.1) *by* $y(\cdot; t_0, x, u)$ *and we call it the* trajectory *of the system corresponding to the initial condition* $y(t_0) = x$ *and to the control* u.

The word "control" is used sometimes to denote elements $v \in U$ and sometimes for functions $u : [t_0, \infty[\to U$; the meaning should always be clear from the context.

We restrict ourselves to autonomous control systems where $f(x, u)$ does not depend on t; this is done for the sake of simplicity since the results we present can be easily extended to the nonautonomous case.

We now list some basic assumptions on our control system which will be made in most of the results of this chapter.

(H0) The control set U is compact.
(H1) There exists $K_1 > 0$ such that $|f(x_2, u) - f(x_1, u)| \leq K_1|x_2 - x_1|$, for all $x_1, x_2 \in \mathbb{R}^n$, $u \in U$.
(H2) f_x exists and is continuous; in addition, there exists $K_2 > 0$ such that $||f_x(x_2, u) - f_x(x_1, u)|| \leq K_2|x_2 - x_1|$, for all $x_1, x_2 \in \mathbb{R}^n$, $u \in U$.

It is well known from the theory of ordinary differential equations that assumption (H1) ensures the existence of a unique global solution to the state equation (7.1) for any choice of t_0, x and u. Observe also that assumptions (H0) and (H1), together with the continuity of f, imply

$$|f(x, u)| \leq C + K_1|x|, \qquad x \in \mathbb{R}^n, \ u \in U, \tag{7.2}$$

where $C = \max_{u \in U} |f(0, u)|$. In addition, one can obtain the following estimates on the trajectories which will be repeatedly used in our analysis.

Lemma 7.1.2 *Let t_0, t_1 be given, with $t_0 < t_1$.*

(i) Let f satisfy (H0) *and* (H1). *Then, for any $r > 0$ there exists $R > 0$ such that*

$$|y(t; t_0, x, u)| \leq R, \qquad \forall t \in [t_0, t_1] \tag{7.3}$$

for all controls $u : [t_0, t_1] \rightarrow U$ and for all $x \in B_r$.
(ii) Let f satisfy (H1). *Then, there exists $c > 0$ such that*

$$|y(t; t_0, x_0, u) - y(t; t_0, x_1, u)| \leq c|x_0 - x_1|, \qquad \forall t \in [t_0, t_1] \tag{7.4}$$

for all $u : [t_0, t_1] \rightarrow U$ and $x_0, x_1 \in \mathbb{R}^n$.
(iii) If f satisfies (H1) *and* (H2) *then the constant c in* (ii) *can be chosen in such a way that*

$$\left| y(t; t_0, x_0, u) + y(t; t_0, x_1, u) - 2y\left(t; t_0, \frac{x_0 + x_1}{2}, u\right) \right| \leq c|x_0 - x_1|^2 \tag{7.5}$$

also holds, for all $u : [t_0, t_1] \rightarrow U$, $x_0, x_1 \in \mathbb{R}^n$ and $t \in [t_0, t_1]$.

Proof — Let us set for simplicity $y(\cdot) := y(\cdot; t_0, x, u)$. We have, by (7.2),

$$|y(t)| \leq |x| + \int_{t_0}^{t} (C + K_1|y(s)|)\, ds$$

$$\leq |x| + C(t_1 - t_0) + K_1 \int_{t_0}^{t} |y(s)|\, ds.$$

Then Gronwall's inequality (see Theorem A. 4.3) implies that

$$|y(t)| \leq [|x| + C(t_1 - t_0)]e^{K_1(t_1 - t_0)}, \qquad t \in [t_0, t_1],$$

and (7.3) follows. To prove (7.4), we set, for $i = 0, 1$, $y_i(t) := y(t; t_0, x_i, u)$. We have, by (H1), that

$$|y_0(t) - y_1(t)| = \left| x_0 - x_1 + \int_{t_0}^{t} [f(y_0(s), u(s)) - f(y_1(s), u(s))]\, ds \right|$$

$$\leq |x_0 - x_1| + K_1 \int_{t_0}^{t} |y_0(s) - y_1(s)|\, ds.$$

Thus (7.4) also follows from Gronwall's inequality. Let us now assume that property (H2) holds. First we observe that

$$\left| f(x_0, u) + f(x_1, u) - f\left(\frac{x_0 + x_1}{2}, u\right)\right|$$

$$= \left| \int_0^1 f_x\left(\frac{x_0 + x_1}{2} + t\frac{x_1 - x_0}{2}, u\right) \cdot \frac{x_1 - x_0}{2}\, dt\right.$$

$$\left. - \int_0^1 f_x\left(x_0 + t\frac{x_1 - x_0}{2}, u\right) \cdot \frac{x_1 - x_0}{2}\, dt\right|$$

$$\le \int_0^1 K_2 \left|\frac{x_0 + x_1}{2} - x_0\right| \left|\frac{x_1 - x_0}{2}\right| dt = \frac{K_2}{4}|x_1 - x_0|^2$$

for all $x_0, x_1 \in \mathbb{R}^n$, $u \in U$. Therefore

$$|f(x_0, u) + f(x_1, u) - 2f(x_2, u)|$$

$$\le \left| f(x_0, u) + f(x_1, u) - 2f\left(\frac{x_0 + x_1}{2}, u\right)\right|$$

$$+ 2\left| f\left(\frac{x_0 + x_1}{2}, u\right) - f(x_2, u)\right|$$

$$\le \frac{K_2}{4}|x_0 - x_1|^2 + K_1|x_0 + x_1 - 2x_2|$$

for all $x_0, x_1, x_2 \in \mathbb{R}^n$, $u \in U$. Let us now define y_0, y_1 as before and set in addition $y_2(t) := y(t; t_0, (x_0 + x_1)/2, u)$. We find that

$$|y_0(t) + y_1(t) - 2y_2(t)|$$

$$= \left| \int_{t_0}^t [f(y_0(s), u(s)) + f(y_1(s), u(s)) - 2f(y_2(s), u(s))]\, ds\right|$$

$$\le \frac{K_2}{4}\int_{t_0}^t |y_0(s) - y_1(s)|^2\, ds + K_1\int_{t_0}^t |y_0(s) + y_1(s) - 2y_2(s)|\, ds$$

$$\le \frac{c^2(t_1 - t_0)K_2}{4}|x_0 - x_1|^2 + K_1\int_{t_0}^t |y_0(s) + y_1(s) - 2y_2(s)|\, ds,$$

where we have applied (7.4). Using Gronwall's inequality again we obtain (7.5). ∎

An *optimal control problem* consists of choosing the control strategy u in the state equation (7.1) in order to minimize a given functional. We now introduce the so-called *Mayer problem*, while other types of control problems will be treated later.

Let $g : \mathbb{R}^n \to \mathbb{R}$ be a continuous function and let $T > 0$. For any $(t, x) \in [0, T] \times \mathbb{R}^n$, we consider the following problem:

(MP) minimize $g(y(T; t, x, u))$ over all control strategies $u : [t, T] \to U$.

Definition 7.1.3 *A control $u : [t, T] \to U$ such that the infimum in* (MP) *is attained is called* optimal *for problem* (MP) *with initial point* (t, x). *The corresponding solution* $y(\cdot) = y(\cdot; t, x, u)$ *of the state equation is called an* optimal trajectory *or a* minimizer.

The function g is called the *final cost* of the Mayer problem. An optimal control problem where one is only interested in the behavior of the trajectories up to a given finite time, like the one considered here, is called a problem with *finite time horizon*.

Let us study the existence of optimal controls for the Mayer problem.

Theorem 7.1.4 *Let f satisfy* (H0), (H1), *let $g \in C(\mathbb{R}^n)$, and let the set*

$$f(x, U) := \{f(x, u) \; : \; u \in U\} \tag{7.6}$$

be convex for all $x \in \mathbb{R}^n$. Then, for any $(t, x) \in [0, T] \times \mathbb{R}^n$, there exists an optimal control for problem (MP).

The theorem will follow as a corollary of two results which are fundamental in optimal control theory. The first result, often called *Filippov's lemma*, states that the control system (7.1) admits an equivalent formulation as a differential inclusion. Let us set for simplicity $\mathcal{F}(x) = f(x, U)$. We say that an absolutely continuous function $y : [t_0, t_1] \to \mathbb{R}^n$ is a solution of the differential inclusion

$$y' \in \mathcal{F}(y) \tag{7.7}$$

if it satisfies $y'(t) \in f(y(t), U)$ for $t \in [t_0, t_1]$ a.e. Clearly, if y is a solution of the state equation (7.1) for some control strategy u, then it is also a solution of the differential inclusion (7.7). The next result says that the converse is also true.

Theorem 7.1.5 (Filippov's Lemma) *Let $y : [t_0, t_1] \to \mathbb{R}^n$ be a solution of the differential inclusion* (7.7). *Then there exists a measurable function $u : [t_0, t_1] \to U$ such that $y'(t) = f(y(t), u(t))$ for $t \in [t_0, t_1]$ a.e..*

Proof — Let $y : [t_0, t_1] \to \mathbb{R}^n$ be a solution of the differential inclusion (7.7). Let us set, for all $t \in [t_0, t_1]$ such that $y'(t)$ exists,

$$\mathcal{U}(t) = \{u \in U \; : \; y'(t) = f(y(t), u)\}.$$

By definition, $\mathcal{U}(t)$ is nonempty for a.e. $t \in [t_0, t_1]$. To prove the theorem we have to show the existence of a measurable function $u : [t_0, t_1] \to U$ such that $u(t) \in \mathcal{U}(t)$ for a.e. $t \in [t_0, t_1]$. For this purpose we shall use a measurable selection theorem recalled in the appendix (see Theorem A. 5.2) and an approximation procedure.

By Lusin's theorem there exists, for every $\varepsilon > 0$, a compact set $K \subset [t_0, t_1]$ such that the measure of $[t_0, t_1] \setminus K$ is less than ε and such that the restriction of $y'(t)$ on K is continuous. By the arbitrariness of ε, we can also find a sequence $\{K_i\}$ of disjoint compact subsets of $[t_0, t_1]$, such that the restriction of $y'(t)$ on K_i is continuous and that $\cup_{i=1}^{\infty} K_i$ covers $[t_0, t_1]$ up to a set of measure zero.

Now, if we set for a given i

$$\mathcal{U}_i(t) = \begin{cases} \mathcal{U}(t) & t \in K_i \\ \emptyset & \text{otherwise} \end{cases}$$

we easily find that $\mathcal{U}_i$ is a multifunction with closed graph (see Section A. 5 for the definition). Hence, by Proposition A. 5.3 and Theorem A. 5.2, there exists a measurable selection of $\mathcal{U}_i$, that is, a measurable function $u_i : K_i \to U$ such that $u_i(t) \in \mathcal{U}(t)$ for all $t \in K_i$.

Now let us fix any $\bar{u} \in U$ and define

$$u(t) = \begin{cases} u_i(t) & t \in K_i \text{ for some } i \\ \bar{u} & t \notin \cup_{i=1}^{\infty} K_i. \end{cases}$$

It is easily checked that $u(\cdot)$ is measurable and $u(t) \in \mathcal{U}(t)$ for $t \in [t_0, t_1]$ almost everywhere, and so the theorem is proved. ∎

The other basic result needed for the proof of Theorem 7.1.4 is the following compactness property for the trajectories of the control system.

Theorem 7.1.6 *Assume that* (H0), (H1) *hold and that* $f(x, U)$ *is convex for all* $x \in \mathbb{R}^n$. *Let* $\{y_k\}$ *be a sequence of trajectories of* (7.1) *in some given interval* $[t_0, t_1]$, *that is,* $y_k(\cdot) = y(\cdot; t_0, x_k, u_k)$ *for some* $x_k \in \mathbb{R}^n$ *and* $u_k : [t_0, t_1] \to U$. *If the trajectories* y_k *are uniformly bounded, then there exists a subsequence* $\{y_{k_h}\}$ *converging uniformly to an arc* $\bar{y} : [t_0, t_1] \to \mathbb{R}^n$ *which is also a trajectory of* (7.1).

Proof — Let $\{y_k\}$ be a uniformly bounded sequence of trajectories. Since f satisfies (7.2) and $y_k' = f(y_k, u_k)$, the trajectories y_k are uniformly Lipschitz continuous. Hence, by the Ascoli–Arzelà theorem, there exists a subsequence, which we still denote by $\{y_k\}$, converging uniformly to some Lipschitz arc $\bar{y} : [t_0, t_1] \to \mathbb{R}^n$.

To prove the theorem we need to show that $\bar{y}$ is a trajectory of the system. By Filippov's Lemma, it suffices to show that

$$\bar{y}'(t) \in \mathcal{F}(\bar{y}(t)), \qquad t \in [t_0, t_1] \text{ a.e..} \tag{7.8}$$

In the following computations we consider only points in a neighborhood of the trajectory $\bar{y}$, which is bounded, and so we can assume that $|f(x, u)|$ is bounded by some constant M_f. By Rademacher's theorem, $\bar{y}$ is differentiable a.e. in $[t_0, t_1]$. Let t be a point of differentiability for $\bar{y}$. For a fixed $\varepsilon > 0$, let us set $\mathcal{F}_\varepsilon(x) = f(x, U) + \overline{B_\varepsilon(0)}$. Such a set is convex since we are assuming that $f(x, U)$ is convex. By (H1) we find

$$|f(y_k(s)), u_k(s)) - f(\bar{y}(t), u_k(s))| \le K_1 |y_k(s) - \bar{y}(t)|$$
$$\le K_1(|y_k(s) - y_k(t)| + |y_k(t) - \bar{y}(t)|)$$
$$\le K_1(M_f |s - t| + ||y_k - \bar{y}||_\infty).$$

Hence, $f(y_k(s)), u_k(s)) \in \mathcal{F}_\varepsilon(\bar{y}(t))$ if $|s - t|$ is small and k is large. Therefore, if $|h|$ is small and k is large we find, using Theorem A. 1.27,

$$\frac{y_k(t+h) - y_k(t)}{h} = \frac{1}{h} \int_t^{t+h} f(y_k(s), u_k(s)) \, ds \in \mathcal{F}_\varepsilon(\bar{y}(t)).$$

If we let $k \to \infty$ we obtain that $h^{-1}[\bar{y}(t+h) - \bar{y}(t)] \in \mathcal{F}_\varepsilon(\bar{y}(t))$, and so, letting $h \to 0$, $\bar{y}'(t) \in \mathcal{F}_\varepsilon(\bar{y}(t))$. Since $\varepsilon > 0$ is arbitrary, we conclude that $\bar{y}'(t) \in \mathcal{F}(\bar{y}(t))$ at any point t where $\bar{y}$ is differentiable, and so $\bar{y}$ is a trajectory of our system. ∎

Remark 7.1.7 It is clear from the proof that in the statement of Theorem 7.1.6 it suffices to assume that $f(x, U)$ is convex for all $x \in C$, where C is a closed set containing all the trajectories y_k. Also, assumption (H1) can be weakened by requiring f to be bounded and Lipschitz continuous with respect to x in all sets of the form $B_R \times U$.

Remark 7.1.8 It is easy to show that, without the convexity of the sets $f(x, U)$, the compactness of trajectories may fail. Consider for instance the one-dimensional system $y' = u$, $u \in U$, where $U = \{-1, 1\}$. Then one can consider controls $u_k :$ $[0, 1] \to U$ taking alternately the values 1 and -1 on intervals of length $1/k$. If we set $y_k(\cdot) = y(\cdot; 0, 0, u_k)$, we easily find that $y_k \to \bar{y}$ uniformly, where $\bar{y} \equiv 0$. However, $\bar{y}$ is not a trajectory of the system, since $\bar{y}' \equiv 0 \notin U$.

Proof of Theorem 7.1.4. Let us fix a point $(t, x) \in [0, T[\times \mathbb{R}^n$. We can find a sequence of controls $u_k : [t, T] \to U$ such that, setting $y_k(\cdot) := y(\cdot; t, x, u_k)$, we have

$$\lim_{k \to \infty} g(y_k(T)) = \inf_{u:[t,T] \to U} g(y(T; t, x, u)). \tag{7.9}$$

Such a sequence $\{u_k\}$ is called a *minimizing sequence*. We want to show that a subsequence of $\{y_k\}$ converges to an optimal trajectory.

By estimate (7.3), all trajectories of system (7.1) starting at (t, x) are uniformly bounded in $[t, T]$. Therefore, we can apply Theorem 7.1.6 to conclude that a subsequence of $\{y_k\}$ converges uniformly to a function $\bar{y} : [t, T] \to \mathbb{R}^n$ which is a trajectory of our system. We have $\bar{y}(t) = x$ since $y_k(t) = x$ for all k. In addition, since g is continuous, we deduce from (7.9)

$$g(\bar{y}(T)) = \lim_{k \to \infty} g(y_k(T)) = \inf_{u:[t,T] \to U} g(y(T; t, x, u)),$$

showing that $\bar{y}$ is optimal. ∎

7.2 The value function

We now apply to the Mayer problem the dynamic programming approach as we have done in the previous chapter for the calculus of variations. Throughout the section we assume that hypotheses (H0) and (H1) are satisfied. Unless explicitly stated, we do not assume the convexity of the sets $f(x, U)$, which ensures the existence of optimal trajectories since such a property is not needed for the results of this section. We begin by introducing the value function of our problem.

Definition 7.2.1 *Given* $(t, x) \in [0, T] \times \mathbb{R}^n$, *we define*

$$V(t, x) = \inf\{g(y(T; t, x, u)) \; : \; u : [t, T] \to U \text{ measurable }\}.$$

The function V is called the value function *of the control problem* (MP).

From the continuity of g and estimate (7.3) we see that V is finite everywhere. In addition, V satisfies the *dynamic programming principle* similar to the one we have seen in the calculus of variations (see Theorem 1.2.2).

Theorem 7.2.2 *For any given* $(t, x) \in \,]0, T[$ *and any* $s \in [t, T]$ *we have*

$$V(t, x) = \inf_{u:[t,s] \to U} V(s, y(s; t, x, u)). \tag{7.10}$$

In addition, a control $u : [t, T] \to U$ *is optimal for* (t, x) *if and only if*

$$V(t, x) = V(s, y(s; t, x, u)) \quad \text{for all } s \in [t, T]. \tag{7.11}$$

Proof — Let us take any $s \in [t, T]$, $u \in L^1([t, s], U)$ and denote for simplicity $x_1 = y(s; t, x, u)$. For any given $\varepsilon > 0$, we can find $v \in L^1([s, T], U)$ such that

$$g(y(T; s, x_1, v)) \leq V(s, x_1) + \varepsilon.$$

Let us now define $w \in L^1([t, T], U)$ as

$$w(\tau) = \begin{cases} u(\tau) & \tau \in [t, s] \\ v(\tau) & \tau \in \,]s, T]. \end{cases}$$

Then, by the semigroup property, $y(T; t, x, w) = y(T; s, x_1, v)$. Therefore

$$V(t, x) \leq g(y(T; t, x, w)) = g(y(T; s, x_1, v)) \leq V(s, x_1) + \varepsilon.$$

By the arbitrariness of ε we deduce

$$V(t, x) \leq V(s, x_1) = V(s, y(s; t, x, u))$$

and since u was also arbitrary, we obtain that $V(t, x)$ is not greater than the right-hand side in (7.10).

To prove the converse inequality, let us fix $\varepsilon > 0$ and let $w \in L^1([t, T], U)$ be such that
$$V(t, x) \geq g(y(T; t, x, w)) - \varepsilon.$$

Let us call u, v the restrictions of w to the intervals $[t, s]$ and $[s, T]$, respectively. If we again set $x_1 = y(s; t, x, u)$, we have

$$V(s, x_1) \leq g(y(T; s, x_1, v)) = g(y(T; t, x, w)) \leq V(t, x) + \varepsilon.$$

Since ε is arbitrary, this completes the proof of (7.10). The proof of (7.11) can be obtained by similar arguments, which are left to the reader. ∎

From the previous theorem we deduce that, for any control strategy $u : [t, T] \to U$,

$$V(t, x) \leq V(s, y(s; t, x, u)), \qquad \forall s \in \,]t, T], \qquad (7.12)$$

with equality for every s if and only if u is optimal for (t, x). Hence, if y is any trajectory of the system, the function $s \to V(s, y(s))$ is nondecreasing, and it is constant if and only if y is optimal.

We begin our analysis of the regularity of the value function by proving the Lipschitz continuity.

Theorem 7.2.3 *Let f satisfy* (H0), (H1) *and let $g \in \mathrm{Lip}_{loc}(\mathbb{R}^n)$. Then $V \in \mathrm{Lip}_{loc}([0, T] \times \mathbb{R}^n)$.*

Proof — Given $r > 0$, we want to prove that V is Lipschitz continuous in $[0, T] \times B_r$. We first observe that, by estimate (7.3), there exists $R > 0$ such that all admissible trajectories starting from points (t, x) with $x \in B_r$ stay inside B_R. We then denote by K_g the Lipschitz constant of g in B_R and by M_f the supremum of $|f(x, u)|$ for $(x, u) \in B_R \times U$.

Let us first estimate the variation of V between two points with the same t coordinate. We take $x_1, x_2 \in B_r$ and $t \in [0, T]$, and suppose for instance that $V(t, x_1) \leq V(t, x_2)$. For any given $\varepsilon > 0$ we can find, by the definition of V, a control $u : [t, T] \to U$ such that

$$g(y(T; t, x_1, u)) \leq V(t, x_1) + \varepsilon.$$

Let us denote for simplicity $\bar{x}_1 = y(T; t, x_1, u)$, $\bar{x}_2 = y(T; t, x_2, u)$. Then, by (7.4), we have $|\bar{x}_2 - \bar{x}_1| \leq c|x_2 - x_1|$ for some $c > 0$ independent of x_1, x_2, t. We find that

$$\begin{aligned} V(t, x_2) &\leq g(\bar{x}_2) \leq g(\bar{x}_1) + K_g |\bar{x}_2 - \bar{x}_1| \\ &\leq V(t, x_1) + \varepsilon + c K_g |x_2 - x_1|. \end{aligned}$$

Since $\varepsilon > 0$ is arbitrary, we obtain

$$|V(t, x_2) - V(t, x_1)| \leq c K_g |x_2 - x_1|. \qquad (7.13)$$

The above estimate holds for $x_1, x_2 \in B_r$. Actually, by our choice of R, it also holds if x_1, x_2 belong to trajectories which have started from B_r at a previous time, i.e., are of the form $x = y(t; \bar{x}, \bar{t}, u)$ for some $\bar{x} \in B_r$ and $\bar{t} \in [0, t]$.

Let us now consider two arbitrary points $(t_1, x_1), (t_2, x_2) \in [0, T] \times B_r$ and suppose for instance $t_1 < t_2$. For any $\varepsilon > 0$ we can find, by (7.10), a control $u : [t_1, t_2] \to U$ such that, setting $\bar{x} = y(t_2; t_1, x_1, u)$, we have

$$0 \leq V(t_2, \bar{x}) - V(t_1, x_1) \leq \varepsilon. \qquad (7.14)$$

On the other hand we have, by (7.13),

$$\begin{aligned} |V(t_2, \bar{x}) - V(t_2, x_2)| &\leq c K_g (|\bar{x} - x_1| + |x_1 - x_2|) \\ &\leq c K_g (M_f |t_2 - t_1| + |x_1 - x_2|). \end{aligned}$$

By the arbitrariness of $\varepsilon > 0$ in (7.14) we obtain our assertion.   ∎

Let us now introduce the *hamiltonian function* associated with (MP), defined as

$$H(x, p) = \max_{u \in U} -p \cdot f(x, u).$$

Observe that $H(x, p)$ is the support function of the convex hull of $f(x, U)$ evaluated at $-p$. This implies in particular that H is convex and homogeneous of degree one in the p variable. Therefore, it is not differentiable at $p = 0$ except in trivial cases. The smoothness of H for $p \neq 0$ depends on the properties of the set $f(x, U)$ and will be investigated later. In general, we can say that H is locally Lipschitz continuous; more precisely, using (7.2) and (H1), we easily obtain

$$|H(x, p) - H(x, q)| \leq \max_{u \in U} |f(x, u) \cdot (p - q)| \leq (C + K_1|x|)|p - q|$$

$$|H(x, p) - H(y, p)| \leq \max_{u \in U} |f(x, u) - f(y, u)||p| \leq K_1|x - y||p|,$$

for all $x, y, p, q \in \mathbb{R}^n$.

The next result shows that the value function satisfies a Hamilton–Jacobi equation whose hamiltonian is the function H introduced above.

Theorem 7.2.4 *Under the hypotheses of the previous theorem, the value function* V *is the unique viscosity solution of the problem*

$$\begin{cases} -\partial_t V(t, x) + H(x, \nabla V(t, x)) = 0, & (t, x) \in \,]0, T[\, \times \mathbb{R}^n \\ V(T, x) = g(x) & x \in \mathbb{R}^n. \end{cases} \quad (7.15)$$

Proof — Let us first show that a backward Cauchy problem of the form (7.15) is well posed. To this purpose, suppose that V is a viscosity solution to this problem, and set $U(t, x) = V(T - t, x)$. Then, it is easily checked that $(p_t, p_x) \in D^+U(t, x)$ (resp. $D^-U(t, x)$) if and only if $(-p_t, p_x) \in D^+V(T - t, x)$ (resp. $D^-V(T - t, x)$). Therefore, U is a viscosity solution to

$$\begin{cases} \partial_t U(t, x) + H(x, \nabla U(t, x)) = 0, & (t, x) \in \,]0, T[\, \times \mathbb{R}^n \\ U(0, x) = g(x) & x \in \mathbb{R}^n. \end{cases}$$

Since the comparison principle holds for this problem (see Theorem 5.2.12), an analogous property holds for problem (7.15) and we have, in particular, uniqueness of the viscosity solution in the class of continuous functions.

We now proceed to check that the value function V is a viscosity solution of (7.15). Let us take $(t, x) \in \,]0, T[\, \times \mathbb{R}^n$ and $(p_t, p_x) \in D^+V(t, x)$. Let M_f be the supremum of $|f|$ over $B_R(x) \times U$, with B_R large enough to contain all admissible trajectories starting from (t, x). Then we have, for any control $u : [t, T] \to U$,

$$|y(\sigma; t, x, u) - x| \leq M_f(\sigma - t), \qquad \sigma > t. \quad (7.16)$$

Let us take an arbitrary element $v \in U$ and let us set $y(\cdot) = y(\cdot; t, x, v)$ where v means the constant control strategy equal to v. We find, by (7.16) and (H1), that

$$y(s) = x + \int_t^s f(y(\sigma), v) d\sigma$$

$$= x + f(x, v)(s - t) + \int_t^s [f(y(\sigma), v) - f(x, v)] d\sigma$$

$$= x + f(x, v)(s - t) + o(s - t), \qquad \text{as } s \downarrow t.$$

This implies, by the definition of $D^+ V$,

$$V(s, y(s)) \leq V(t, x) + (p_t + f(x, v) \cdot p_x)(s - t) + o(s - t).$$

On the other hand we have $V(s, y(s)) \geq V(t, x)$ for all $s > t$, by the dynamic programming principle (7.12). We deduce that

$$-p_t - f(x, v) \cdot p_x \leq 0,$$

which implies, by the arbitrariness of $v \in U$, that

$$-p_t + H(x, p_x) \leq 0,$$

which proves that u is a viscosity subsolution of the equation.

To prove that u is a supersolution, let us take $(p_t, p_x) \in D^- V(t, x)$ and let us fix $\varepsilon > 0$. Using (7.16) and the definition of $D^- V$ it is easy to see that if we choose $\delta > 0$ small enough, we have

$$\frac{V(t + \delta, y(t + \delta)) - V(t, x) - p_t \delta - p_x \cdot (y(t + \delta) - x)}{\delta} \geq -\varepsilon$$

for any $y(\cdot) = y(\cdot; t, x, u)$ trajectory starting from (t, x). On the other hand, by properties (7.16), (H1) and by the definition of H,

$$p_x \cdot (y(t + \delta) - x) = \int_t^{t+\delta} p_x \cdot f(y(\sigma), u(\sigma)) d\sigma$$

$$\geq \int_t^{t+\delta} p_x \cdot f(x, u(\sigma)) d\sigma - \delta^2 |p_x| M_f K_1$$

$$\geq -\delta H(x, p_x) - \delta^2 |p_x| M_f K_1.$$

Thus we obtain

$$\frac{V(t + \delta, y(t + \delta)) - V(t, x)}{\delta} \geq p_t - H(x, p_x) - \delta |p_x| M_f K_1 - \varepsilon$$

for any trajectory $y(\cdot) = y(\cdot; t, x, u)$ starting from (t, x). Now, by (7.10), we can find a trajectory $y(\cdot)$ such that $V(t + \delta, y(t + \delta)) - V(t, x) \leq \delta^2$. We conclude that

$$p_t - H(x, p_x) \leq \delta(1 + |p_x| M_f K_1) + \varepsilon.$$

By letting $\varepsilon, \delta \to 0$ we obtain that $-p_t + H(x, p_x) \geq 0$. Hence, V is a viscosity supersolution of the equation.   ∎

Using similar arguments we can derive an additional inequality at the points lying along an optimal trajectory and different from the endpoints.

Theorem 7.2.5 *Let f satisfy hypotheses* (H0) *and* (H1). *Let $y : [t, T] \to \mathbb{R}^n$ be an optimal trajectory for a point $(t, x) \in [0, T] \times \mathbb{R}^n$. Then, for any $\tau \in {]}t, T[$, we have*

$$-p_t + H(y(\tau), p_x) = 0, \qquad \forall (p_t, p_x) \in D^+ V(\tau, y(\tau)). \qquad (7.17)$$

Proof — Let us take $\tau \in [t, T[$. We already know from the previous proof that the left-hand side of (7.17) is nonpositive, so it suffices to prove that, if $\tau > t$, the converse inequality also holds. First we observe that, by the dynamic programming principle,

$$V(\tau, y(\tau)) = V(\tau - h, y(\tau - h)), \qquad 0 \leq h \leq \tau - t.$$

On the other hand, if $(p_t, p_x) \in D^+ V(\tau, y(\tau))$, we have

$$V(\tau - h, y(\tau - h)) - V(\tau, y(\tau)) \leq -p_t h - p_x \cdot (y(\tau) - y(\tau - h)) + o(h).$$

Therefore we find that

$$
\begin{aligned}
0 &\leq -p_t h - p_x \cdot (y(\tau) - y(\tau - h)) + o(h) \\
&= -p_t h - \int_{\tau - h}^{\tau} p_x \cdot f(y(\sigma), u(\sigma)) \, d\sigma + o(h) \\
&= -p_t h - \int_{\tau - h}^{\tau} p_x \cdot f(y(\tau), u(\sigma)) \, d\sigma + o(h) \\
&\leq h[-p_t + H(y(\tau), p_x)] + o(h),
\end{aligned}
$$

which yields the conclusion.   ∎

From the above result we can derive a bound on the size of the superdifferential of V along an optimal trajectory. In the next proof we use the set $\nabla^+ V(t, x)$; we recall that such a set is defined as the superdifferential of $x \mapsto V(t, x)$ and that if V is semiconcave, it coincides with the vectors which are space components of elements of $D^+ V(t, x)$ (see Lemma 3.3.16).

Theorem 7.2.6 *Under the hypotheses of the previous theorem, let $y : [t, T] \to \mathbb{R}^n$ be an optimal trajectory for a point $(t, x) \in [0, T] \times \mathbb{R}^n$. Given $\tau \in {]}t, T[$, let $v \in \{1, \ldots, n\}$ be such that any normal cone to the convex hull of the set $f(y(\tau), U)$ has dimension less than or equal to v. Then $\dim D^+ V(\tau, y(\tau)) \leq v$.*

Proof — If $\tau \in {]}t, T[$, we know from the previous theorem that $p \in \mathbb{R}^n$ belongs to $\nabla^+ V(\tau, y(\tau))$ if and only if $(H(y(\tau), p), p) \in D^+ V(\tau, y(\tau))$. In particular, $\nabla^+ V(\tau, y(\tau))$ and $D^+ V(\tau, y(\tau))$ have the same dimension.

Let us set $V = -\text{co } f(y(\tau), U)$, $M = \nabla^+ V(\tau, y(\tau))$. We claim that V and M satisfy property (ii) of Lemma A. 1.24. To see this, observe first that $\sigma_V(\cdot) = H(y(\tau), \cdot)$. Then, given $p_0, p_1 \in M$ and $t \in [0, 1]$, we have $(\sigma_V(p_i), p_i) \in D^+ V(\tau, y(\tau))$ for $i = 0, 1$, and so

$$(t\sigma_V(p_0) + (1-t)\sigma_V(p_1), tp_0 + (1-t)p_1) \in D^+ V(\tau, y(\tau))$$

by the convexity of the superdifferential. Again by the previous theorem, this implies that $\sigma_V(tp_0 + (1-t)p_1) = t\sigma_V(p_0) + (1-t)\sigma_V(p_1)$, and this proves property (ii) of Lemma A. 1.24. That lemma then implies that M is contained in some normal cone to the set V, and therefore $\dim D^+ V(\tau, y(\tau)) \leq \nu$. ∎

Remark 7.2.7 The previous theorem gives an upper bound on the size of the superdifferential of V along an optimal trajectory. When $f(x, U)$ is smooth, we can take $\nu = 1$ and obtain that the superdifferential has dimension at most one. We will see in the following (Corollary 7.3.5 and Theorem 7.3.16) that under further assumptions one can refine this estimate and prove that V is smooth at the interior points of an optimal trajectory. On the other hand, simple counterexamples, like the one after Theorem 7.3.16, show that such a property cannot be expected in full generality for the Mayer problem. This behavior is different from the problem in the calculus of variations considered in the previous chapter, where the value function is differentiable along an optimal trajectory (see Theorem 6.3.6 and Corollary 6.4.10).

Let us now study the semiconcavity of the value function.

Theorem 7.2.8 *Let f satisfy* (H0), (H1), (H2) *and let* $g \in \text{SCL}_{loc}(\mathbb{R}^n)$. *Then* $V \in \text{SCL}_{loc}([0, T] \times \mathbb{R}^n)$.

Proof — We assume for simplicity that there always exists an optimal control for problem (MP). If this is not the case, one can use an approximation procedure, as in the proof of Theorem 7.2.3.

For a given $r > 0$, we will prove the semiconcavity of V in $[0, T] \times B_r$. Let $R > 0$ be such that all trajectories of (7.1) starting from points in $[0, T] \times B_r$ stay in $[0, T] \times B_R$. Let us denote by M_f the supremum of $|f(x, u)|$ on $B_R \times U$, by K_g and by α_g the Lipschitz constant and the semiconcavity constant of g in B_R, respectively.

We first prove the semiconcavity inequality for a triple of points with the same time component. Given x, h such that $x \pm h \in B_r$ and given $t \in [0, T]$, let $u : [t, T] \to U$ be an optimal control for the point (t, x). Let us set for simplicity

$$y(\cdot) = y(\cdot; t, x, u), \qquad y_-(\cdot) = y(\cdot; t, x - h, u), \qquad y_+(\cdot) = y(\cdot; t, x + h, u).$$

Then, by Lemma 7.1.2, we have

$$|y_+(T) - y_-(T)| \leq c|h|, \qquad |y_+(T) + y_-(T) - 2y(T)| \leq c|h|^2$$

for some constant $c > 0$ which does not depend on h. It follows, using the definition of V, the optimality of u, the Lipschitz continuity and semiconcavity of g,

$$V(t, x + h) + V(t, x - h) - 2V(t, x)$$
$$\leq g(y_+(T)) + g(y_-(T)) - 2g(y(T))$$
$$= g(y_+(T)) + g(y_-(T)) - 2g\left(\frac{y_+(T) + y_-(T)}{2}\right)$$
$$+2g\left(\frac{y_+(T) + y_-(T)}{2}\right) - 2g(y(T))$$
$$\leq \alpha_g |y_+(T) - y_-(T)|^2 + K_g |y_+(T) + y_-(T) - 2y(T)|$$
$$\leq (\alpha_g c^2 + K_g c)|h|^2, \tag{7.18}$$

which proves the semiconcavity inequality in this case.

Let us now consider x, h and t, τ such that $x \pm h \in B_r$ and such that $0 \leq t - \tau \leq t + \tau \leq T$. Let $u : [t, T] \to U$ be an optimal control for (t, x). We define

$$\bar{u}(s) = u\left(\frac{t + \tau + s}{2}\right), \qquad s \in [t - \tau, t + \tau]$$

and we set

$$y(s) = y(s; t, x, u), \qquad s \in [t, t + \tau],$$
$$\bar{y}(s) = y(s; t - \tau, x - h, \bar{u}), \qquad s \in [t - \tau, t + \tau].$$

Then we have, for all $s_1 \in [t - \tau, t + \tau]$, $s_2 \in [t, t + \tau]$,

$$|\bar{y}(s_1) - y(s_2)| = \left|-h + \int_{t-\tau}^{s_1} f(\bar{y}(s), \bar{u}(s))\, ds - \int_t^{s_2} f(y(s), u(s))\, ds\right|$$
$$\leq |h| + (s_1 - t + \tau)M_f + (s_2 - t)M_f$$
$$\leq |h| + 3M_f \tau. \tag{7.19}$$

Let us set $\bar{x} = \bar{y}(t + \tau)$, $\hat{x} = y(t + \tau)$. We have

$$|x + h - \bar{x}| = \left|2h + \int_{t-\tau}^{t+\tau} f(\bar{y}(s), \bar{u}(s))\, ds\right| \leq 2|h| + 2M_f \tau. \tag{7.20}$$

In addition, using the definition of $\bar{u}$, hypothesis (H1) and estimate (7.19),

$$|x + h + \bar{x} - 2\hat{x}|$$
$$= \left|\int_{t-\tau}^{t+\tau} f(\bar{y}(s), \bar{u}(s))\, ds - 2\int_t^{t+\tau} f(y(s), u(s))\, ds\right|$$
$$= \left|2\int_t^{t+\tau} [f(\bar{y}(2s - t - \tau), u(s)) - f(y(s), u(s))]\, ds\right|$$
$$\leq 2\int_t^{t+\tau} K_1 |\bar{y}(2s - t - \tau) - y(s)|\, ds$$
$$\leq 2K_1 \tau(|h| + 3M_f \tau). \tag{7.21}$$

From the dynamic programming principle and the optimality of u we deduce that

$$V(t + \tau, x + h) + V(t - \tau, x - h) - 2V(t, x)$$
$$\leq V(t + \tau, x + h) + V(t + \tau, \bar{x}) - 2V(t + \tau, \hat{x})$$
$$= V(t + \tau, x + h) + V(t + \tau, \bar{x}) - 2V\left(t + \tau, \frac{x + h + \bar{x}}{2}\right)$$
$$+ 2V\left(t + \tau, \frac{x + h + \bar{x}}{2}\right) - 2V(t + \tau, \hat{x}).$$

Using (7.18) and (7.20) we obtain

$$V(t + \tau, x + h) + V(t + \tau, \bar{x}) - 2V\left(t + \tau, \frac{x + h + \bar{x}}{2}\right)$$

$$\leq (\alpha_g c^2 + K_g c)\frac{|x + h - \bar{x}|^2}{4} \leq (\alpha_g c^2 + K_g c)(|h| + M_f \tau)^2.$$

In addition, if we denote by K_V the Lipschitz constant of V in $[0, T] \times B_R$ we obtain, using (7.21) that

$$2V\left(t + \tau, \frac{x + h + \bar{x}}{2}\right) - 2V(t + \tau, \hat{x})$$
$$\leq K_V|x + h + \bar{x} - 2\hat{x}| \leq 2K_1 K_V \tau(|h| + 3M_f \tau).$$

From the above estimates we conclude that there is a constant C_r depending only on r such that

$$V(t + \tau, x + h) + V(t - \tau, x - h) - 2V(t, x) \leq C_r(\tau^2 + |h|^2). \qquad \blacksquare$$

Remark 7.2.9 Using similar arguments one can treat the case when g is semiconcave with a general modulus ω, finding that V is semiconcave with a modulus of the form $\bar{\omega}(r) = c_1\omega(c_2 r)$ (see [32]). Observe also that the above semiconcavity result cannot be deduced from the general results of Chapter 4 about the semiconcavity of solutions of Hamilton–Jacobi equations because H is not strictly convex and in general not C^1.

Example 7.2.10 Let us consider the state equation $y' = u$, where $u \in U = B_1$. Then it is easy to see that the value function is

$$V(x, t) = \min\{g(z) : |z - x| \leq T - t\}.$$

The corresponding Hamilton–Jacobi equation is $u_t + |\nabla u| = 0$. Let us consider some specific examples in one-dimensional space where the value function can be computed explicitly. In view of the analysis of optimality conditions that will be done in the next section, it is interesting to analyze the singularities of the value function in these particular cases and to see whether they are related with the non-uniqueness of the optimal trajectories.

(i) Suppose that the final cost $g : \mathbb{R} \to \mathbb{R}$ is strictly monotone increasing. Then, for any $(t, x) \in [0, T] \times \mathbb{R}$, the optimal control is the constant one $u \equiv -1$, and so the unique optimal trajectory is $y(s) = x - s + t$. The value function is $V(t, x) = g(x + t - T)$. Observe that it has the same regularity of the final cost; this shows that the semiconcavity assumption on g in Theorem 7.2.8 is necessary and that there is no longer the gain of semiconcavity as in the problem of the calculus of variations studied in the previous chapter (see Theorem 6.4.3). Notice also that, if $\bar{x} \in \mathbb{R}$ is a point of nondifferentiability for g, then $y(s) = \bar{x} - s + T$ is an optimal trajectory consisting of singular points of V; thus, one cannot expect in general the value function of a Mayer problem to be differentiable along optimal trajectories. Finally, let us remark that the optimal trajectory is unique for any choice of initial condition, and thus the uniqueness of the optimal trajectory is not related to the differentiability of V in this case.

(ii) Suppose that $g : \mathbb{R} \to \mathbb{R}$ is an even function, smooth, with $x g'(x) < 0$ for all $x \neq 0$. Then it is easily seen that for a given (t, x) with $x \neq 0$, the optimal trajectory is unique and is given by $y(s) = x + \mathrm{sgn}\,(x)(s - t)$, while for points of the form $(t, 0)$, the optimal trajectories are $y(s) = \pm(s - t)$. It follows that

$$V(t, x) = \begin{cases} g(x + t - T) & x \leq 0 \\ g(x + T - t) & x \geq 0. \end{cases}$$

It is easily seen that V is not differentiable at the points of the form $(t, 0)$ with $t < T$, which are also the points for which the optimal trajectory is not unique. Theorem 7.2.8 ensures that V be semiconcave; this can be also seen directly since V is the minimum of the two smooth functions $g(x + t - T)$ and $g(x - t + T)$. Observe that the solution is singular for times arbitrarily close to the final time T. Such a behavior cannot occur in the problem of the calculus of variations of the previous chapter; in that case, as remarked after Definition 6.3.4, there are no singular points in a neighborhood of the initial time. This behavior is interesting also from the point of view of the associated Hamilton–Jacobi equation: one has a smooth final value, but the corresponding viscosity solution is nonsmooth even for times close to the terminal time. This is a consequence of the nonsmoothness of the hamiltonian, since otherwise the local existence theorem for classical solutions of Hamilton–Jacobi equations would hold (see Theorem 5.1.1).

(iii) Let us consider the final cost $g(x) = \cos x$, and let us assume $T > \pi$. It suffices to describe the behavior of the system for $x \in [-\pi, \pi]$. Let us first consider a time $t \in [T - \pi, T[$. We have various cases. If $x = 0$, then there are two optimal trajectories, namely $y(s) = \pm(s - t)$. If $0 < |x| \leq \pi + t - T$, then the unique optimal trajectory is $y(x) = x + \mathrm{sgn}\,(x)(s - t)$. If $\pi + t - T < |x| \leq \pi$, then any admissible trajectory with $y(t) = x$ and $y(T) = \pm\pi$ is optimal; there exist infinitely many such trajectories. Analogously, if we consider points (t, x) with $t < T - \pi$, then any of the infinite admissible trajectories ending at $\pm\pi$ are optimal. We deduce that the value function has the following form.

$$V(t, x) = \begin{cases} \cos(x + t - T) & \text{if } T - \pi \leq t \leq T \\ & \text{and } T - \pi - t \leq x \leq 0 \\ \cos(x + T - t) & \text{if } T - \pi \leq t \leq T \\ & \text{and } 0 \leq x \leq -T + \pi + t \\ -1 & \text{otherwise.} \end{cases}$$

Thus, V is singular along the segment $\{(t, 0) \ : \ t \in]T - \pi, T[\}$, which consists of points where the optimal trajectory is not unique. On the other hand, it is identically equal to -1 for $t \leq T - \pi$. Therefore, it is singular in a certain time interval but is becomes smooth as one proceeds backwards in time. Observe that all points (t, x) with $t < T - \pi$ are points where V is differentiable (V is even constant in a neighborhood) but admit infinitely many optimal trajectories. ∎

Although the natural property to be expected in minimization problems is semiconcavity, there are special cases where one can prove that the value function of a control problem is semiconvex. One example is the following, dealing with the Mayer problem for a linear system with convex final cost.

Theorem 7.2.11 Let $f(x, u) = Ax + Bu$, where A, B are $n \times n$ and $n \times m$ matrices, respectively. Suppose that $U \subset \mathbb{R}^m$ is convex and compact and that the final cost $g : \mathbb{R}^n \to \mathbb{R}$ is a convex function. Then the value function V of problem (MP) satisfies the following:

(i) $V(t, \cdot)$ is convex in $\mathbb{R}^n$ for all $t \in [0, T]$;
(ii) V as a function of (t, x) is locally semiconvex with a linear modulus in $[0, T] \times \mathbb{R}^n$.

Proof — For given $x, h \in \mathbb{R}^n$ and $t \in [0, T[$, let $u_-, u_+ : [t, T] \to \mathbb{R}^n$ be optimal controls for the points $(t, x - h)$, $(t, x + h)$, respectively. Let us define $\bar{u}(s) := (u_-(s) + u_+(s))/2$. Then $\bar{u}(s) \in U$ since we are assuming that U is convex. Let us set for simplicity $\bar{y}(\cdot) = y(\cdot; t, x, \bar{u})$, $y_\pm(\cdot) = y(\cdot; t, x \pm h, u_\pm)$. Thanks to the linearity of the system we have

$$\bar{y}(s) = \frac{1}{2}[y_-(s) + y_+(s)], \qquad s \in [t, T].$$

Therefore, by the optimality of u_-, u_+ and by the convexity of g,

$$2V(t, x) - V(t, x - h) - V(t, x + h)$$
$$\leq 2g(\bar{y}(T)) - g(y_-(T)) - g(y_+(T)) \leq 0,$$

showing that $V(t, \cdot)$ is convex.

Let us now prove (ii). For any fixed any $r > 0$ we can find $R > 0$ such that all trajectories of the system starting from points in B_r stay inside B_R up to time T. By Theorem 7.2.3, V is Lipschitz continuous in $[0, T] \times B_R$; let us denote by K_V its Lipschitz constant.

Let us take $x, h \in \mathbb{R}^n$ and $t, \tau \geq 0$ such that $x \pm h \in B_r$ and $0 \leq t - \tau < t + \tau \leq T$. Let $u_- : [t - \tau, T] \to U$ be an optimal control for $(t - \tau, x - h)$. We define

$$\bar{u}(s) = u_-(2s - t - \tau), \qquad s \in [t, t + \tau].$$

Let us set

$$y_-(\cdot) = y(\cdot; t - \tau, x - h, u_-), \qquad \bar{y}(\cdot) = y(\cdot; t, x, \bar{u}).$$

We have, by the dynamic programming principle, that

$$V(t - \tau, x - h) = V(t + \tau, y_-(t + \tau)), \quad V(t, x) \leq V(t + \tau, \bar{y}(t + \tau)). \quad (7.22)$$

In addition, setting $M_R = \max\{|Ax + Bu| : x \in B_R, u \in U\}$, we have

$$|y_-(s_1) - \bar{y}(s_2)| \leq |h| + 3\tau M_R, \qquad s_1 \in [t - \tau, t + \tau], s_2 \in [t, t + \tau]$$

and therefore

$$x + h + y_-(t + \tau) - 2\bar{y}(t + \tau) = \int_{t-\tau}^{t+\tau} y'_-(s)\, ds - 2 \int_t^{t+\tau} \bar{y}'(s)\, ds$$

$$= \int_{t-\tau}^{t+\tau} (Ay_-(s) + Bu_-(s))\, ds - 2 \int_t^{t+\tau} (A\bar{y}(s) + Bu_-(2s - t - \tau))\, ds$$

$$= \int_{t-\tau}^{t+\tau} A\left(y_-(s) - \bar{y}\left(\frac{s + t + \tau}{2}\right)\right) ds$$

$$\leq 2\tau \|A\|(|h| + 3\tau M_R).$$

We conclude, using part (i) and (7.22), that

$$2V(t, x) - V(t - \tau, x - h) - V(t + \tau, x + h)$$
$$\leq 2V(t + \tau, \bar{y}(t + \tau)) - V(t + \tau, y_-(t + \tau)) - V(t + \tau, x + h)$$
$$= 2V(t + \tau, \bar{y}(t + \tau)) - 2V\left(t + \tau, \frac{y_-(t + \tau) + x + h}{2}\right)$$
$$+ 2V\left(t + \tau, \frac{y_-(t + \tau) + x + h}{2}\right)$$
$$- V(t + \tau, y_-(t + \tau)) - V(t + \tau, x + h)$$
$$\leq K_V |y_-(t + \tau) + x + h - 2\bar{y}(t + \tau)|$$
$$\leq 2K_V \|A\|(\tau |h| + 3M_R \tau^2),$$

and so V is semiconvex with a linear modulus.   ∎

Let us point out that, even for a linear system, the semiconvexity of the final cost does not imply that the value function is semiconvex. For instance, in Examples 7.2.10(ii) and (iii) the final cost is smooth, hence semiconvex; on the other hand, the value function is semiconcave and not smooth, and so it is not semiconvex.

Combining the previous theorem with the semiconcavity result of Theorem 7.2.8 and recalling Corollary 3.3.8 we obtain the following.

Corollary 7.2.12 *Under the hypotheses of Theorem 7.2.11, if we assume in addition that $g \in \mathrm{SCL}_{loc}(\mathbb{R}^n)$, then $V \in C^{1,1}_{loc}([0, T] \times \mathbb{R}^n)$.*

Let us remark that in the above corollary g is implicitly required to be of class $C^{1,1}$, since it is assumed to be both convex and semiconcave. We see that, for a linear system with smooth convex cost, the value function has the same regularity of the data; we do not have the formation of singularities which is otherwise usual for the value function of a control problem.

7.3 Optimality conditions

A fundamental result in optimal control theory is the so-called *Pontryagin maximum principle*; for the Mayer problem studied here it can be stated in the following way.

Theorem 7.3.1 *Let the control system (f, U) satisfy (H0), (H1) and let $g \in C(\mathbb{R}^n)$. Suppose also that f_x exists and is continuous with respect to x. Given $(t, x) \in [0, T] \times \mathbb{R}^n$, let $u : [t, T] \to U$ be an optimal control for problem (MP) with initial point (t, x) and let $y(\cdot) = y(\cdot; t, x, u)$ be the corresponding optimal trajectory. For any $q \in D^+g(y(T))$, let $p : [t, T] \to \mathbb{R}^n$ be the solution of the equation*

$$\begin{cases} p'(s) = -f_x^T(y(s), u(s))\, p(s), & s \in [t, T]\ a.e. \\ p(T) = q. \end{cases} \tag{7.23}$$

Then, for a.e. $s \in [t, T]$, $p(s)$ satisfies

$$-f(y(s), u(s)) \cdot p(s) \geq -f(y(s), v) \cdot p(s), \qquad \forall v \in U. \tag{7.24}$$

Remark 7.3.2 By f_x^T in (7.23) we denote the transpose of the jacobian of f with respect to x. Thus, the equation may be rewritten component-wise as

$$p_i'(s) = -\sum_{j=1}^n \frac{\partial f_j}{\partial x_i}(y(s), u(s))\, p_j(s).$$

We observe that (7.23) is the adjoint linearized equation associated with (7.1) (see Appendix A. 4).

Proof — Let us fix any $\bar{s} \in \,]t, T[$ which is a Lebesgue point for the function $s \to f(y(s), u(s))$, that is, a value such that

$$\lim_{h \downarrow 0} \frac{1}{h} \int_{\bar{s}-h}^{\bar{s}+h} |f(y(s), u(s)) - f(y(\bar{s}), u(\bar{s}))|\, ds = 0.$$

It is well known from measure theory (see e.g., [72, p. 44]) that all $s \in \,]t, T[$ have this property except for a set of measure zero.

Let us fix any $v \in U$ and define, for $\varepsilon > 0$ small,

$$u_\varepsilon(s) = \begin{cases} u(s) & s \in [t, T] \setminus [\bar{s} - \varepsilon, \bar{s}] \\ v & s \in [\bar{s} - \varepsilon, \bar{s}]. \end{cases} \tag{7.25}$$

Let us set $y_\varepsilon(s) = y(s; t, x, u_\varepsilon)$ and $\bar{x} = y(\bar{s})$. Since our arguments are of a local nature, we can assume that $|f|$ is bounded by some constant M_f. We first observe that

$$|y_\varepsilon(s) - \bar{x}| \leq |y_\varepsilon(s) - y(\bar{s} - \varepsilon)| + |y(\bar{s} - \varepsilon) - \bar{x}|$$
$$\leq 2M_f\varepsilon, \qquad \forall s \in [\bar{s} - \varepsilon, \bar{s}],$$

since $y(\bar{s} - \varepsilon) = y_\varepsilon(\bar{s} - \varepsilon)$. Therefore, using also the property that $\bar{s}$ is a Lebesgue point for $s \to f(y(s), u(s))$,

$$y_\varepsilon(\bar{s}) - y(\bar{s}) = \int_{\bar{s}-\varepsilon}^{\bar{s}} [f(y_\varepsilon(s), v) - f(y(s), u(s))]\, ds$$
$$= \int_{\bar{s}-\varepsilon}^{\bar{s}} [f(\bar{x}, v) - f(y(s), u(s))]\, ds + o(\varepsilon)$$
$$= \varepsilon[f(\bar{x}, v) - f(\bar{x}, u(\bar{s}))] + o(\varepsilon).$$

By well-known results on ordinary differential equations (see Theorem A. 31) we deduce, for $s \geq \bar{s}$,

$$y_\varepsilon(s) = y(s) + \varepsilon w(s) + o(\varepsilon) \tag{7.26}$$

where w is the solution of the linearized problem

$$\begin{cases} w'(s) = f_x(y(s), u(s))\, w(s) & s \geq \bar{s} \\ w(\bar{s}) = f(\bar{x}, v) - f(\bar{x}, u(\bar{s})). \end{cases}$$

If we take $q \in D^+g(y(T))$ and define p by solving (7.23), we obtain that $p(s) \cdot w(s)$ is constant for $s \in [\bar{s}, T]$. Since y is an optimal trajectory, we have $g(y_\varepsilon(T)) \geq g(y(T))$ for all $\varepsilon > 0$ and thus we obtain

$$0 \leq g(y_\varepsilon(T)) - g(y(T)) \leq q \cdot [y_\varepsilon(T) - y(T)] + o(\varepsilon)$$
$$= \varepsilon q \cdot w(T) + o(\varepsilon) = \varepsilon p(T) \cdot w(T) + o(\varepsilon) = \varepsilon p(\bar{s}) \cdot w(\bar{s}) + o(\varepsilon)$$

which implies that

$$0 \leq p(\bar{s}) \cdot w(\bar{s}) = p(\bar{s}) \cdot [f(y(\bar{s}), v) - f(y(\bar{s}), u(\bar{s}))].$$

Since this holds for all $v \in U$ and for $\bar{s} \in \,]t, T[$ a.e., the theorem is proved. ∎

The previous result is called a "maximum principle" because of inequality (7.24), which shows that the quantity $-f(y(s), v) \cdot p(s)$ is maximized for $v = u(s)$. The terminal condition for $p(T)$ in (7.23) is called the *transversality condition*. Any arc $p(\cdot)$ satisfying the adjoint equation (7.23) together with the transversality condition and inequality (7.24) is called a *dual arc* or a *co-state* associated with the optimal pair (u, y). The Pontryagin maximum principle is a necessary condition for the optimality of trajectories which can be regarded as a generalization of the Euler–Lagrange equations in the calculus of variations. In fact, in the cases where property (7.24) determines uniquely $u(s)$ as a function of $y(s)$, $p(s)$, equations (7.1) and (7.23) become a first order system in the pair (y, p), similar to the hamiltonian system (6.13) in the calculus of variations. The analogy will be made precise for certain classes of problems we consider later (see Corollary 7.3.7).

The previous theorem gives the existence of a dual arc associated with y if $D^+g(y(T))$ is nonempty; such a property is ensured if g, for instance, is differentiable or semiconcave. It is possible to give more refined versions of the maximum principle showing that a dual arc exists also if g is less regular, e.g., locally Lipschitz continuous (see for instance [55, Theorem 4.9.1]).

An important property satisfied by a dual arc in connection with the value function is the inclusion described in the next theorem. We recall that the symbol $\nabla^+ V(t, x)$ denotes the superdifferential of $x \rightarrow V(t, x)$ (see Definition 3.28).

Theorem 7.3.3 *Let f, U, g be as in the previous theorem, let (u, y) be an optimal pair for the point $(t, x) \in [0, T] \times \mathbb{R}^n$, and let $p : [t, T] \rightarrow \mathbb{R}^n$ be a dual arc associated with (u, y). Then*

$$p(s) \in \nabla^+ V(s, y(s)), \qquad \forall s \in [t, T].$$

Proof — For simplicity of notation we prove the assertion only in the case $s = t$, the general case being entirely analogous. Let us fix any $h \in \mathbb{R}^n$, $|h| = 1$. For any $\varepsilon > 0$ let us set $y_\varepsilon(\cdot) = y(\cdot; t, x + \varepsilon h, u)$. Then

$$y_\varepsilon(s) = y(s) + \varepsilon w(s) + o(\varepsilon),$$

where w is the solution of the linearized problem

$$\begin{cases} w'(s) = f_x(y(s), u(s))w(s) & s \geq t \\ w(t) = h. \end{cases}$$

By definition of V we have $g(y_\varepsilon(T)) \geq V(t, x + \varepsilon h)$ for all $\varepsilon > 0$. As in the proof of the maximum principle, we find that $g(y_\varepsilon(T)) - g(y(T)) \leq \varepsilon p(s) \cdot w(s) + o(\varepsilon)$, where $p(s) \cdot w(s)$ is constant for $s \in [t, T]$. Therefore

$$V(t, x + \varepsilon h) - V(t, x) \leq g(y_\varepsilon(T)) - g(y(T))$$
$$= \varepsilon p(t) \cdot w(t) + o(\varepsilon) = p(t) \cdot \varepsilon h + o(\varepsilon),$$

where the remainder term $o(\varepsilon)$ can be taken independent of h. Since h can be any unit vector, this proves that $p(t) \in \nabla^+ V(t, x)$. ∎

The previous result can be interpreted as an invariance property of the superdifferential of V with respect to the adjoint equation. The subdifferential satisfies an analogous property, with a reversed time direction, as the next result shows.

Theorem 7.3.4 *Let f, U, g be as in the previous theorem, let (u, y) be an optimal pair for the point $(t_0, x_0) \in [0, T] \times \mathbb{R}^n$ and let $p : [t_0, T] \to \mathbb{R}^n$ be any solution of the adjoint equation*

$$p'(s) = -f_x^T(y(s), u(s))p(s), \qquad s \in [t_0, T].$$

Suppose that $p(t_0) \in \nabla^- V(t_0, x_0)$. Then

$$p(t) \in \nabla^- V(t, y(t)), \qquad t \in [t_0, T].$$

Proof — Let $t \in [t_0, T]$ and let h be any unit vector. Let $w : [t_0, t] \to \mathbb{R}^n$ be the solution of

$$w'(s) = f_x(y(s), u(s))w(s), \qquad w(t) = h.$$

For any $\varepsilon > 0$, let us set $y_\varepsilon(\cdot) = y(\cdot; t_0, x_0 + \varepsilon w(t_0), u)$. Then

$$y_\varepsilon(s) = y(s) + \varepsilon w(s) + o(\varepsilon), \qquad s \in [t_0, t].$$

By the dynamic programming principle, the hypothesis that $p(t_0)$ belongs to $\nabla^- V(t_0, x_0)$ and the property that $p \cdot w$ is constant,

$$V(t, y_\varepsilon(t)) \geq V(t_0, x_0 + \varepsilon w(t_0)) \geq V(t_0, x_0) + \varepsilon p(t_0) \cdot w(t_0) + o(\varepsilon)$$
$$= V(t, y(t)) + \varepsilon p(t) \cdot w(t) + o(\varepsilon).$$

On the other hand, by the Lipschitz continuity of V,

$$V(t, y_\varepsilon(t)) = V(t, y(t) + \varepsilon w(t)) + o(\varepsilon) = V(t, y(t) + \varepsilon h) + o(\varepsilon),$$

and so we conclude that

$$V(t, y(t) + \varepsilon h) - V(t, y(t)) \geq \varepsilon p(t) \cdot w(t) + o(\varepsilon).$$

By the arbitrariness of h, we deduce that $p(t) \in \nabla^- V(t, y(t))$. ∎

If we are in the hypotheses of the semiconcavity theorem, then the subdifferential of V is nonempty only at the differentiability points of V. Thus we obtain the following result (see also Remark 7.2.7).

Corollary 7.3.5 *Suppose that properties (H0), (H1), (H2) hold and $g \in SCL_{loc}(\mathbb{R}^n)$. Let (u, y) be an optimal pair for a point (t_0, x_0) and let p be a dual arc associated to (u, y). If V is differentiable at (t_0, x_0), then it is differentiable at $(t, y(t))$ for all $t \in [t_0, T]$ and*

$$DV(t, y(t)) = (H(p(t)), p(t)), \qquad t \in [t_0, T]. \tag{7.27}$$

Proof — Since V is differentiable at (t_0, x_0) we obtain from Theorem 7.3.3 that

$$\nabla^+ V(t_0, x_0) = \nabla^- V(t_0, x_0) = \{p(t_0)\}.$$

Then Theorem 7.3.4 ensures that $p(t) \in \nabla^- V(t, y(t))$ for all $t \in [t_0, T]$. On the other hand, we know from Theorem 7.2.8 that V is semiconcave. In particular, $V(t, \cdot)$ is semiconcave, and so $p(t) \in \nabla^- V(t, y(t))$ implies that $V(t, \cdot)$ is differentiable at x and that $\nabla V(t, y(t)) = p(t)$. Then Lemma 3.3.16 implies that $D^+ V(t, y(t))$ is a segment with endpoints $(\lambda_1, p(t))$, $(\lambda_2, p(t))$ for some $\lambda_1 \leq \lambda_2$. Since $D^+ V$ is the convex hull of $D^* V$ we have

$$(\lambda_i, p(t)) \in D^* V(t, y(t)), \qquad i = 1, 2.$$

Since all elements of $D^* V$ satisfy the Hamilton–Jacobi equation (7.15), we deduce that $\lambda_1 = \lambda_2 = H(p(t))$, showing that $D^+ V(t, y(t))$ is a singleton. Hence V is differentiable at $(t, y(t))$ and (7.27) holds.   ∎

If the hamiltonian associated to our problem is suitably smooth, it is possible to derive further optimality conditions and to establish some relations between the optimal trajectories for a given point (t, x) and the generalized gradients of V at (t, x). As we have already observed, we cannot expect the hamiltonian to be smooth everywhere, since it will be nondifferentiable at the points with $p = 0$. Let us focus our attention on the case where these are the only singularities of the hamiltonian. More precisely, we assume that:

(H3) The hamiltonian H belongs to $C^{1,1}_{loc}(\mathbb{R}^n \times (\mathbb{R}^n \setminus \{0\}))$.

A study of the regularity of the support functions to a convex set shows that such a property is satisfied, for instance, if the sets $\{f(x, U) : x \in \mathbb{R}^n\}$ are a family of uniformly convex sets of class C^2 (see Theorem A. 1.22 and Definition A. 1.23).

Let us first derive an expression for the derivatives of H.

Theorem 7.3.6 *If* (H3) *holds, then we have, for any* (x, p) *with* $p \neq 0$,

$$H_x(x, p) = -f_x^T(x, u^*(x, p))p, \qquad H_p(x, p) = -f(x, u^*(x, p)), \qquad (7.28)$$

where $u^*(x, p) \in U$ *is any vector such that*

$$-f(x, u^*) \cdot p = \max_{u \in U} -f(x, u) \cdot p.$$

Proof — For given x, p, let us set

$$U^*(x, p) = \{u^* \in U : -f(x, u^*) \cdot p = \max_{u \in U} -f(x, u) \cdot p\}.$$

Since H is the maximum of smooth functions, its differentiability can be analyzed using Theorem 3.4.4 on marginal functions. We obtain that H is differentiable at (x, p) if and only if the sets

$$\{-f_x^T(x, u^*)p : u^* \in U^*(x, p)\}, \qquad \{-f(x, u^*) : u^* \in U^*(x, p)\} \qquad (7.29)$$

are singletons, and in that case they coincide with $\{H_x(x, p)\}$ and $\{H_p(x, p)\}$, respectively. Therefore, if (H3) holds the derivatives of H are given by (7.28).   ∎

Corollary 7.3.7 *Suppose that hypotheses* (H0)–(H3) *are satisfied. Let* (u, y) *be an optimal pair for the point* $(t, x) \in [0, T] \times \mathbb{R}^n$ *and let* $p : [t, T] \to \mathbb{R}^n$ *be a dual arc associated with* (u, y) *such that* $p(\bar{s}) \neq 0$ *for some* $\bar{s} \in [t, T]$. *Then* $p(s) \neq 0$ *for all* $s \in [t, T]$ *and* (y, p) *solves the system*

$$\begin{cases} y'(s) = -H_p(y(s), p(s)) \\ p'(s) = H_x(y(s), p(s)) \end{cases} \qquad s \in [t, T]. \tag{7.30}$$

Therefore, y and p are of class C^1.

Proof — Since the equation satisfied by $p(\cdot)$ is linear, if $p(\bar{s}) \neq 0$ for some $\bar{s}$, then $p(s) \neq 0$ for all s. Therefore the pair (y, p) remains in the set where H is differentiable. Recalling the expressions (7.28) for the derivative of H and inequality (7.24), which can be restated as $u(t) = u^*(y(t), p(t))$, we obtain our assertion. ∎

Remark 7.3.8 Observe that system (7.30) is the characteristic system associated to the Hamilton–Jacobi equation (7.15). Thus the two notions, a priori independent, of optimal trajectory for the control problem and of characteristic curve for the Hamilton–Jacobi equation turn out to be equivalent, at least formally. Similar properties hold for problems of the calculus of variations (see Theorem 6.4.6) and for the control problems which will be studied in what follows. ∎

We have seen in the previous chapter (see Theorem 6.4.9) that in the case of the calculus of variations the optimal trajectories starting from a point (t, x) are in one-to-one correspondence with the reachable gradients of the value function at (t, x). The same statement does not hold in general for the Mayer problem (see Example 7.2.10) but we can obtain some results which are much in the same spirit, and are stated below in Theorems 7.3.9 to 7.3.14 (see also Corollary 7.3.18). Roughly speaking, the difficulty here is that the hamiltonian H is singular when $p = 0$, and this forces us to treat separately the case where $0 \in D^*V(t, x)$. From now on we make the following assumptions, in addition to (H0)–(H3):

(H4) $f(x, U)$ is convex for all $x \in \mathbb{R}^n$;
(H5) $g \in C^1(\mathbb{R}^n) \cap SCL_{loc}(\mathbb{R}^n)$.

Theorem 7.3.9 *Assume properties* (H0)–(H5). *Let* V *be differentiable at* $(t, x) \in [0, T[\times \mathbb{R}^n$, *with* $DV(t, x) \neq 0$. *Consider the pair* (y, p) *which solves system* (7.30) *with initial conditions* $y(t) = x$, $p(t) = \nabla V(t, x)$. *Then* y *is an optimal trajectory for* (t, x), p *is a dual arc associated with* y *and* $p(s) = \nabla V(s, y(s))$ *for all* $s \in [t, T]$. *In addition,* y *is the unique optimal trajectory starting at* (t, x) *and* p *is the unique dual arc associated with* y.

Proof — Observe first that the hypothesis $DV(t, x) \neq 0$ implies that $\nabla V(t, x) = 0$; to see this, recall that, by Theorem 7.2.4, we have $u_t(t, x) = H(x, \nabla V(t, x))$, and that $H(x, 0) = 0$.

By Theorem 7.1.4 there exists an optimal trajectory y starting from (t, x). In addition, $D^+ g(y(T))$ is nonempty by assumption (H5) and so Theorem 7.3.1 implies that there exists a dual arc p associated with y. Since V is differentiable at (t, x), Corollary 7.3.5 implies that V is differentiable at $(s, y(s))$ and that $p(s) = \nabla V(s, y(s))$ for all $s \in [t, T]$. In particular, $p(t) = \nabla V(t, x) \neq 0$. By the previous corollary, the pair (y, p) coincides with the solution of system (7.30) with initial conditions $y(t) = x$, $p(t) = \nabla V(t, x)$. We can repeat the argument for any optimal trajectory, finding that it must coincide with y. The dual arc associated with y is also unique, by Corollary 7.3.5.   ∎

Theorem 7.3.10 *Let properties* (H0)–(H5) *hold. Given a point* $(t, x) \in [0, T[\times \mathbb{R}^n$ *and a vector* $\bar{p} = (\bar{p}_t, \bar{p}_x) \in D^* V(t, x)$ *such that* $\bar{p} \neq 0$, *let us associate with* $\bar{p}$ *the pair* (y, p) *which solves system* (7.30) *with initial conditions* $y(t) = x$, $p(t) = \bar{p}_x$. *Then* y *is an optimal trajectory for* (t, x), p *is a dual arc associated with* y *and* $p(s) \in \nabla^* V(s, y(s))$ *for all* $s \in [t, T]$. *The map from* $D^* V(t, x)$ *to the set of optimal trajectories from* (t, x) *defined in this way is injective.*

Proof — We recall that every $(p_t, p_x) \in D^* V(t, x)$ satisfies the Hamilton–Jacobi equation $p_t + H(x, p_x) = 0$; therefore, p_t is uniquely determined by p_x and $(p_t, p_x) = (0, 0)$ if and only if $p_x = 0$, by the properties of H.

Let us now take $(t, x) \in [0, T[\times \mathbb{R}^n$ and a vector $\bar{p} = (\bar{p}_t, \bar{p}_x) \in D^* V(t, x)$ with $\bar{p}_x \neq 0$. Then there exists a sequence $\{(t_k, x_k)\}$ of points where V is differentiable, such that

$$\lim_{k \to \infty} (t_k, x_k) = (t, x), \qquad \lim_{k \to \infty} \nabla V(t_k, x_k) = \bar{p}_x.$$

Since $\bar{p}_x \neq 0$, we have $\nabla V(t_k, x_k) \neq 0$ for k large enough. By the previous theorem, there exists a unique trajectory $y_k : [t, T] \to \mathbb{R}^n$ optimal for (t_k, x_k) and it is obtained by solving (7.30) with initial conditions $y(t_k) = x_k$, $p(t_k) = \nabla V(t_k, x_k)$. By continuous dependence, the pairs (y_k, p_k) converge to (y, p) obtained solving (7.30) with the initial conditions $y(t) = x$, $p(t) = \bar{p}_x$. Since V is continuous we have

$$g(y(T)) = \lim_{k \to \infty} g(y_k(T)) = \lim_{k \to \infty} V(x_k, t) = V(x, t),$$

which means that y is optimal for (t, x). In addition, for any $s \in]t, T]$,

$$p(s) = \lim p_k(s) = \lim \nabla V(s, y_k(s)) \in \nabla^* V(s, y(s)).$$

In particular, this shows that p is the dual arc associated with y, with final value $p(T) = Dg(y(T))$.

Let us prove the injectivity of the map. Suppose that we have two elements $\bar{p}, \bar{q} \in D^* V(t, x)$ yielding two optimal pairs $(y, p), (\hat{y}, q)$ such that $y \equiv \hat{y}$. Since g is differentiable, we have $p(T) = Dg(y(T)) = Dg(\hat{y}(T)) = q(T)$. But then the pairs (y, p) and $(\hat{y}, q)$ satisfy system (7.30) with the same terminal conditions, and therefore coincide. In particular $\bar{p} = \bar{q}$, which proves that our map is injective.   ∎

In the previous theorem the assumption that $g \in C^1$ plays an important role and is part of the statement would be false otherwise. This can be easily seen from

Example 7.2.10(i), where the optimal trajectory is unique even at those points which are singular for V, and thus the map defined in the previous theorem is not injective.

Let us now consider the case where $0 \in D^*V(t, x)$.

Theorem 7.3.11 *Assume* (H0)–(H5) *and let* $(t, x) \in [0, T[\times \mathbb{R}^n$ *be such that* $0 \in D^*V(t, x)$. *Then there exists* $y : [t, T] \to \mathbb{R}^n$ *optimal trajectory for* (t, x) *such that* $Dg(y(T)) = 0$.

Proof — Since $0 \in D^*V(t, x)$, we can find a sequence $\{(t_k, x_k)\}$ such that V is differentiable at $\{(t_k, x_k)\}$ and

$$\lim_{k \to \infty} (t_k, x_k) = (t, x), \qquad \lim_{k \to \infty} DV(t_k, x_k) = 0.$$

Let y_k be an optimal trajectory for (t_k, x_k) and let p_k be an associated dual arc. By Theorem 7.1.6 we can assume, after possibly passing to a subsequence, that $y_k \to y$ uniformly, where y is an admissible trajectory for (t, x). Since

$$g(y(T)) = \lim_{k \to \infty} g(y_k(T)) = \lim_{k \to \infty} V(x_k, t_k) = V(x, t)$$

we see that y is optimal. In addition, we have $p_k(t_k) = \nabla V(t_k, x_k) \to 0$. Since the equation satisfied by p_k is linear, with a coefficient which is uniformly bounded by assumption (H1), this implies that $p_k(T) \to 0$ as well, and so $Dg(y(T)) = \lim Dg(y_k(T)) = \lim p_k(T) = 0$. ∎

Observe that the above result is weaker than the one in Theorem 7.3.10 because it only gives the existence of y without saying that y is the C^1 solution of an ordinary differential system with suitable initial conditions. However, this is in the nature of the problem we are considering: if we look back at Example 7.2.10(iii), we see that $\nabla V(t, x) = 0$ at all points with $t \le T - \pi$, but the optimal trajectories have a different structure at every point and there is no way of obtaining them only from the knowledge of the differential of V.

Corollary 7.3.12 *Assume* (H0)–(H5) *and suppose also that* $Dg(x) \ne 0$ *for all* $x \in \mathbb{R}^n$. *Then* $0 \notin D^*V(t, x)$ *for all* $(t, x) \in [0, T] \times \mathbb{R}^n$.

Corollary 7.3.13 *Assume* (H0)–(H5) *and suppose also that* $Dg(x) \ne 0$ *for all* $x \in \mathbb{R}^n$. *Then* V *is differentiable at* (t, x) *if and only if there exists a unique optimal trajectory starting at* (t, x).

Proof — We recall that V is semiconcave by Theorem 7.2.8 and therefore V is differentiable at (t, x) if and only if $D^*V(t, x)$ is a singleton. Thus, the statement follows from the previous corollary and from Theorems 7.3.9, 7.3.10. ∎

If the hypothesis $Dg \ne 0$ is not satisfied, we can prove the following weaker version of the previous corollary.

Theorem 7.3.14 *Assume properties* (H0)–(H5). *Then, for any* $(t, x) \in [0, T[\times \mathbb{R}^n$ *the following holds: if V is not differentiable at* (t, x), *then there exists more than one optimal trajectory for problem* (MP) *with initial point* (t, x). *On the other hand, if V is differentiable at* (t, x) *with nonzero gradient, there exists a unique optimal trajectory.*

Proof — Suppose that V is not differentiable at a point $(t, x) \in [0, T[\times \mathbb{R}^n$. Since V is semiconcave, we can find two distinct elements $\bar{p}, \tilde{p} \in D^*V(t, x)$. Then we can apply Theorems 7.3.10 or 7.3.11 to find two optimal trajectories $\bar{y}, \tilde{y}$ associated with $\bar{p}, \tilde{p}$. If $\bar{p}, \tilde{p}$ are both nonzero, the two trajectories are distinct by Theorem 7.3.10. They are distinct also if one of the two vectors, say for instance $\bar{p}$, is zero, since in that case we have $Dg(\bar{y}(T)) \neq 0$ and $Dg(\tilde{y}(T)) = 0$. This proves the first part of the theorem. The second follows from Theorem 7.3.9. ∎

The previous theorem leaves open the possibility that the optimal trajectory is not unique at a point where V is differentiable with zero gradient. Such a behavior can actually occur, as we have seen in Example 7.2.10(iii).

We conclude our discussion of the Mayer problem by giving two results about the propagation of singularities and the regularity along optimal trajectories. We will need the following further assumption on f:

(H6) for all $x \in \mathbb{R}^n$, $f(x, U)$ is not a singleton and, if $n > 1$, it has a C^1 boundary.

Observe that such a property implies that any normal cone to $f(x, U)$ is a half-line starting from 0.

Theorem 7.3.15 *Assume that properties* (H0)–(H6) *are satisfied. If V is not differentiable at a point* $(t, x) \in]0, T[\times \mathbb{R}^n$, *then there exists a Lipschitz singular arc for V starting from* (t, x) *as in Theorem 4.2.2.*

Proof — Let $(t, x) \in]0, T[\times \mathbb{R}^n$ be a singular point for V. We suppose that

$$\partial D^+V(t, x) = D^*V(t, x) \tag{7.31}$$

and we want to derive a contradiction. First observe that we have

$$-p_t + H(x, p_x) = 0, \qquad \forall (p_t, p_x) \in \partial D^+V(t, x), \tag{7.32}$$

since the same property holds for the elements of $D^*V(t, x)$. We claim that the dimension of $D^+V(t, x)$ is strictly less than $n + 1$. If it is not so, then $D^+V(t, x)$ has nonempty interior. Taking (p_t, p_x) in the interior of $D^+V(t, x)$, we have $\{\lambda : (\lambda, p_x) \in D^+V(t, x)\} = [\lambda_1, \lambda_2]$ with $\lambda_1 < \lambda_2$. But then (λ_1, p_x) and (λ_2, p_x) both belong to $\partial D^+V(t, x)$, in contradiction with (7.32). Hence, $D^+V(t, x)$ cannot have dimension $n + 1$. This implies that $D^+V(t, x) = \partial D^+V(t, x) = D^*V(t, x)$; in particular, (7.32) holds for all $(p_t, p_x) \in D^+V(t, x)$.

Arguing as in the proof of Theorem 7.2.6, we find that $\nabla^+V(t, x)$ is contained in some normal cone to the set $f(x, U)$. From assumption (H6) we deduce that $\nabla^+V(t, x)$ is either a singleton or a segment contained in a half-line starting from the origin. Since V is semiconcave and nondifferentiable at (t, x), $D^+V(t, x)$ is not a singleton. Then (7.32) implies that $\nabla^+V(t, x)$ is also not a singleton. Hence, we can find two nonzero elements $p_x, \tilde{p}_x \in \nabla^+V(t, x)$ such that $\tilde{p}_x = \lambda p_x$ for some $\lambda > 0, \lambda \neq 1$. Let us denote by (y, p) (resp. by $(\tilde{y}, \tilde{p})$) the solution of system (7.30) with initial conditions $y(t) = x$, $p(t) = p_x$ (resp. $\tilde{y}(t) = x$, $\tilde{p}(t) = \tilde{p}_x$). Since $D^+V(t, x) = D^*V(t, x)$, Theorem 7.3.10 implies that both y and $\tilde{y}$ are optimal trajectories for (t, x) and that $p, \tilde{p}$ are associated dual arcs. In particular, since g is differentiable, we have at the final time the relations

$$p(T) = Dg(y(T)), \qquad \tilde{p}(T) = Dg(\tilde{y}(T)). \tag{7.33}$$

On the other hand, since the equation for p is invariant under multiplication for a positive λ, we have that $\tilde{y} \equiv y$, $\tilde{p} \equiv \lambda p$. But this contradicts (7.33). The contradiction shows that our initial assumption (7.31) cannot be satisfied. Therefore $\partial D^+V(t, x) \setminus D^*V(t, x) \neq \emptyset$, and we can conclude by Theorem 4.2.2. ∎

Theorem 7.3.16 *Suppose that properties* (H0)–(H6) *are satisfied. Given* $(t, x) \in [0, T[\times \mathbb{R}^n$ *such that* $0 \notin D^*V(t, x)$, *let* $y : [t, T] \to \mathbb{R}^n$ *be an optimal trajectory for* (t, x). *Then* V *is differentiable at all points of the form* $(s, y(s))$, *with* $t < s \leq T$.

Proof — Since $0 \notin D^*V(t, x)$, it is easy to see that $0 \notin D^*V(s, y(s))$ for s in a right neighborhood of t. It suffices to show that V is differentiable at $(s, y(s))$ for these values of s; then V will be differentiable also at later times by Corollary 7.3.5.

Let us argue by contradiction and suppose that $D^+V(\bar{s}, y(\bar{s}))$ is not a singleton for some $\bar{s}$ chosen as above. Set for simplicity $\bar{x} = y(\bar{s})$. Then, $D^*V(\bar{s}, \bar{x})$ contains at least two elements. We recall that different elements of $D^*V(\bar{s}, \bar{x})$ have different space components, as a consequence of Theorem 7.2.4. Let p be the dual arc associated with y such that $p(T) = Dg(y(T))$. We know from Theorem 7.3.3 that $p(\bar{s}) \in \nabla^+V(\bar{s}, \bar{x})$. Since $D^*V(\bar{s}, \bar{x})$ is not a singleton, there is at least one vector $\hat{p} \in \nabla^+V(\bar{s}, \bar{x})$ which is different from $p(\bar{s})$ and is the space component of an element of $D^*V(\bar{s}, \bar{x})$. We have seen in the proof of Theorem 7.2.6 that $\nabla^+V(\bar{s}, \bar{x})$ is contained in some normal cone to the convex set $f(\bar{x}, U)$. Thus, we deduce from (H6) that $\nabla^+V(\bar{s}, \bar{x})$ is contained in a half-line starting from the origin; it cannot contain the origin by the assumption that $0 \notin D^*V(\bar{t}, \bar{x})$. It follows that $\hat{p}$ and $p(\bar{s})$ are both nonzero and satisfy $\hat{p} = \lambda p(\bar{s})$ for some $\lambda > 0, \lambda \neq 1$. If we set $p_1(s) = \lambda p(s)$ for all $s \in [\bar{s}, T]$, we have that the pair (y, p_1) solves system (7.30) with conditions $y(\bar{s}) = \bar{x}$, $p_1(\bar{s}) = \hat{p}$. By Theorem 7.3.10, p_1 is a dual arc associated with y in $[\bar{s}, T]$ and satisfies $p_1(T) = Dg(y(T))$. This is a contradiction, since $p_1(T) = \lambda p(T) \neq p(T) = Dg(y(T))$. Thus, V must be differentiable at $(\bar{s}, y(\bar{s}))$. ∎

Example 7.3.17 Let us show that the hypothesis that $0 \notin D^*V(t, x)$ in the previous theorem is essential. Let $n = 1$, $U = [-1, 1]$, $f(x, u) = u$. We take as our final cost a function g with the following properties:

(i) $g(x) = 0$ for all $x \le -2$
(ii) $g(x) = -x$ for all $x \ge 0$
(iii) $g \ge 0$ in $[-2, 0]$.

We assume that the values of g in $[-2, 0]$ are such that g is smooth in $\mathbb{R}$. We consider the Mayer problem (MP) with $T = 2$. Although we have not completely specified g, our requirements suffice to characterize $V(t, x)$ for points (t, x) with $t \in [0, 1]$. In fact, let us observe that the points reachable by an admissible trajectory starting at a given point (t, x) are those of the interval $[x - 2 + t, x + 2 - t]$. If $t \in [0, 1]$ and $x \le t - 2$ we have $x - 2 + t \le 2t - 4 \le -2$ and $x + 2 - t \le 0$, and so it is easily seen that the minimum of g on $[x - 2 + t, x + 2 - t]$ is 0. On the other hand, if $x \ge t - 2$, we find that the optimal trajectory is the rightmost one ending at $x + 2 - t$. It follows that, for $t \in [0, 1]$,

$$V(t, x) = \begin{cases} 0 & x \le t - 2 \\ t - 2 - x & x \ge t - 2. \end{cases}$$

Observe that the segment $y(s) = s - 2$ is an optimal trajectory for the point $(0, -2)$. However, the points of this trajectory are singular for all $s \in [0, 1]$ since $\nabla^+ V(s, y(s)) = [-1, 0]$ for all $s \in [0, 1]$. ∎

Corollary 7.3.18 *Assume properties* (H0)–(H6). *If* $(t, x) \in [0, T[\times \mathbb{R}^n$ *is such that* $0 \notin D^* V(t, x)$, *then the map defined in Theorem 7.3.10 gives a a one-to-one correspondence between* $D^* V(t, x)$ *and the optimal trajectories of problem* (MP) *with initial point* (t, x).

Proof — We only need to show that the map given by Theorem 7.3.10 is surjective. To see this, let y be any optimal trajectory for (t, x) and let p be a dual arc associated with y. Then, by Theorems 7.3.3 and 7.3.16, V is differentiable at $(s, y(s))$ and satisfies $DV(s, y(s)) = (H(p(s)), p(s))$ for all $s \in]t, T]$. It follows that $p(t) = \lim_{s \to t} p(s)$ is the space component of an element of $D^* V(t, x)$, and this proves our assertion. ∎

7.4 The Bolza problem

In this section we consider another kind of optimal control problem with finite time horizon. As in the Mayer problem, we are given a control system (f, U), a function $g \in C(\mathbb{R}^n)$ and a time $T > 0$; in addition, a function $L \in C(\mathbb{R}^n \times U)$ is assigned, called *running cost*. For any $(t, x) \in [0, T] \times \mathbb{R}^n$, we consider the functional

$$J_{t,x}(u) = \int_t^T L(y(s), u(s)) \, ds + g(y(T)), \quad \text{where } y(\cdot) = y(\cdot; t, x, u) \quad (7.34)$$

and the control problem

(BP) minimize $J_{t,x}(u)$ over all controls $u : [t, T] \to U$.

Such a control problem is called of *Bolza type*.

Remark 7.4.1 Clearly, the Mayer problem can be regarded as a special case of the Bolza problem, corresponding to the choice $L(x, u) \equiv 0$. On the other hand, it is also possible to transform a Bolza problem into a Mayer problem by adding an additional space variable. We can set in fact $Y = (y_0, y) \in \mathbb{R} \times \mathbb{R}^n$ and define a new control system on $\mathbb{R} \times \mathbb{R}^n$ whose state equation is $Y' = \tilde{f}(Y, u)$, where $\tilde{f} : \mathbb{R} \times \mathbb{R}^n \times U \to \mathbb{R} \times \mathbb{R}^n$ is defined by

$$\tilde{f}(Y, u) = (L(y, u), f(y, u)).$$

Then it is easily seen that an arc $y : [t, T] \to \mathbb{R}^n$ is optimal for the Bolza problem with initial point (t, x) if and only if the arc Y corresponding to the same control u is optimal for the Mayer problem with initial point $(t, (0, x))$ and final cost $\tilde{g} : \mathbb{R} \times \mathbb{R}^n \to \mathbb{R}$ given by $\tilde{g}(x_0, x) = x_0 + g(x)$.

Remark 7.4.2 Observe that the problem in the calculus of variations considered in the previous chapter is also a special case of the Bolza problem, where the state equation is $y' = u$ and the control set is $U = \mathbb{R}^n$. We are studying the problems under slightly different hypotheses (here we are taking $L(x, u)$ independent of t and we have a final cost rather than an initial one) but these differences are not substantial. Observe that a control set of the form $U = \mathbb{R}^n$ does not satisfy the compactness assumption (H0), which we are often requiring in this chapter; however, as we will see in Theorem 7.4.6 (see also Theorem 6.1.2(iii)) if L is coercive the problem behaves as if the control space were bounded.

We assume throughout that L satisfies the following.

(L1) For any $R > 0$ there exists γ_R such that $|L(x_2, u) - L(x_1, u)| \leq \gamma_R |x_2 - x_1|$, for all $x_1, x_2 \in B_R, u \in U$.

For some results we also need the following.

(L2) For any $x \in \mathbb{R}^n$, the following set is convex:

$$\mathcal{L}(x) := \{(\lambda, v) \in \mathbb{R}^{n+1} : \exists u \in U \text{ such that } v = f(x, u), \lambda \geq L(x, u)\}.$$

(L3) For any $R > 0$ there exists λ_R such that

$$L(x, u) + L(y, u) - 2L\left(\frac{x+y}{2}, u\right) \leq \lambda_R |x - y|^2, \qquad x, y \in B_R, u \in U.$$

Remark 7.4.3 Assumption (L2) is related to the lower semicontinuity of the integral term in the functional J and its role will be clear in Theorem 7.4.4. It is easy to see that (L2) implies the convexity of $f(x, U)$ but it is in general a stronger assumption. The two properties are equivalent if $L = L(x)$. If f is linear with respect to u, then (L2) is satisfied if U is convex and L is convex with respect to u. If f is invertible with respect to u and if $h_x : f(x, U) \to U$ denotes its inverse, then (L2) is satisfied if and only if, for all x, $f(x, U)$ is convex and $v \to L(x, h_x(v))$ is a convex function of $v \in f(x, U)$.

We now study the existence of optimal trajectories for the Bolza problem. The main step is contained in the next theorem, which is similar to the compactness Theorem 7.1.6, but describes in addition the behavior of an integral functional with respect to the uniform convergence of trajectories.

Theorem 7.4.4 *Assume that* (H0), (H1), (L1), (L2) *hold. Let* $\{y_k\}$ *be a sequence of trajectories of* (7.1) *in some given interval* $[t_0, t_1]$, *that is,* $y_k(\cdot) = y(\cdot; t_0, x_k, u_k)$ *for some* $x_k \in \mathbb{R}^n$ *and* $u_k : [t_0, t_1] \rightarrow U$. *If the trajectories* y_k *are uniformly bounded, then there exists a subsequence* $\{y_{k_h}\}$ *converging uniformly to an arc* $\bar{y}$ *which is a trajectory of* (7.1) *associated with some control* $\bar{u}$ *and satisfies*

$$\int_{t_0}^{t_1} L(\bar{y}(t), \bar{u}(t)) \, dt \leq \liminf_{k \to \infty} \int_{t_0}^{t_1} L(y_k(t), u_k(t)) \, dt. \tag{7.35}$$

Proof — By assumption all trajectories y_k are contained in some ball B_R for a suitable $R > 0$. Since (H0) holds, we can find M such that $L(x, u) \leq M$ for all $(x, u) \in B_R \times U$.

Let us first extract a subsequence such that the lim inf in (7.35) becomes a limit. Here and in the following we denote the subsequences by the same subscript k as the original sequence.

We now consider a control system $(\hat{U}, \hat{f})$ where the state space has dimension $n + 1$ and the control space has dimension $m + 1$. As a new control space we take $\hat{U} = U \times [0, 1]$ and we denote the controls by $\hat{u} = (u, u_0)$, with $u \in U$ and $u_0 \in [0, 1]$. We denote the points in $\mathbb{R}^{n+1}$ by $\hat{x} = (x, x_0)$, where $x \in \mathbb{R}^n$, $x_0 \in \mathbb{R}$. We consider the state equation

$$\begin{cases} y' = f(y, u) \\ y_0' = u_0 M + (1 - u_0) L(y, u) \end{cases} \tag{7.36}$$

corresponding to $\hat{f}(\hat{x}, \hat{u}) = (f(x, u), Mu_0 + (1 - u_0)L(x, u))$. Thus, the first n component of $\hat{y}$ evolve according to the original system.

Given any $\hat{x} = (x, x_0)$ such that $|x| < R$, we have $L(x, u) < M$ and therefore

$$\hat{f}(\hat{x}, \hat{U}) = \{(f(x, u), Mu_0 + (1 - u_0)L(x, u)) \; : \; u \in U, u_0 \in [0, 1]\}$$
$$= \{(f(x, u), z) \; : \; u \in U, L(x, u) \leq z \leq M\}$$
$$= \{(y, y_0) \in \mathcal{L}(x) \; : \; y_0 \leq M\},$$

where $\mathcal{L}(x)$ is the set introduced in (L2). We deduce that $\hat{f}(\hat{x}, \hat{U})$ is convex for all $\hat{x} = (x, x_0) \in B_R \times \mathbb{R}$, since is the intersection of convex sets.

Let us now denote by $\hat{y}_k$ the trajectories of (7.36) corresponding to the initial conditions $(x_k, 0)$ and to the controls $(u_k, 0)$. We obtain that $\hat{y}_k = (y_k, z_k)$, where the y_k's are our original trajectories and

$$z_k(t) = \int_{t_0}^{t} L(y_k(s), u_k(s)) \, ds, \qquad t \in [t_0, t_1].$$

Since the trajectories $\hat{y}_k$ are contained in $B_R \times \mathbb{R}$, we can apply Theorem 7.1.6 (see also Remark 7.1.7). We obtain that there exist $\bar{y}, \bar{z}$ such that $y_k \to \bar{y}, z_k \to \bar{z}$ uniformly in $[t_0, t_1]$. In addition, $(\bar{y}, \bar{z})$ is a trajectory of (7.36) and so there exist $\bar{u} : [t_0, t_1] \to U$ and $\bar{u}_0 : [t_0, t_1] \to [0, 1]$ such that

$$\bar{y}' = f(\bar{y}, \bar{u}), \qquad \bar{z}' = \bar{u}_0 M + (1 - \bar{u}_0) L(\bar{y}, \bar{u}).$$

The first equation says that $\bar{y}$ is a trajectory of the original system (7.1) corresponding to the control $\bar{u}$. The second equation implies that $\bar{z}' \geq L(\bar{y}, \bar{u})$ and therefore

$$\int_{t_0}^{t_1} L(\bar{y}(t), \bar{u}(t)) \, dt$$
$$\leq \int_{t_0}^{t_1} \bar{z}'(t) \, dt = \bar{z}(t_1) - \bar{z}(t_0) = \lim_{k \to \infty} [z_k(t_1) - z_k(t_0)]$$
$$= \lim_{k \to \infty} \int_{t_0}^{t_1} L(y_k(t), u_k(t)) \, dt.$$

Thus $\bar{y}$ satisfies the desired properties.   ∎

Theorem 7.4.5 *Assume that hypotheses* (H0), (H1), (L1), (L2) *hold and that g is continuous. Then, for any* $(t, x) \in [0, T] \times \mathbb{R}^n$, *there exists an optimal control for problem* (BP).

Proof — We follow the same method of Theorem 7.1.4. We consider a minimizing sequence $\{u_k\}$ and the associated trajectories $\{y_k\}$. We apply Theorem 7.4.4 to find that a subsequence of y_k converges to an admissible trajectory $\bar{y}$. Using (7.35) we obtain that $\bar{y}$ is optimal.   ∎

Let us now consider the case when the control set U is unbounded (for instance, $U = \mathbb{R}^m$) and thus assumption (H0) is not satisfied. We assume instead that the running cost L is coercive with respect to u. More precisely, we replace assumption (H0) with the following ones:

(H*) There exists K_0 such that

$$|f(x, u)| \leq K_0(1 + |x| + |u|), \qquad \forall x \in \mathbb{R}^n, \, u \in U.$$

(L*) There exists $l_0 \geq 0$ and a function $l : [0, \infty[\to [0, \infty[$ with $l(r)/r \to +\infty$ as $r \to +\infty$ and such that

$$L(x, u) \geq l(|u|) - l_0, \qquad \forall x \in \mathbb{R}^n, \, u \in U.$$

Theorem 7.4.6 *Let* (f, U) *be a control system satisfying* (H*), (H1). *Let* $T > 0$, *let* $g : \mathbb{R}^n \to \mathbb{R}$ *be locally Lipschitz and bounded from below and let* $L \in C(\mathbb{R}^n \times U)$ *satisfy* (L*), (L1). *Then, for any* $R > 0$ *there exists* $\mu_R > 0$ *with the following property: given* $(t, x) \in [0, T] \times B_R$, *if we set* $\mathcal{M}_R = \{u : [t, T] \to U : \|u\|_\infty \leq \mu_R\}$ *we have*

$$\inf_{u \in L^1([t, T], U)} J_{t,x}(u) = \inf_{u \in \mathcal{M}_R} J_{t,x}(u).$$

Proof — Let $\{u_k\} \subset L^1([t, T], U)$ be such that

$$\lim_{k \to \infty} J_{t,x}(u_k) = \inf_{u \in L^1([t,T],U)} J_{t,x}(u) := \lambda$$

and let us set $y_k = y(\cdot; t, x, u_k)$. We have

$$\int_t^T l(|u_k(s)|) \, ds \leq \int_t^T L(y_k(s), u_k(s)) \, ds + l_0 T$$
$$\leq J_{t,x}(u_k) + l_0 T - \inf g.$$

By (L*) there exists M_0 such that $l(r) > r$ for all $r > M_0$. Then we have, for k large enough,

$$\|u_k\|_1 \leq M_0 T + \int_{\{|u| > M_0\}} l(|u_k(s)|) \, ds$$
$$\leq M_0 T + \lambda + 1 + l_0 T - \inf g.$$

Thus, if we set $M = M_0 T + \lambda + 1 + l_0 T - \inf g$, we see that the infimum of $J_{t,x}(u)$ does not change if it is taken only over the controls u with $\|u\|_1 < M$.

Let us now consider any control $u \in L^1([t, T])$ satisfying the bound $\|u\|_1 \leq M$ and set for simplicity $y(\cdot) := y(\cdot; t, x, u)$. We find, using (H*),

$$|y(s)| \leq |x| + K_0 \int_t^s (1 + |u(\tau)| + |y(\tau)|) \, d\tau$$
$$\leq R + K_0(T + M) + \int_t^s |y(\tau)| \, d\tau, \qquad s \in [t, T].$$

Thus, using Gronwall's inequality,

$$|y(s)| \leq (R + K_0(T + M)) e^{K_0 T}, \qquad s \in [t, T]. \tag{7.37}$$

Let us set for simplicity $R^* = (R + K_0(T + M)) e^{K_0 T}$. We now define, for a given $\mu > 0$,

$$I_\mu = \{s \in [t, T] : |u(s)| > \mu\},$$
$$u_\mu(s) = \begin{cases} u(s) & \text{if } |u(s)| \leq \mu \\ 0 & \text{if } |u(s)| > \mu. \end{cases}$$

We assume for simplicity that $0 \in U$, so that u_μ is an admissible control; otherwise 0 can be replaced by any fixed element of U, with slight changes in the following computations. Our aim is to show that, since L is superlinear in u, the "truncated" control u_μ yields a lower value of the functional than the original control u if μ is chosen suitably large.

We set $y_\mu(\cdot) := y(\cdot; t, x, u_\mu)$. Then we have, taking into account (H1), (H*) and (7.37),

$$|y_\mu(s) - y(s)| \le \left| \int_t^T [f(y_\mu, u_\mu) - f(y, u)] \, d\tau \right|$$

$$\le \int_t^T |f(y_\mu, u_\mu) - f(y, u_\mu)| \, d\tau + \int_t^T |f(y, u_\mu) - f(y, u)| \, d\tau$$

$$\le K_1 \int_t^T |y_\mu(\tau) - y(\tau)| \, d\tau + K_0 \int_{I_\mu} [2 + 2|y(\tau)| + |u(\tau)|] \, d\tau$$

$$\le K_1 \int_t^T |y_\mu(\tau) - y(\tau)| \, d\tau + 2(1 + R^*) K_0 \int_{I_\mu} (1 + |u(\tau)|) \, d\tau.$$

By Gronwall's inequality

$$|y_\mu(s) - y(s)| \le \alpha_1 e^{K_1 s} \int_{I_\mu} (1 + |u(\tau)|) \, d\tau,$$

where $\alpha_1 = 2(1 + R^*) K_0$. Therefore, if we denote by K_g the Lipschitz constant of g over B_{R^*},

$$|g(y_\mu(T)) - g(y(T))| \le K_g |y_\mu(T) - y(T)|$$

$$\le K_g \alpha_1 e^{K_1 T} \left(|I_\mu| + \int_{I_\mu} |u(\tau)| \, d\tau \right). \tag{7.38}$$

Similarly, using (L1), (L*) and (7.37) we find

$$\int_t^T [L(y_\mu(s), u_\mu(s)) - L(y(s), u(s))] \, ds$$

$$\le \gamma_{R^*} \int_t^T |y_\mu(s) - y(s)| \, ds + \int_{I_\mu} [L(y(s), 0) - L(y(s), u(s))] \, ds$$

$$\le \gamma_{R^*} \alpha_1 T e^{K_1 T} \left(|I_\mu| + \int_{I_\mu} |u(\tau)| \, d\tau \right)$$

$$+ \alpha_2 |I_\mu| - \int_{I_\mu} l(|u(s)|) \, ds, \tag{7.39}$$

with $\alpha_2 = L(0, 0) + R^* \gamma_{R^*} + l_0$. Let us assume $\mu \ge 1$, so that $|I_\mu| \le \int_{I_\mu} |u|$. Then we obtain from (7.38) and (7.39)

$$J_{t,x}(u_\mu) - J_{t,x}(u) \le \int_{I_\mu} [\alpha_3 |u(s)| - l(|u(s)|)] \, ds$$

with $\alpha_3 = 2(\gamma_{R^*} T + K_g) \alpha_1 e^{K_1 T} + \alpha_2$. Let us now choose $\mu_R > 1$ in such a way that $l(r) \ge \alpha_3 r$ for all $r \ge \mu_R$. Then $J_{t,x}(u_{\mu_R}) \le J_{t,x}(u)$ for all controls u with $\|u\|_1 \le M$. Since the bound $\|u\|_1 \le M$ is satisfied along a minimizing sequence, we have proved that the infimum of the functional remains the same if the functional is restricted over controls u with $\|u\|_\infty \le \mu_R$. ∎

The previous result shows that the control problem under consideration is equivalent to one with compact control space. Thus, we can apply Theorem 7.4.5 and obtain the following existence result.

Corollary 7.4.7 *Let the hypotheses of the previous theorem be satisfied, and let* (L2) *also hold. Then there exists an optimal control for problem* (BP) *for any initial condition* $(t, x) \in [0, T] \times \mathbb{R}^n$.

Example 7.4.8 Let us consider the linear control system (f, U) with $U = \mathbb{R}^m$ and $f(x, u) = Ax + Bu$, where A, B are $n \times n$ and $n \times m$ matrices. Given $T > 0$, we want to minimize the functional

$$J(y) = \int_t^T \{\langle My(s), y(s)\rangle + \langle Nu(s), u(s)\rangle\} \, ds + \langle Py(T), y(T)\rangle$$

over all trajectories $y(\cdot) = y(\cdot; t, x, u)$ starting from a given point $(t, x) \in [0, T] \times \mathbb{R}^n$. Here M, P and N are $n \times n$ and $m \times m$ symmetric matrices, respectively. This is a Bolza problem, corresponding to the running cost $L(x, u) = \langle Mx, x\rangle + \langle Nu, u\rangle$ and final cost $g(x) = \langle Px, x\rangle$. It is a classical problem in control theory and is called the *linear-quadratic regulator*. If M, P are nonnegative definite and N is positive definite, then this problem satisfies the assumptions of the previous corollary.

Let us introduce the value function for the Bolza problem.

Definition 7.4.9 *Given* $(t, x) \in [0, T] \times \mathbb{R}^n$, *we define*

$$V(t, x) = \inf\{J_{t,x}(u) \ : \ u : [t, T] \to U \text{ measurable }\}.$$

The function V is called the value function *of the control problem* (BP).

As in Theorem 7.2.2 one can prove that the value function satisfies the *dynamic programming principle*: for any given $(t, x) \in {]0, T[}$ and any $s \in [t, T]$ we have

$$V(t, x) = \inf_{u:[t,s] \to U} V(s, y(s; t, x, u)) + \int_t^s L(y(\tau; t, x, u), u(\tau)) \, d\tau. \quad (7.40)$$

As in the Mayer problem, the value function is Lipschitz continuous and semiconcave under suitable assumptions on the data, as the following results show.

Theorem 7.4.10 *Let our control system satisfy assumptions* (H0), (H1), (L1) *and let* $g \in \text{Lip}_{loc}(\mathbb{R}^n)$. *Then* $V \in \text{Lip}_{loc}([0, T] \times \mathbb{R}^n)$.

Proof — It is possible to repeat the proof of Theorem 7.2.3 with some slight modifications in the estimates due to additional terms containing the running cost L. We use instead the transformation of a Bolza problem into a Mayer problem, which allows us to deduce the result directly from Theorem 7.2.3.

As described in Remark 7.4.1, we transform our original problem into a problem in $\mathbb{R}^{n+1}$. We denote the generic point of $\mathbb{R}^{n+1}$ by (x_0, x), with $x_0 \in \mathbb{R}$ and $x \in \mathbb{R}^n$. The dynamics is given by $\tilde{f}(x_0, x, u) = (L(x, u), f(x, u))$, while the final cost is

$\tilde{g}(x_0, x) = x_0 + g(x)$. It is easy to check that, if we denote by $\tilde{V}(t, x_0, x)$ the value function of the Mayer problem for system $(\tilde{f}, U)$ with final cost $\tilde{g}$, then

$$\tilde{V}(t, x_0, x) = x_0 + V(t, x), \qquad t \in [0, T], \ (x_0, x) \in \mathbb{R}^{n+1}, \tag{7.41}$$

where V is the value function of our original Bolza problem.

Using our assumptions we easily find that system $(\tilde{f}, U)$ satisfies (H0) and that $\tilde{g} \in \mathrm{Lip}_{loc}(\mathbb{R}^{n+1})$. Assumption (H1) may not be satisfied since the Lipschitz requirement on L in (L1) is only local. However, taking into account the a priori bound on the trajectories (7.3), we see that the value function in a given compact set does not change if we modify $L(x, u)$ for large x in order to have global Lipschitz continuity. Thus, we can assume that the system $(\tilde{f}, U)$ satisfies (H1) as well. We then apply Theorem 7.2.3 to this system with final cost $\tilde{g}$. We obtain that $\tilde{V}$ is locally Lipschitz and so, by (7.41), the same holds for V. ∎

Theorem 7.4.11 *Let assumptions* (H0), (H1), (H2), (L1), (L3) *be satisfied and let* $g \in \mathrm{SCL}_{loc}(\mathbb{R}^n)$. *Then* $V \in \mathrm{SCL}_{loc}([0, T] \times \mathbb{R}^n)$.

Proof — It is not convenient to use the transformation of the previous theorem, since the resulting system would not satisfy (H2). However, the result can be proved by a procedure which is very similar to the one of Theorem 7.2.8 (see also Theorem 6.4.1).

Let $r > 0$ be fixed and let $R > 0$ be such that all trajectories starting from B_r at a time $t \in [0, T]$ stay inside B_R in $[t, T]$. Let us first consider a triple of points with the same time coordinate. Given x, h such that $x \pm h \in B_r$ and given $t \in [0, T]$, let $u : [t, T] \to U$ be an optimal control for the point (t, x). Let us set

$$y(\cdot) = y(\cdot; t, x, u), \qquad y_-(\cdot) = y(\cdot; t, x - h, u), \qquad y_+(\cdot) = y(\cdot; t, x + h, u).$$

Then, by Lemma 7.1.2, we have

$$|y_+(s) - y_-(s)| \leq c|h|, \qquad |y_+(s) + y_-(s) - 2y(s)| \leq c|h|^2 \tag{7.42}$$

for some constant $c > 0$. It follows, using the definition of V and the optimality of u, that

$$V(t, x + h) + V(t, x - h) - 2V(t, x)$$
$$\leq g(y_+(T)) + g(y_-(T)) - 2g(y(T)) \tag{7.43}$$
$$+ \int_t^T (L(y_+(s), u(s)) + L(y_-(s), u(s)) - 2L(y(s), u(s))) \, ds.$$

As in (7.18) we find that

$$g(y_+(T)) + g(y_-(T)) - 2g(y(T)) \leq C|h|^2.$$

Here and in the following we denote by C any constant depending only on r. We also find, by (L1) and (L3),

$$L(y_+(s), u(s)) + L(y_-(s), u(s)) - 2L(y(s), u(s))$$

$$= L(y_+(s), u(s)) + L(y_-(s), u(s)) - 2L\left(\frac{y_-(s) + y_+(s)}{2}, u(s)\right)$$

$$+ 2L\left(\frac{y_-(s) + y_+(s)}{2}, u(s)\right) - 2L(y(s), u(s))$$

$$\leq \lambda_R |y_+(s) - y_-(s)|^2 + \gamma_R |y_+(s) - y_-(s) - 2y(s)|$$

which is also estimated by $C|h|^2$, by virtue of (7.42). Thus (7.43) implies that

$$V(t, x + h) + V(t, x - h) - 2V(t, x) \leq C|h|^2 \qquad (7.44)$$

which proves the semiconcavity inequality in this case.

Let us now consider x, h and t, τ such that $x \pm h \in B_r$ and such that $0 \leq t - \tau \leq t + \tau \leq T$. We take a control $u : [t, T] \to U$ optimal for (t, x) and define

$$\bar{u}(s) = u\left(\frac{t + \tau + s}{2}\right), \qquad s \in [t - \tau, t + \tau],$$

$$y(s) = y(s; t, x, u), \qquad s \in [t, t + \tau],$$

$$\bar{y}(s) = y(s; t - \tau, x - h, \bar{u}), \qquad s \in [t - \tau, t + \tau],$$

$$\bar{x} = \bar{y}(t + \tau), \qquad \hat{x} = y(t + \tau).$$

By the dynamic programming principle

$$V(t + \tau, x + h) + V(t - \tau, x - h) - 2V(t, x)$$

$$\leq V(t + \tau, x + h) + V(t + \tau, \bar{x}) - 2V(t + \tau, \hat{x}) \qquad (7.45)$$

$$+ \int_{t-\tau}^{t+\tau} L(\bar{y}(s), \bar{u}(s)) \, ds - 2 \int_{t}^{t+\tau} L(y(s), u(s)) \, ds.$$

We can proceed as in the proof of Theorem 7.2.8 and obtain from (7.44) that

$$V(t + \tau, x + h) + V(t + \tau, \bar{x}) - 2V(t + \tau, \hat{x}) \leq C(|h|^2 + \tau^2).$$

In addition, as in (7.19), we have that $|\bar{y}(s_1) - y(s_2)| \leq C(|h| + \tau)$ for all $s_1 \in [t - \tau, t + \tau]$, $s_2 \in [t, t + \tau]$. Thus we can estimate the last term in (7.45) using (L1) to find that

$$\int_{t-\tau}^{t+\tau} L(\bar{y}(s), \bar{u}(s)) \, ds - 2 \int_{t}^{t+\tau} L(y(s), u(s)) \, ds$$

$$= 2 \int_{t}^{t+\tau} [L(\bar{y}(2s - t - \tau), u(s)) - L(y(s), u(s))] \, ds$$

$$\leq 2\tau \gamma_R C(|h| + \tau) \leq C'(|h|^2 + \tau^2).$$

Gathering the above estimates, we recover the desired semiconcavity inequality for V.   ∎

Analogous results hold when the control space is not necessarily compact but the running cost is superlinear with respect to u.

Theorem 7.4.12 *Let* (f, U) *be a control system satisfying* (H*) *and* (H1).

(i) *Let* $g : \mathbb{R}^n \to \mathbb{R}$ *be locally Lipschitz and bounded from below, and let* $L \in C(\mathbb{R}^n \times U)$ *satisfy* (L1) *and* (L*). *Then the value function* V *of problem* (BP) *is locally Lipschitz.*

(ii) *Suppose, in addition, that* $g \in \mathrm{SCL}_{loc}(\mathbb{R}^n)$, *f satisfies* (H2) *and* L *satisfies the following: for any* $R > 0$, *there exists* λ_R *such that*

$$L(x, u) + L(y, u) - L\left(\frac{x+y}{2}, u\right) \le \lambda_R |x - y|^2, \quad x, y \in B_R, \ u \in U \cap B_R.$$

Then $V \in \mathrm{SCL}_{loc}([0, T] \times \mathbb{R}^n)$.

Proof — Let us take any $r > 0$; by Theorem 7.4.6 there exists $\mu > 0$ such that for any $(t, x) \in [0, T] \times B_r$, the value of $V(t, x)$ does not change if we replace the control set U by $U \cap B_\mu$. Then the theorem is an immediate consequence of the previous regularity results obtained in the case of U compact. ∎

As for the Mayer problem, one can prove a semiconvexity result for linear systems with convex cost.

Theorem 7.4.13 *Consider a control system* (f, U) *with* U *convex (possibly unbounded) and* $f(x, u) = Ax + Bu$ *with* A, B *matrices. Let* $L : \mathbb{R}^n \times U \to \mathbb{R}$ *and* $g : \mathbb{R}^n \to \mathbb{R}$ *be convex functions. Suppose in addition that* g *is bounded from below and that* L *satisfies* (L*). *Then the value function* V *of the Bolza problem associated with the system is convex with respect to* x *and locally semiconvex with a linear modulus with respect to* (t, x). *If in addition* $g \in C_{loc}^{1,1}$ *and* $L(\cdot, u) \in C_{loc}^{1,1}$ *for all* u, *then* $V \in C_{loc}^{1,1}$.

Proof — We follow the same procedure of the corresponding result for the Mayer problem (Theorem 7.2.11), so we only show how to estimate the additional terms coming from the presence of the running cost L. In the first step, when we estimate the quantity $2V(t, x) - V(t, x - h) - V(t, x + h)$ we find the same terms as in the Mayer problem plus an integral term involving L which has the form

$$\int_t^T \left[2L\left(\frac{y_-(s) + y_+(s)}{2}, \frac{u_-(s) + u_+(s)}{2}\right) \right.$$
$$\left. - L(y_-(s), u_-(s)) - L(y_+(s), u_+(s)) \right] ds$$

which is nonpositive, since L is convex. Thus we conclude also in this case that $V(t, \cdot)$ is a convex function of x.

In the second step, when we estimate $2V(t, x) - V(t - \tau, x - h) - V(t + \tau, x + h)$, the additional term has the form

$$\Lambda := 2 \int_t^{t+\tau} L(\bar{y}(s), u_-(2s - t - \tau)) \, ds - \int_{t-\tau}^{t+\tau} L(y_-(s), u_-(s)) \, ds.$$

We now rescale time in the first integral and use the local Lipschitz continuity of L (which follows from the convexity) to obtain

$$
\begin{aligned}
\Lambda &= \int_{t-\tau}^{t+\tau} \left[L\!\left(\bar{y}\left(\frac{s+t+\tau}{2}\right), u_-(s)\right) - L(y_-(s), u_-(s)) \right] ds \\
&\leq K \int_{t-\tau}^{t+\tau} \left| y\left(\frac{s+t+\tau}{2}\right) - y_-(s) \right| ds \\
&\leq 2K\tau C(|h| + \tau) \leq C'(|h|^2 + \tau^2),
\end{aligned}
$$

and so V is semiconvex with a linear modulus.

If $g, L(\cdot, u) \in C_{loc}^{1,1}$, then they belong to SCL $_{loc}$. We can apply Theorem 7.4.12 to find that V belongs to SCL $_{loc}$, in addition to being convex, and therefore is in $C_{loc}^{1,1}$.

∎

Observe that the linear quadratic regulator considered in Example 7.4.8 satisfies all assumptions of the last part of previous theorem and so the corresponding value function is locally of class $C^{1,1}$.

The value function of a Bolza problem also satisfies a suitable Hamilton–Jacobi equation. The hamiltonian function in this case is defined as

$$
H(x, p) = \max_{u \in U} [-p \cdot f(x, u) - L(x, u)]. \tag{7.46}
$$

Then we have the following result.

Theorem 7.4.14 *Under the hypotheses of Theorem 7.4.10 or 7.4.12(i), the value function V is a viscosity solution of the problem*

$$
\begin{cases}
-\partial_t V(t, x) + H(x, \nabla V(t, x)) = 0, & (t, x) \in \,]0, T[\, \times \mathbb{R}^n \\
V(T, x) = g(x) & x \in \mathbb{R}^n.
\end{cases} \tag{7.47}
$$

Proof — Similar to Theorem 7.2.4 (see also Theorem 6.4.5). ∎

Definition (7.46) can be regarded as a generalization of the Legendre–Fenchel transform which defines the hamiltonian in the calculus of variations. It is easy to see that H is convex in p and locally Lipschitz continuous in x. We have seen for the Mayer problem that the analysis of the optimality conditions is more difficult if the hamiltonian is not differentiable or not strictly convex. For the Bolza problem the properties of the hamiltonian can be very different depending on the properties of f and L. For instance, if $L = L(x)$, then the function L does not influence the minimization in (7.46), and so the behavior of H with respect to p is the same as in the Mayer problem. On the other hand, the presence of a suitable $L = L(x, u)$ can improve the properties of the hamiltonian, and in some cases H can be both differentiable and strictly convex, at least for $|p|$ small.

Example 7.4.15 Consider again the linear quadratic regulator (Example 7.4.8). If the matrix N is positive definite, then an easy computation shows that the Hamiltonian function is

$$H(x, p) = -\langle Ax, p \rangle - \langle Mx, x \rangle + \frac{1}{4}\langle BN^{-1}B^T p, p \rangle.$$

Therefore H is smooth; if $BN^{-1}B^T$ is positive definite, then H is also strictly convex.

Theorem 7.4.16 *Let f, g, L satisfy the assumptions of Theorem 7.4.11 or 7.4.12(ii). Suppose in addition that, for all $x \in \mathbb{R}^n$, $H(x, \cdot)$ is strictly convex. Let $y : [t, T] \to \mathbb{R}^n$ be an optimal trajectory for a point $(t, x) \in [0, T] \times \mathbb{R}^n$. Then V is differentiable at $(\tau, y(\tau))$ for all $\tau \in]t, T[$.*

Proof — As in the case of a Mayer problem (see Theorem 7.2.5) we can prove that if $y : [t, T] \to \mathbb{R}^n$ is an optimal trajectory, then we have

$$-p_t + H(y(\tau), p_x) = 0, \qquad \forall \tau \in]t, T[, \ (p_t, p_x) \in D^+V(\tau, y(\tau)).$$

Since $H(y(\tau), \cdot)$ is strictly convex and $D^+V(\tau, y(\tau))$ is a convex set, the above equality implies that $D^+V(\tau, y(\tau))$ is a singleton. Since V is semiconcave, this implies that V is differentiable at $(\tau, y(\tau))$. ∎

We now give the maximum principle and the co-state inclusion for the Bolza problem. Here and in the rest of the section we consider the case where the control set U is compact; all results have straightforward extensions in the case where assumption (H0) is replaced by (H*) and (L*).

Theorem 7.4.17 *Let f, L satisfy hypotheses (H0), (H1), (L1) and let $g \in C(\mathbb{R}^n)$. Suppose in addition that f_x, L_x exist and are continuous with respect to x. Given $(t, x) \in [0, T] \times \mathbb{R}^n$, let $u : [t, T] \to U$ be an optimal control for problem (BP) with initial point (t, x) and let $y(\cdot) = y(\cdot; t, x, u)$ be the corresponding optimal trajectory. For a given $q \in D^+g(y(T))$, let $p : [t, T] \to \mathbb{R}^n$ be the solution of the equation*

$$\begin{cases} p'(s) = -f_x^T(y(s), u(s)) \, p(s) - L_x(s, y(s), u(s)), & s \in [t, T] \ a.e. \\ p(T) = q. \end{cases} \tag{7.48}$$

Then, $p(s)$ satisfies, for $s \in [t, T]$ a.e.,

$$-f(y(s), u(s)) \cdot p(s) - L(y(s), u(s)) \geq -f(y(s), v) \cdot p(s) - L(y(s), v)$$

for all $v \in U$. In addition,

$$p(s) \in \nabla^+V(s, y(s)) \qquad \forall s \in [t, T].$$

Proof — The strategy of proof is similar to the one used for the Mayer problem. We fix $\bar{s} \in]t, T[$ which is a Lebesgue point for the functions $s \to f(y(s), u(s))$ and $s \to L(y(s), u(s))$. We then take an arbitrary $v \in U$ and define, for $\varepsilon > 0$ small, the perturbed controls

$$u_\varepsilon(s) = \begin{cases} u(s) & s \in [t, T] \setminus [\bar{s} - \varepsilon, \bar{s}] \\ v & s \in [\bar{s} - \varepsilon, \bar{s}] \end{cases} \tag{7.49}$$

and set $y_\varepsilon(s) = y(s; t, x, u_\varepsilon)$ and $\bar{x} = y(\bar{s})$. As in the proof of Theorem 7.3.1, we find

$$y_\varepsilon(s) = y(s) + \varepsilon w(s) + o(\varepsilon), \qquad s \geq \bar{s}, \tag{7.50}$$

where w is the solution of the linearized problem

$$\begin{cases} w'(s) = f_x(y(s), u(s)) \, w(s) & s \geq \bar{s} \\ w(\bar{s}) = f(\bar{x}, v) - f(\bar{x}, u(\bar{s})). \end{cases}$$

If we take $q \in D^+ g(y(T))$ and define p by solving (7.48), we obtain

$$\frac{d}{ds} \langle p(s), w(s) \rangle = -\langle L_x(y(s), u(s)), w(s) \rangle, \tag{7.51}$$

and therefore

$$\int_{\bar{s}}^T \langle L_x(y(s), u(s)), \, w(s) \rangle \, ds = p(\bar{s}) \cdot w(\bar{s}) - p(T) \cdot w(T)$$

$$= p(\bar{s}) \cdot [f(\bar{x}, v) - f(\bar{x}, u(\bar{s}))] - q \cdot w(T).$$

Therefore, using the fact that $\bar{s}$ is a Lebesgue point for L,

$$\int_t^T [L(y_\varepsilon(s), u_\varepsilon(s)) - L(y(s), u(s))] \, ds$$

$$= \int_{\bar{s}-\varepsilon}^{\bar{s}} [L(y_\varepsilon(s), v) - L(y(s), u(s))] \, ds$$

$$+ \int_{\bar{s}}^T [L(y_\varepsilon(s), u(s)) - L(y(s), u(s))] \, ds$$

$$= \varepsilon[L(\bar{x}, v) - L(\bar{x}, u(\bar{s}))] + \varepsilon \int_{\bar{s}}^T \langle L_x(y(s), u(s)), \, w(s) \rangle \, ds + o(\varepsilon)$$

$$= \varepsilon[L(\bar{x}, v) - L(\bar{x}, u(\bar{s})) + p(\bar{s}) \cdot [f(\bar{x}, v) - f(\bar{x}, u(\bar{s}))] - q \cdot w(T)].$$

Now since $q \in D^+ g(y(T))$ we have

$$g(y_\varepsilon(T)) - g(y(T)) \leq q \cdot (y_\varepsilon(T) - y(T)) = \varepsilon q \cdot w(T) + o(\varepsilon).$$

Thus we conclude, using the optimality of u,

$$0 \leq \int_t^T [L(y_\varepsilon(s), u_\varepsilon(s)) - L(y(s), u(s))]\, ds + g(y_\varepsilon(T)) - g(y(T))$$
$$\leq \varepsilon[L(\bar{x}, v) - L(y(\bar{s}), u(\bar{s}))] + \varepsilon p(\bar{s}) \cdot [f(\bar{x}, v) - f(\bar{x}, u(\bar{s}))] + o(\varepsilon).$$

Since this holds for all $v \in U$ and for $\bar{s} \in \,]t, T[$ a.e., the first assertion is proved.

To prove the inclusion in the superdifferential, let us fix any $h \in \mathbb{R}^n$, $|h| = 1$ and let us set, for $\varepsilon > 0$ $y_\varepsilon(\cdot) = y(\cdot; t, x + \varepsilon h, u)$. Then

$$y_\varepsilon(s) = y(s) + \varepsilon w(s) + o(\varepsilon),$$

where w is the solution of the linearized problem

$$\begin{cases} w'(s) = f_x(y(s), u(s))w(s) & s \geq t \\ w(t) = h. \end{cases}$$

With computations similar to the first part of the proof we find that

$$\int_t^T [L(y_\varepsilon(s), u(s)) - L(y(s), u(s))]\, ds$$
$$= \varepsilon \int_t^T \langle L_x(y(s), u(s)), w(s)\rangle\, ds + o(\varepsilon)$$
$$= \varepsilon\, [p(t) \cdot h - q \cdot w(T)] + o(\varepsilon).$$

Since $q \in D^+ g(y(T))$ we have also in this case

$$g(y_\varepsilon(T)) - g(y(T)) \leq q \cdot (y_\varepsilon(T)) - y(T)) = \varepsilon q \cdot w(T) + o(\varepsilon),$$

and so we conclude, by the optimality of u, that

$$V(t, x + \varepsilon h) - V(t, x)$$
$$\leq \int_t^T [L(y_\varepsilon(s), u(s)) - L(y(s), u(s))]\, ds + g(y_\varepsilon(T)) - g(y(T))$$
$$\leq \varepsilon p(t) \cdot h + o(\varepsilon).$$

Such an estimate holds for any unit vector h, with $o(\varepsilon)$ independent of h. Thus we have proved that $p(t) \in \nabla^+ V(t, x)$. The proof that $p(s) \in \nabla^+ V(s, y(s))$ for a general $s \in [t, T]$ is entirely analogous. ∎

Corollary 7.4.18 *Let the hypotheses of the previous theorem be satisfied, and suppose in addition that $H \in C^{1,1}_{loc}(\mathbb{R}^n \times \mathbb{R}^n)$. Let (u, y) be an optimal pair for the point $(t, x) \in [0, T] \times \mathbb{R}^n$ and let $p : [t, T] \to \mathbb{R}^n$ be a dual arc associated with (u, y). Then (y, p) solves the system*

$$\begin{cases} y'(s) = H_x(y(s), p(s)) \\ p'(s) = -H_p(y(s), p(s)) \end{cases} \qquad s \in [t, T]. \qquad (7.52)$$

As a consequence, y, p are of class C^1.

Proof — We are assuming that H is smooth; from Theorem 3.4.4 on marginal functions we deduce that the derivatives of H are given by

$$H_x(x, p) = -f_x^T(x, u^*(x, p))p - L_x(x, u^*(x, p)),$$
$$H_p(x, p) = -f(x, u^*(x, p))$$

(7.53)

where $u^*(x, p)$ is any element of U such that

$$-f(x, u^*) \cdot p - L(x, u^*) = \max_{u \in U} -f(x, u) \cdot p - L(x, u).$$

Then the assertion follows from the maximum principle. ∎

Example 7.4.19 Unlike the Mayer problem, one can give nontrivial examples of Bolza problems where the hamiltonian H is everywhere differentiable, as required in the above corollary. This was the case in the linear quadratic regulator (see Examples 7.4.8 and 7.4.15). We give here another such example, where the control space is bounded. Let $U = \bar{B}_r$ and let

$$f(x, u) = \sigma(x)u + h(x),$$
$$L(x, u) = l(x) + \frac{1}{2}|u|^2,$$

where $\sigma(\cdot)$ is a matrix with $C^{1,1}$ entries, h and l are of class $C^{1,1}$. It is easy to see that for all $(x, p) \in \mathbb{R}^n \times \mathbb{R}^n$ there exists a unique $u^* = u^*(x, p)$ such that the infimum in (7.46) is attained, given by

$$u^*(x, p) = \begin{cases} -\sigma^T(x)p & \text{if} & |\sigma^T(x)p| \leq r \\ -r\dfrac{\sigma^T(x)p}{|\sigma^T(x)p|} & \text{if} & |\sigma^T(x)p| > r. \end{cases}$$

Thus the hamiltonian is given by

$$H(x, p) = \begin{cases} \dfrac{|\sigma^T(x)p|^2}{2} - l(x) - h(x) \cdot p & \text{if} \quad |\sigma^T(x)p| \leq r \\ r|\sigma^T(x)p| - \dfrac{r^2}{2} - l(x) - h(x) \cdot p & \text{if} \quad |\sigma^T(x)p| > r. \end{cases}$$

It is easily checked that $H \in C_{loc}^{1,1}(\mathbb{R}^n \times \mathbb{R}^n)$.

Theorem 7.4.20 *Suppose that hypotheses* (H0)–(H2) *and* (L1)–(L3) *are satisfied. Assume in addition that* $g \in SCL_{loc}(\mathbb{R}^n)$, *that* L_x *exists and is continuous with respect to* x, *that* $H \in C_{loc}^{1,1}(\mathbb{R}^n \times \mathbb{R}^n)$ *and is strictly convex. Given* $(t, x) \in [0, T[\times \mathbb{R}^n$ *and* $\bar{p} = (\bar{p}_t, \bar{p}_x) \in D^*V(t, x)$, *let us associate with* $\bar{p}$ *the pair* (y, p) *which solves system* (7.52) *with initial conditions* $y(t) = x$, $p(t) = \bar{p}_x$. *Then* y *is an optimal trajectory for* (t, x). *The map from* $D^*V(t, x)$ *to the set of optimal trajectories defined in this way is one-to-one.*

We skip the proof, which is analogous to the one of Theorem 6.4.9. Observe that since we are assuming here that H is differentiable everywhere, we do need to treat separately the case where $0 \in D^*V(t, x)$, as we did for the Mayer problem.

Bibliographical notes

For a general introduction to optimal control theory the reader may consult for instance [111, 26, 54, 53, 28], while a detailed treatment of the dynamic programming approach can be found in [110, 20] and in the references given after Chapter 1.

Most of the results of Sections 7.1 and 7.2 are classical in control theory and in dynamic programming. Let us mention that the existence result Theorem 7.1.4 is due to Filippov [77]. The semiconcavity of the value function in the case of a Mayer problem (see Theorem 7.2.8) was first studied in [33].

The maximum principle (Theorem 7.3.1) is a classical result in control theory and is originally due to Pontryagin (see e.g., [118]). There are versions of this result in more general settings, e.g., in the presence of constraints on the trajectories or on the endpoints. The property that the dual arc is related to the gradient of the value function in some suitable sense is also well known (see e.g., [57]). The other optimality conditions in Section 7.3 and the results about the propagation of singularities are taken from [33, 47, 4, 129]. Further results about the singular set are given in [2]. The results of Section 7.4 are mainly an adaptation to the Bolza problems of techniques of [46, 33]. A semiconcavity result for the Bolza problem under more general hypotheses can be found in [51]. Many of the results of this chapter have also been extended to some infinite dimensional control systems in [31, 34, 35, 4].

We have not considered here control problems with state constraints; for such problems, semiconcavity is harder to prove and cannot be expected in such general hypotheses as in the cases we have treated here. A semiconcavity result for a particular infinite horizon problem with space constraints has been obtained in [40].

Let us mention here some important topics related to control theory that are not treated in this book. In the study of stochastic optimal control problems one can also introduce the dynamic programming approach (see [80, 81]); the value function in this case satisfies a Hamilton–Jacobi equation of second order, usually degenerate elliptic or parabolic. Some semiconcavity results are available also in this case (see e.g., [95]), but the understanding of the regularity of the solution is far from being as good as in the first order case. Another interesting case is the one of differential games (see [20] and the references therein), where the dynamic programming approach leads to first-order Hamilton–Jacobi equations where the hamiltonian is nonconvex. Here virtually nothing is known about the structure of singularities of the solution. Without convexity of the hamiltonian, semiconcavity cannot be expected, so the analysis should be focused on weaker regularity properties.

8

Control Problems with Exit Time

The control problems considered in this chapter are called with *exit time* because the terminal time of the trajectories is not fixed, but it is the first one at which they reach a given target set. A typical example is the *minimum time problem*, where one wants to steer a point to the target in minimal time. It is interesting to observe that the distance function can be regarded as the value function of a particular minimum time problem, and so the properties of the distance function may serve as a guideline for the analysis of the general case.

Our analysis of exit time problems will be similar to that of the previous chapters about finite horizon problems. We first study the existence of optimal controls; then we introduce the value function, give results about its Lipschitz continuity and semiconcavity, analyze the optimality conditions and the properties of optimal trajectories. The variable terminal time introduces some additional difficulties; another new feature of these problems is given by the presence of the target set, whose properties play an important role in the analysis.

The chapter is structured as follows. In Section 8.1 we introduce exit time problems and give a result about the existence of optimal controls. In Section 8.2 we consider the value function and study its regularity properties. We show that the Lipschitz continuity of the value function is ensured by suitable compatibility conditions on the final cost and on the behavior of the dynamics along the boundary of the target set. Then we prove a semiconcavity result for the value function; an important hypothesis here is that the target set satisfies an interior sphere condition, analogous to the one considered in Chapter 2 on the analysis of the distance function. In Section 8.3 we give a semiconvexity result for the minimum time function of a linear system with convex target; as a corollary, we prove differentiability of this function for some classes of linear systems. Finally, Section 8.4 is devoted to the analysis of optimality conditions. We give the Pontryagin maximum principle in the case of a smooth target and prove the co-state inclusion. Under additional hypotheses on the system, we can show that the trajectories are solutions of the associated hamiltonian system and are in one-to-one correspondence with the reachable gradients of the value function.

8.1 Optimal control problems with exit time

Let us consider again a control system (f, U) as in the previous chapter and the associated state equation

$$\begin{cases} y'(t) = f(y(t), u(t)), & t \geq 0 \text{ a.e.} \\ y(0) = x, \end{cases} \tag{8.1}$$

where $u \in L^1([0, T_u], U)$ is a control defined up to some finite time $T_u > 0$ depending on u. For the reader's sake, we quote from the previous chapter some assumptions about the system which we will often require in our following analysis.

(H0) The control set U is compact.
(H1) There exists $K_1 > 0$ such that $|f(x_2, u) - f(x_1, u)| \leq K_1 |x_2 - x_1|$, for all $x_1, x_2 \in \mathbb{R}^n$, $u \in U$.
(H2) f_x exists and is continuous; in addition, there exists $K_2 > 0$ such that $\|f_x(x_2, u) - f_x(x_1, u)\| \leq K_2 |x_2 - x_1|$, for all $x_1, x_2 \in \mathbb{R}^n$, $u \in U$.

We always assume that our system satisfies (H1), ensuring the existence and uniqueness of the solution of (8.1). Since the initial time of the system will be 0 throughout the chapter, we denote the solution of (8.1) simply by $y^{x,u}(\cdot)$.

In addition, we assume that a closed set $K \subset \mathbb{R}^n$ is given, and is called the *target*. For a given trajectory $y = y^{x,u}$ of the system we set

$$\tau(x, u) = \min\{t \geq 0 : y^{x,u}(t) \in K\},$$

with the convention that $\tau(x, u) = +\infty$ if $y^{x,u}(t) \notin K$ for all $t \in [0, T_u]$. We call $\tau(x, u)$ the *exit time* of the trajectory. If $\tau(x, u) < +\infty$ we set for simplicity

$$y_\tau^{x,u} := y^{x,u}(\tau(x, u))$$

to denote the point where the trajectory reaches the target. We denote by $\mathcal{R}$ the set of all x such that $\tau(x, u) < +\infty$ for some control u and call $\mathcal{R}$ the *controllable set*.

In this chapter we consider optimal control problems where the functional to minimize is no longer defined on a fixed time interval, but depends on the exit time $\tau(x, u)$. A simple and important example consists of the minimization of the exit time itself, and is called the *minimum time problem*:

(MTP) given $x \in \mathcal{R}$, minimize $\tau(x, u)$ over all measurable $u : [0, T_u] \to U$.

A control u achieving the minimum in (MTP) and the associated trajectory $y = y^{x,u}$ are called *time-optimal*. The corresponding value function is called the *minimum time function* and is defined by

$$T(x) = \inf\{\tau(x, u) : u \in L^1([0, T_u], U)\} \qquad x \in \mathcal{R}. \tag{8.2}$$

Example 8.1.1 If we consider the system $y' = u$ with $u \in U = B_1$, then it is easy to see that $\mathcal{R} = \mathbb{R}^n$ and $T(x) = d_{\mathcal{K}}(x)$, the distance function from $\mathcal{K}$. The optimal trajectories are the line segments joining a given point x to a projection of x on $\mathcal{K}$. Therefore an optimal trajectory exists for any x, and is unique if and only if the projection of x on $\mathcal{K}$ is unique; we know from Corollary 3.4.5 that this condition is in turn equivalent to the differentiability of $d_{\mathcal{K}}$ at x. Thus, also in this case we see that the uniqueness of the optimal trajectory is related to the differentiability of the value function, as for finite horizon control problems. Later in this chapter we shall describe a class of exit time problems satisfying the same property.

We can consider more general control problems depending on the exit time of the trajectory. In fact, let us suppose that two continuous functions $L : \mathbb{R}^n \times U \to \mathbb{R}$ (called *running cost*) and $g : \mathbb{R}^n \to \mathbb{R}$ (the *terminal cost*) are given, with L positive and g bounded from below. We can consider the functional

$$J(x, u) = \int_0^{\tau(x,u)} L(y^{x,u}(s), u(s))ds + g(y_\tau^{x,u})$$

and study the following control problem:

(ETP) given $x \in \mathcal{R}$, minimize $J(x, u)$ over all $u \in L^1([0, T_u], U)$.

We call a problem of this form an *exit time problem*. Observe that the minimum time problem is a special case of (ETP), corresponding to the choices $L \equiv 1$ and $g \equiv 0$. If $x \in \mathcal{R}$ there exists at least one control u such that $J(x, u) < +\infty$; on the other hand, our assumptions on g, L imply that $J(x, u) \geq \inf g$ for all u. Therefore $\inf_u J(x, u) \neq \pm\infty$ for all $x \in \mathcal{R}$.

For future reference, let us recall some assumptions on L which have been introduced in the previous chapter and which will be often needed in our later analysis.

(L1) For any $R > 0$ there exists γ_R such that $|L(x_2, u) - L(x_1, u)| \leq \gamma_R |x_2 - x_1|$, for all $x_1, x_2 \in B_R, u \in U$.

(L2) For any $x \in \mathbb{R}^n$, the following set is convex:

$$\mathcal{L}(x) := \{(\lambda, v) \in \mathbb{R}^{n+1} : \exists u \in U \text{ such that } v = f(x, u), \lambda \geq L(x, u)\}.$$

(L3) For any $R > 0$ there exists λ_R such that

$$L(x, u) + L(y, u) - 2L\left(\frac{x+y}{2}, u\right) \leq \lambda_R |x - y|^2, \qquad x, y \in B_R, u \in U.$$

A control u and the corresponding trajectory $y^{x,u}$ are called *optimal* for the point x if u is such that the infimum in (ETP) is attained. The value function of the problem is defined by

$$V(x) = \inf\{J(x, u) : u \in L^1([0, T_u], U)\}, \qquad x \in \mathcal{R}. \tag{8.3}$$

For these problems the *dynamic programming principle* takes the following form: for any $x \in \mathbb{R}^n$ and $u : [0, T_u] \to U$ such that $\tau(x, u) < \infty$.

$$V(x) \le \int_0^t L(y^{x,u}(s), u(s))ds + V(y^{x,u}(t)), \qquad \forall t \in [0, \tau(x, u)], \qquad (8.4)$$

with equality if u is optimal. This property can be proved by the same arguments of the previous chapters (Theorem 1.2.2 or 7.2.2).

Let us study the existence of optimal trajectories. We first consider the case of the minimum time problem.

Theorem 8.1.2 *Assume that* (H0), (H1) *hold and that the set* $f(x, U)$ *is convex for all* $x \in \mathbb{R}^n$. *Then, for any* $x \in \mathcal{R}$, *there exists a control which is time optimal for* x.

Proof — Let us consider a sequence $u_k : [0, T_k] \to U$ such that $\tau(x, u_k) \to T(x)$ and let us set $y_k(\cdot) = y^{x,u_k}(\cdot)$. Let us fix $\bar{T} > T(x)$; we can assume that all controls are defined up to time $\bar{T}$ by defining them in an arbitrary way for $t \in [T_k, \bar{T}]$ if $T_k < \bar{T}$.

The trajectories y_k are uniformly bounded in the interval $[0, \bar{T}]$ by (7.3) and have bounded Lipschitz constants by (7.2). Theorem 7.1.6 ensures that after possibly passing to a subsequence, $y_k \to \bar{y}$ uniformly, where $\bar{y} = y^{x,\bar{u}}$ for some control $\bar{u} : [0, \bar{T}] \to U$.

In addition we have

$$\begin{aligned} |\bar{y}(T(x)) - y_k(\tau(x, u_k))| &\le |\bar{y}(T(x)) - y_k(T(x))| \\ &\quad + |y_k(T(x)) - y_k(\tau(x, u_k))| \\ &\le \|\bar{y} - y_k\|_\infty + \mathrm{Lip}(y_k)|T(x) - \tau(x, u_k)| \end{aligned} \qquad (8.5)$$

and the right-hand side tends to 0 as $k \to \infty$. By definition $y_k(\tau(x, u_k)) \in \mathcal{K}$ for all k; since $\mathcal{K}$ is closed, we deduce that $\bar{y}(T(x)) \in \mathcal{K}$, that is, $\bar{y}$ is a time optimal trajectory. ∎

On the other hand, as soon as a nonconstant final cost g is present in the problem, the above hypotheses no longer ensure the existence of an optimal trajectory, as shown by the following example.

Example 8.1.3 Consider the control system in $\mathbb{R}^2$ with $f(x, u) = u$, $u \in U := B_1$. Then the admissible trajectories are those with speed at most one. Consider the target $\mathcal{K}$ consisting of the two points $(0, 0)$ and $(1, 0)$. Let $L \equiv 1$ and let g be such that $g(0, 0) = 0$, $g(1, 0) = 2$. Consider problem (ETP) with initial point $(2, 0)$. By taking trajectories from $(2, 0)$ to $(0, 0)$ not passing through $(1, 0)$ one can make the functional arbitrarily close to 2, but not exactly 2. The trajectories ending at $(1, 0)$ do not improve the performance since they yield a value of the functional which is at least 3. Therefore no trajectory achieves the infimum of the functional.

Let us note another feature of this example. We can easily compute the value function and find that $V(x, y) = \sqrt{x^2 + y^2}$ for all $(x, y) \ne (1, 0)$; on the other hand, since $(1, 0) \in \mathcal{K}$ we have $V(1, 0) = g(1, 0) = 2$, and thus $V(x, y) \not\to V(1, 0)$ as $(x, y) \to (1, 0)$. ∎

Let us show that an existence result for problem (ETP) can be obtained if suitable assumptions are added.

Theorem 8.1.4 *Assume properties* (H0), (H1), (L1) *and* (L2). *Suppose in addition that there exist constants* $N, G, \alpha > 0$ *such that*

$$\begin{cases} L(x, u) \geq \alpha, & |f(x, u)| \leq N, \quad x \in \mathbb{R}^n, u \in U, \\ |g(x) - g(y)| \leq G|x - y|, & x, y \in \partial \mathcal{K}, \\ \alpha > NG. \end{cases} \qquad (8.6)$$

Then, for any initial point $x \in \mathcal{R}$, *there exists an optimal control for problem* (ETP).

Remark 8.1.5 Assumption (8.6) rules out the behavior of the previous example. Roughly speaking, it implies that it is not convenient to choose a long trajectory if there is a short one arriving at the target. In fact, going from one point of the target x_1 to another one x_2 over a time interval of length T increases the L-term in the functional by at least αT, while the g-term can decrease by no more than $G|x_2 - x_1| \leq GNT$. Since $\alpha > GN$, the functional increases. Thus, if one can reach the target at a point x_1 which is very close to the initial one, it is not convenient to take another trajectory ending at a point x_2 which is far away (as it happened in the previous example, with $x_1 = (1, 0)$ and $x_2 = (0, 0)$). We will see in the following that (8.6), together with other assumptions, ensures also the continuity along the boundary of the target of the value function (another property violated in the example). For a detailed discussion of these matters, including sharper formulations of (8.6), see [110, Ch. 5] and [20, Sect IV.3.1].

Remark 8.1.6 Let us mention that if the final cost g is constant (as in the case of the minimum time problem), it is no longer necessary to require the global boundedness of f in assumption (8.6). In other words, the results we prove in the following assuming (8.6) are also valid if g is constant under the only hypothesis that $L \geq \alpha > 0$. The proofs require some easy adaptation which we leave to the reader.

Proof of Theorem 8.1.4 — Let $u_k : [0, T_k] \to U$ be a minimizing sequence. We set for simplicity $\tau_k := \tau(x, u_k)$, $y_k(\cdot) = y^{x, u_k}(\cdot)$, $z_k = y_k(\tau_k)$. Let us first show that the times $\{\tau_k\}$ are uniformly bounded. We have

$$|z_k - z_1| \leq |z_k - x| + |x - z_1| \leq N(\tau_k + \tau_1)$$

and so

$$J(x, u_k) \geq \alpha \tau_k + g(z_k) \geq \alpha \tau_k + g(z_1) - G|z_k - z_1| \\ \geq \alpha \tau_k + g(z_1) - GN(\tau_k + \tau_1).$$

Since $J(x, u_k) \to V(x)$ and $\alpha - GN > 0$, the sequence $\{\tau_k\}$ must be bounded.

Let us set $T^* = \max \tau_k$. We can assume that u_k, y_k are defined in $[0, T^*]$ by setting u_k equal to some constant control $\bar{u}$ for $t \geq \tau_k$. By passing to a subsequence, we can also assume that τ_k converges to some $\bar{\tau}$.

We now apply Theorem 7.4.4 to the trajectories y_k. We obtain that, up to a subsequence, they converge uniformly to a trajectory $\bar{y}$ corresponding to some control $\bar{u}$. In addition,

$$\int_0^\tau L(\bar{y}(t), \bar{u}(t))\, dt \le \liminf_{k\to\infty} \int_0^\tau L(y_k(t), u_k(t))\, dt$$

for all $\tau \in [0, T^*]$. Observe that, since L is locally bounded,

$$\left| \int_{\bar{\tau}}^{\tau_k} L(y_k(t), u_k(t))\, dt \right| \le \sup L |\bar{\tau} - \tau_k| \to 0$$

and therefore

$$\int_0^{\bar{\tau}} L(\bar{y}(t), \bar{u}(t))\, dt \le \liminf_{k\to\infty} \int_0^{\bar{\tau}} L(y_k(t), u_k(t))\, dt$$
$$= \liminf_{k\to\infty} \int_0^{\tau_k} L(y_k(t), u_k(t))\, dt.$$

Moreover

$$|g(y_k(\tau_k)) - g(\bar{y}(\bar{\tau}))| \le G(|y_k(\tau_k) - y_k(\bar{\tau})| + |y_k(\bar{\tau}) - \bar{y}(\bar{\tau})|) \to 0.$$

Since by assumption

$$V(x) = \lim_{k\to\infty} \left(\int_0^{\tau_k} L(y_k(t), u_k(t))\, dt + g(y_k(\tau_k)) \right),$$

we conclude that

$$V(x) \ge \int_0^{\bar{\tau}} L(\bar{y}(t), \bar{u}(t))\, dt + g(\bar{y}(\bar{\tau})). \tag{8.7}$$

Arguing as in (8.5) we find that $y(\bar{\tau}) \in \mathcal{K}$, which implies that $\tau(x, \bar{u}) \le \bar{\tau}$. Suppose that $\tau(x, \bar{u}) = \bar{\tau}$. Then inequality (8.7) becomes $V(x) \ge J(x, \bar{u})$, that is, $\bar{u}$ is optimal for x. To conclude the proof it suffices therefore to exclude that $\tau(x, \bar{u}) < \bar{\tau}$. Let us set for simplicity $t^* = \tau(x, \bar{u})$ and suppose that $t^* < \bar{\tau}$. We have, by definition of V,

$$V(x) \le \int_0^{t^*} L(\bar{y}(t), \bar{u}(t))\, dt + g(\bar{y}(t^*)).$$

Taking into account (8.7) we deduce

$$0 \ge \int_{t^*}^{\bar{\tau}} L(\bar{y}(t), \bar{u}(t))\, dt + g(\bar{y}(\bar{\tau})) - g(\bar{y}(t^*))$$
$$\ge \alpha(\bar{\tau} - t^*) - G|\bar{y}(\bar{\tau}) - \bar{y}(t^*)|$$
$$\ge \alpha(\bar{\tau} - t^*) - GN(\bar{\tau} - t^*),$$

contradicting (8.6). Thus we cannot have $\tau(x, \bar{u}) < \bar{\tau}$ and this completes the proof that $\bar{y}$ is optimal. ∎

With similar arguments we can show that the uniform limit of optimal trajectories is an optimal trajectory, provided the value function V is continuous (sufficient conditions for the continuity of V will be given in the next section).

Theorem 8.1.7 *Let the assumptions of Theorem 8.1.4 hold, and suppose in addition that V is continuous in $\mathcal{R}$. Let $\{y_k\} = \{y^{x_k, u_k}\}$ be a family of optimal trajectories converging uniformly to an arc y over a time interval $[0, T]$ with $T \geq \liminf \tau(x_k, u_k)$. Then y is also an optimal trajectory.*

Proof — By Theorem 7.1.6, the arc y is of the form $y = y^{x,u}$, for some $x \in \mathbb{R}^n$ and $u : [0, T] \to U$. After possibly extracting a subsequence, we can assume that $\tau(x_k, u_k) \to \tau$ for some $\tau \leq T$. We easily obtain that $y(\tau) \in \mathcal{K}$ and so $\tau(x, u) \leq \tau$. Arguing as in the proof of Theorem 8.1.4 and using the continuity of V we obtain

$$V(x) = \lim_{k \to \infty} V(x_k) = \lim_{k \to \infty} \left(\int_0^{\tau(x_k, u_k)} L(y_k(t), u_k(t)) \, dt + g(y_\tau^{x_k, u_k}) \right)$$

$$\geq \int_0^\tau L(y(t), u(t)) \, dt + g(y(\tau)).$$

We can now deduce that $\tau(x, u) = \tau$, since if $\tau(x, u) < \tau$ we find a contradiction, as in the proof of Theorem 8.1.4. Hence the right-hand side in the above inequality coincides with $J(x, u)$; this implies that we have an equality and that u is optimal for x. ∎

As in the previous chapters, one can show that the value function V of an exit time problem is a viscosity solution of a suitable partial differential equation. In fact, let us define

$$H(x, p) = \max_{u \in U} [-f(x, u) \cdot p - L(x, u)]. \tag{8.8}$$

Then the *dynamic programming equation* associated to (ETP) is

$$H(x, DV(x)) = 0. \tag{8.9}$$

In particular (see Proposition A. 1.17), for the minimum time problem we obtain the equation

$$\sigma_{\text{co} \, f(x,U)}(-p) - 1 = 0, \tag{8.10}$$

where $\sigma(\cdot)$ denotes the support function and co $f(x, U)$ is the convex hull of $f(x, U)$.

Theorem 8.1.8 *If assumptions (H0), (H1), (L1) hold and if the value function V is continuous in the interior of $\mathcal{R} \setminus \mathcal{K}$, then it is a viscosity solution of equation (8.9) in this set.*

Proof — Let us take x in the interior of $\mathcal{R} \setminus \mathcal{K}$ and $p \in D^+V(x)$. For any given $v \in U$, let us consider the trajectory $y(\cdot)$ starting from x and corresponding to the constant control v. Then we have $y(t) = x + f(x, v)t + o(t)$ as $t \to 0$ and therefore, since $p \in D^+V(x)$,

$$V(y(t)) - V(x) \leq p \cdot (y(t) - x) + o(|y(t) - x|) = p \cdot f(x, v)t + o(t).$$

On the other hand, by the dynamic programming principle (8.4),

$$V(x) \leq V(y(t)) + \int_0^t L(y(s), v)\, ds = V(y(t)) + tL(x, v) + o(t).$$

It follows that
$$-p \cdot f(x, v) - L(x, v) \leq 0.$$

Since $v \in U$ is arbitrary, we conclude that $H(x, p) \leq 0$.

Let us now take $p \in D^-V(x)$ and $\varepsilon > 0$. For any given $t \in {]}0, T(x){[}$, let $u(\cdot)$ be a control such that $J(x, u) \leq V(x) + \varepsilon t$. Then, since the restriction of $y^{x,u}$ to the interval $[t, \tau(x, u)]$ is an admissible trajectory for $y(t)$, we find that

$$\begin{aligned}
V(y(t)) &\leq \int_t^{\tau(x,u)} L(y^{x,u}(s), u(s))\, ds + g(y_\tau^{x,u}) \\
&= J(x, u) - \int_0^t L(y^{x,u}(s), u(s))\, ds \\
&\leq V(x) + \varepsilon t - \int_0^t L(y^{x,u}(s), u(s))\, ds.
\end{aligned}$$

On the other hand, since $p \in D^-V(x)$,

$$\begin{aligned}
V(y(t)) - V(x) &\geq p \cdot (y(t) - x) + o(t) \\
&= \int_0^t p \cdot f(y^{x,u}(s), u(s))\, ds + o(t).
\end{aligned}$$

Hence

$$\begin{aligned}
0 &\leq \int_0^t [-p \cdot f(y^{x,u}(s), u(s)) - L(y^{x,u}(s), u(s))]\, ds + \varepsilon t + o(t) \\
&= \int_0^t [-p \cdot f(x, u(s)) - L(x, u(s))]\, ds + \varepsilon t + o(t) \\
&\leq tH(x, p) + \varepsilon t + o(t).
\end{aligned}$$

Therefore $H(x, p) \geq -\varepsilon$. By the arbitrariness of ε, this completes the proof. ∎

If we consider points along optimal trajectories different from the endpoints we can prove that the elements of D^+V satisfy also the reverse inequality. In fact the following result holds, whose proof is omitted since it is similar to that of Theorem 7.2.5.

Proposition 8.1.9 *Let properties* (H0), (H1), (L1) *be satisfied and let y be an optimal trajectory reaching the target at time τ. Then we have*

$$H(y(t), p) = 0, \qquad \forall p \in D^+V(y(t)), \ \forall t \in \,]0, \tau[\,.$$

We conclude the section recalling that the value function of problem (ETP) is characterized by the property of being a viscosity solution of (8.9). It is remarkable that in the next statement the a priori knowledge of the controllable set $\mathcal{R}$ is not required; in fact it is a uniqueness result for the pair $(V, \mathcal{R})$ rather than for V alone. We will see in the following section sufficient conditions ensuring that $V, \mathcal{R}$ satisfy the properties required on w, Ω in the statement below.

Theorem 8.1.10 *Let properties* (H0), (H1), (L1) *hold, and assume that $\partial \mathcal{K}$ is compact, that g is continuous and that $L \geq \alpha > 0$. Let $\Omega \subset \mathbb{R}^n$ and $w \in C(\Omega)$ satisfy the following properties.*

(i) Ω is open and contains $\mathcal{K}$;
(ii) w is bounded from below and is a viscosity solution of (8.9) in $\mathcal{R} \setminus \mathcal{K}$;
(iii) $w(x) = g(x)$ for all $x \in \partial \mathcal{K}$, $\lim_{x \to \bar{x}} w(x) = +\infty$ for all $\bar{x} \in \partial \mathcal{R}$.

Then $\Omega = \mathcal{R}$ and $w = V$, where V is the value function of problem (ETP).

Proof — See [20, Cor. IV.4.3]. ∎

8.2 Lipschitz continuity and semiconcavity

We investigate now the hypotheses under which the value function of an optimal control problem with exit time is Lipschitz continuous and semiconcave. It will turn out that these properties are related not only to the regularity of the data, but also to the smoothness of the target and to the behavior of the control system near the target.

We begin with some properties of the minimum time function which will be useful also in the analysis of general exit time problems. For given $\delta > 0$, let us introduce the notation

$$\mathcal{K}_\delta := \{x \in \mathbb{R}^n \ : \ d_\mathcal{K}(x) \leq \delta\}.$$

The next result shows that a Lipschitz estimate on the minimum time function T in a neighborhood of the target is sufficient to deduce the local Lipschitz continuity of T in the whole controllable set $\mathcal{R}$, as well as other properties of $\mathcal{R}$.

Theorem 8.2.1 *Let properties* (H0), (H1) *be satisfied. Suppose that for any $R > 0$ there exist $\delta, k > 0$ such that*

$$\mathcal{K}_\delta \cap B_R \subset \mathcal{R} \quad \text{and} \quad T(x) \leq k d_\mathcal{K}(x), \quad x \in \mathcal{K}_\delta \cap B_R. \tag{8.11}$$

Then

(i) the controllable set $\mathcal{R}$ is open;

(ii) for any $\bar{x} \in \partial\mathcal{R}$ we have $\lim_{x \to \bar{x}} T(x) = +\infty$;

(iii) the minimum time function T is locally Lipschitz continuous on $\mathcal{R}$.

Proof — Given $x_0 \in \mathcal{R} \setminus \mathcal{K}$ and $\varepsilon > 0$, let u be a control with $\tau(x_0, u) < T(x_0) + \varepsilon$. Let us set for simplicity $\tau_0 = \tau(x_0, u)$. By estimate (7.4) there exists $c > 0$ depending only on τ_0 such that

$$|y^{x_1,u}(t) - y^{x_2,u}(t)| \le c|x_1 - x_2|, \qquad 0 \le t \le 2\tau_0 + 1, \qquad (8.12)$$

for all $x_1, x_2 \in \mathbb{R}^n$. Thus, for a given $\rho > 0$, we find

$$|y^{x,u}(\tau_0) - y^{x_0,u}(\tau_0)| \le c\rho, \qquad \forall x \in B_\rho(x_0).$$

Since $y^{x_0,u}(\tau_0) \in \mathcal{K}$, this implies that $d_\mathcal{K}(y^{x,u}(\tau_0)) \le c\rho$. Let δ, k be such that (8.11) holds with $R = |y^{x_0,u}(\tau_0)| + 1$. If we choose ρ such that $\rho \le \min\{1/c, \delta/c\}$, we obtain that $y^{x,u}(\tau_0) \in \mathcal{K}_\delta \cap B_R$ for all $x \in B_\rho(x_0)$. Then, by the dynamic programming principle and estimate (8.11),

$$T(x) \le \tau_0 + T(y^{x,u}(\tau_0)) \le \tau_0 + kd_\mathcal{K}(y^{x,u}(\tau_0)) \le T(x_0) + \varepsilon + ck\rho$$

for all $x \in B_\rho(x_0)$. Thus, we obtain a bound on T in a neighborhood of x_0; this can be done for any $x_0 \in \mathcal{R}$, and so we have shown that $\mathcal{R}$ is open and that T is locally bounded in $\mathcal{R}$.

If we suppose in addition that $\varepsilon < 1$ and ρ is such that $ck\rho < T(x_0)$, we have that $T(x) < 2T(x_0) + 1$ for all $x \in B_\rho(x_0)$. Then we can repeat the above argument with an arbitrary pair $x_1, x_2 \in B_{\rho/2}(x_0)$ instead of x_0, x. We take a control u such that $\tau(x_1, u) \le T(x_1) + \varepsilon$. Setting for simplicity $\tau_1 = \tau(x_1, u)$, we obtain from (8.12)

$$d_\mathcal{K}(y^{x_2,u}(\tau_1)) \le |y^{x_2,u}(\tau_1) - y^{x_1,u}(\tau_1)| \le c|x_2 - x_1| \le c\rho \le \delta.$$

Thus, by the dynamic programming principle and (8.11),

$$T(x_2) \le \tau_1 + T(y^{x_2,u}(\tau_1)) \le \tau_1 + kd_\mathcal{K}(y^{x_2,u}(\tau_1))$$
$$\le T(x_1) + ck|x_2 - x_1| + \varepsilon.$$

Since the role of x_1, x_2 can be exchanged, and $\varepsilon > 0$ is arbitrary, we obtain that T is Lipschitz continuous in $B_{\rho/2}(x_0)$.

It remains to prove (ii). For this purpose, suppose that $\{x_n\}$ is any sequence converging to a point $\bar{x} \in \partial\mathcal{R}$. By the previous argument, each x_n is the center of a ball contained in $\mathcal{R}$. If $T(x_n)$ is uniformly bounded, then the radius of the spheres can be taken uniformly bounded from below, in contradiction with the assumption that $x_n \to \bar{x} \in \partial\mathcal{R}$. Thus, $T(x_n) \to +\infty$. ∎

With a similar procedure one can prove that, if the minimum time function satisfies an estimate of the form $T(x) \le m(d_\mathcal{K}(x))$ for $x \in \mathcal{K}_\delta$, where $m(\cdot)$ is any increasing function with $m(r) \to 0$ as $r \to 0$, then (i) and (ii) of the previous theorem still hold and T satisfies an estimate of the form $|T(x) - T(y)| \le c_1 m(c_2|x - y|)$

on any compact subset of $\mathcal{R}$. Thus, the regularity near the target determines the regularity in the whole controllable set (provided the dynamics satisfies (H1)).

We now analyze the question of whether we can find a neighborhood of the target contained in the controllable set $\mathcal{R}$ where estimate (8.11) is satisfied. Easy examples show that such a property cannot be expected without some further assumption of f. Consider for instance the one-dimensional system with state equation $x' = u$, $u \in [0, 1]$ and target $\mathcal{K} = \{0\}$. Then all admissible trajectories are increasing and it is possible to reach the target only starting from the negative axis. Thus, $\mathcal{R} =] - \infty, 0]$ and we see that the controllable set is not a neighborhood of the target. Roughly speaking, this example suggests that one should assume that if x is close to the target, then there exists u such that $f(x, u)$ points in the direction of the target. More precisely, let us introduce the following assumption on the system, which uses the notion of proximal normal (see Definition 3.6.3).

Definition 8.2.2 *We say that the control system (f, U) and the target $\mathcal{K}$ satisfy the Petrov condition if, for any $R > 0$, there exists $\mu > 0$ such that*

$$\min_{u \in U} f(\bar{x}, u) \cdot v < -\mu|v|, \qquad \forall \bar{x} \in \partial\mathcal{K} \cap B_R, \ v \in N_{\mathcal{K}}(\bar{x}). \tag{8.13}$$

The following result shows that the Petrov condition ensures the validity of estimate (8.11) and is actually a necessary and sufficient condition.

Theorem 8.2.3 *Let our system satisfy* (H0), (H1). *Then the two following properties are equivalent:*

(i) for any $R > 0$, there exist $\delta, k > 0$ such that property (8.11) holds;
(ii) for any $R > 0$, there exist $\mu > 0$ such that the Petrov condition (8.13) holds.

Proof — We first show that (ii) implies (i). Let us sketch the strategy of the proof. An immediate consequence of the definition of proximal normal is that if $x \notin \mathcal{K}$ and $\bar{x} \in \mathcal{K}$ is a projection of x onto $\mathcal{K}$, then $v = x - \bar{x} \in N_{\mathcal{K}}(\bar{x})$. On the other hand, the Petrov condition ensures that there exists $\bar{u}$ such that $f(\bar{x}, \bar{u}) \cdot v < -\mu|v|$. If x and $\bar{x}$ are sufficiently close, we have $f(x, \bar{u}) \cdot (\bar{x} - x) \sim f(\bar{x}, \bar{u}) \cdot (\bar{x} - x) > 0$, that is, a trajectory passing through x with speed $f(x, \bar{u})$ is approaching the target. We can use this idea to construct a piecewise constant control strategy steering to the target any point in a suitable neighborhood. This technique is sometimes called "proximal aiming" (see e.g., [55]).

For a fixed $R > 0$, let us take $\mu > 0$ such that

$$\min_{u \in U} f(\bar{x}, u) \cdot v \leq -\mu|v| \tag{8.14}$$

for all $\bar{x} \in \partial\mathcal{K} \cap B_{2R}$ and all v proximal normal to $\mathcal{K}$ at $\bar{x}$. Let us choose M such that

$$\mu/2 \leq M, \qquad |f(x, u)| \leq M, \quad \forall (x, u) \in B_{2R} \times U. \tag{8.15}$$

Let us set

$$\gamma = \sqrt{1 - \left(\frac{\mu}{4M}\right)^2},$$

(8.16)

$$k = \frac{2}{\mu}(1 + \gamma), \qquad \delta = \min\left\{\frac{\mu}{2K_1}, \frac{R}{2(1 + kM)}\right\},$$

(8.17)

where K_1 is the Lipschitz constant of $f(\cdot, u)$ (see (H1)). Let us observe that

$$k(1 - \gamma) = \frac{\mu}{8M^2}.$$

(8.18)

We claim that (8.11) holds with δ, k defined as above. To prove this, let us take an arbitrary $x_0 \in (\mathcal{K}_\delta \setminus \mathcal{K}) \cap B_R$. We will define inductively a sequence $\{x_j\}$ with the properties

$$|x_j - x_0| \le \frac{\mu}{8M} \sum_{h=0}^{j-1} \gamma^h d(x_0),$$

(8.19)

$$d(x_j) \le \gamma^j d(x_0).$$

(8.20)

These inequalities are satisfied for $j = 0$. We define x_{j+1} starting from x_j in the following way. First we observe that by (8.19) and (8.18),

$$|x_j| \le |x_0| + |x_j - x_0| \le R + \delta + \frac{\mu}{8M} \frac{1}{1 - \gamma} d(x_0)$$

$$< R + \delta + kMd(x_0) \le R + (1 + kM)\delta \le \frac{3R}{2}.$$

(8.21)

Let $\bar{x}_j \in \mathcal{K}$ be such that $d(x_j) = |x_j - \bar{x}_j|$. Inequalities (8.21) and (8.20) imply that $\bar{x}_j \in \partial \mathcal{K} \cap B_{2R}$. Setting

$$v_j = x_j - \bar{x}_j,$$

we have that v_j is a proximal normal to $\mathcal{K}$ at $\bar{x}_j$. By (8.14) there exists $u_j \in U$ such that $f(\bar{x}_j, u_j) \cdot v_j \le -\mu|v_j|$. Let us call $y_j(t)$ the trajectory starting at x_j corresponding to the constant control u_j, and let us define $x_{j+1} = y_j(t_j)$, where

$$t_j = \frac{\mu}{8M^2} d(x_j).$$

(8.22)

Let us check that the point x_{j+1} satisfies (8.19) and (8.20). By (8.21), we have $y_j(0) = x_j \in B_{3R/2}$. Thanks to (8.15), we have $|y_j'(t)| \le M$ for all t such that $y_j(t) \in B_{2R}$. In addition, (8.22), (8.20), (8.18) and (8.17) imply that

$$Mt_j \le \frac{\mu d(x_0)}{8M} < kMd(x_0) \le kM\delta < \frac{R}{2}.$$

It follows that

$$y_j(t) \in B_{2R}, \qquad \forall 0 \le t \le t_j.$$

Therefore, using properties (8.14), (H1), (8.15), (8.17) and (8.22), we obtain

$$
\begin{aligned}
\frac{1}{2}\frac{d}{dt}|y_j(t) - \bar{x}_j|^2 &= \langle f(y_j(t), u_j), y_j(t) - \bar{x}_j \rangle \\
&= \langle f(\bar{x}_j, u_j), x_j - \bar{x}_j \rangle + \langle f(y_j(t), u_j), y_j(t) - x_j \rangle \\
&\quad + \langle f(y_j(t), u_j) - f(\bar{x}_j, u_j), x_j - \bar{x}_j \rangle \\
&\le -\mu|x_j - \bar{x}_j| + M|y_j(t) - x_j| + K_1|y_j(t) - \bar{x}_j||x_j - \bar{x}_j| \\
&\le -\mu\, d(x_j) + (K_1\, d(x_j) + M)|y_j(t) - x_j| + K_1 d(x_j)^2 \\
&\le (-\mu + K_1\, d(x_j))\, d(x_j) + (K_1\, d(x_j) + M)Mt \\
&\le -\frac{\mu}{2}d(x_j) + 2M^2 t \le -\frac{\mu}{4}d(x_j)
\end{aligned}
$$

for all $0 \le t \le t_j$. Therefore

$$d^2(y_j(t_j)) \le |y_j(t_j) - \bar{x}_j|^2 \le d^2(x_j) - \frac{\mu}{2}d(x_j)t_j = \gamma^2 d^2(x_j),$$

and this shows that x_{j+1} satisfies (8.20). Property (8.19) also holds, since

$$
\begin{aligned}
|x_{j+1} - x_0| &\le |x_j - x_0| + |x_{j+1} - x_j| \\
&\le \frac{\mu}{8M}\sum_{h=0}^{j-1}\gamma^h d(x_0) + Mt_j \\
&\le \frac{\mu}{8M}\sum_{h=0}^{j}\gamma^h d(x_0).
\end{aligned}
$$

Thus the sequence $\{x_j\}$ satisfies the properties we have claimed. We observe that (8.20) implies that

$$d(x_j) \to 0 \text{ as } j \to +\infty. \tag{8.23}$$

In addition, (8.22), (8.20) and (8.18) imply that

$$\sum_{j=0}^{\infty} t_j \le \frac{\mu}{8M^2}\frac{1}{1-\gamma}d(x_0) = k\, d(x_0). \tag{8.24}$$

We now define a control strategy $\bar{u} : [0, +\infty) \to U$ such that

$$
\begin{cases}
\bar{u}(t) = u_h, & \text{if } \sum_{j=0}^{h-1} t_j \le t < \sum_{j=0}^{h} t_j \text{ for some } h \ge 0 \\
\bar{u}(t) \text{ arbitrary} & \text{if } t \ge \sum_{j=0}^{\infty} t_j.
\end{cases}
$$

Then

$$y^{x_0, \bar{u}}\left(\sum_{h=0}^{j} t_h\right) = x_{j+1}, \qquad j \ge 0.$$

Hence, by (8.23) and (8.24),

$$T(x_0) \le \tau(x_0, \bar{u}) = \sum_{j=0}^{\infty} t_j \le k\, d(x_0),$$

and this proves estimate (8.11).

Let us now prove the converse assertion of the theorem. For a given $R > 0$, let us take any $\bar{x} \in \partial \mathcal{K} \cap B_R$ and $v \in N_{\mathcal{K}}(\bar{x})$. It is not restrictive to suppose $|v| = 1$. By definition of proximal normal there exists $\lambda > 0$ such that

$$B_\lambda(\bar{x} + \lambda v) \cap \mathcal{K} = \emptyset. \tag{8.25}$$

Let k, δ constants such that (8.11) holds with R replaced by $R + 1$. For any $\varepsilon \in\,]0, \delta[$ there exists, by (8.11), a control u_ε such that

$$\tau(\bar{x} + \varepsilon v, u_\varepsilon) < (1 + \varepsilon)k\varepsilon. \tag{8.26}$$

Let us write for simplicity

$$x_\lambda = \bar{x} + \lambda v, \qquad x_\varepsilon = \bar{x} + \varepsilon v, \qquad \tau_\varepsilon = \tau(x_\varepsilon, u_\varepsilon), \qquad y_\varepsilon(\cdot) = y^{x_\varepsilon, u_\varepsilon}(\cdot).$$

Then, by (8.25)

$$|y_\varepsilon(\tau_\varepsilon) - x_\lambda| \ge \lambda,$$

which implies that

$$|y_\varepsilon(\tau_\varepsilon) - x_\lambda|^2 - |y_\varepsilon(0) - x_\lambda|^2 \ge \lambda^2 - (\lambda - \varepsilon)^2 = 2\varepsilon\lambda - \varepsilon^2. \tag{8.27}$$

Let M denote the supremum of f in $B_{R+1} \times U$. If ε is small enough we have $y_\varepsilon(t) \in B_{R+1}$ for $t \in [0, \tau_\varepsilon]$. Therefore

$$\begin{aligned}
\frac{1}{2}\frac{d}{dt}|y_\varepsilon(t) - x_\lambda|^2 &= \langle y_\varepsilon(t) - x_\lambda, f(y_\varepsilon(t), u_\varepsilon(t))\rangle \\
&= \langle \bar{x} - x_\lambda, f(\bar{x}, u_\varepsilon(t))\rangle + \langle y_\varepsilon(t) - \bar{x}, f(\bar{x}, u_\varepsilon(t))\rangle \\
&\quad + \langle y_\varepsilon(t) - x_\lambda, f(y_\varepsilon(t), u_\varepsilon(t)) - f(\bar{x}, u_\varepsilon(t))\rangle \\
&\le -\lambda \inf_{u \in U} v \cdot f(\bar{x}, u) + M|y_\varepsilon(t) - \bar{x}| \\
&\quad + K_1 |y_\varepsilon(t) - x_\lambda||y_\varepsilon(t) - \bar{x}| \\
&\le -\lambda \inf_{u \in U} v \cdot f(\bar{x}, u) + M(\varepsilon + \tau_\varepsilon M) \\
&\quad + K_1(\lambda - \varepsilon + M\tau_\varepsilon)(\varepsilon + \tau_\varepsilon M) \\
&\le -\lambda \inf_{u \in U} v \cdot f(\bar{x}, u) + O(\varepsilon)
\end{aligned}$$

for a.e. $t \in [0, \tau_\varepsilon]$. It follows that

$$|y_\varepsilon(\tau_\varepsilon) - x_\lambda|^2 - |y_\varepsilon(0) - x_\lambda|^2 \le 2\tau_\varepsilon\left[-\lambda \inf_{u \in U} v \cdot f(x, u) + O(\varepsilon)\right].$$

Using (8.27) and (8.26) we obtain

$$2\varepsilon\lambda - \varepsilon^2 \le 2k(1+\varepsilon)\varepsilon[-\lambda \inf_{u\in U} v \cdot f(\bar{x}, u) + O(\varepsilon)],$$

which implies

$$\inf_{u\in U} v \cdot f(\bar{x}, u) \le -\frac{1}{k}.$$

Since v is an arbitrary unit vector, we deduce that (8.13) holds. ∎

We now analyze the Lipschitz continuity of the value function for a general exit time problem of the form (ETP). We begin by showing that as a consequence of the compatibility condition (8.6), the exit time of the trajectories which are optimal (or almost optimal) for problem (ETP) is estimated by the minimum time function T.

Lemma 8.2.4 *Assume properties* (H0), (H1), (L1), (8.6) *and* (8.13). *Then, for any set* $A \subset\subset \mathcal{R}$ *there exists* $\beta > 0$ *with the following property. If* $x \in A$ *there exists* $\varepsilon > 0$ *such that if* u *is a control such that* $J(x, u) \le V(x)+\varepsilon$, *then* $\tau(x, u) \le \beta T(x)$.

Proof — Let $A \subset\subset \mathcal{R}$ be given. Then the minimum time function is bounded from above on A by some constant T^*. Let us denote by B^* the set

$$B^* = \{x \in \mathbb{R}^n \;:\; \text{dist}(x, A) \le N(T^* + 1)\},$$

with N as in (8.6). Let us also choose $M > 0$ such that $L(y, u) \le M$ for any $(y, u) \in B^* \times U$.

Consider now a point $x \in A \setminus \mathcal{K}$ and let $\varepsilon, \eta \in \,]0, 1]$. Then we can find controls v, w such that

$$J(x, v) \le V(x) + \varepsilon, \qquad \tau(x, w) \le T(x) + \eta.$$

Since $\tau(x, w) \le T^* + 1$, we deduce from assumption (8.6) that the trajectory $y^{x,w}$ is contained in B^* for all $t \in [0, \tau(x, w)]$. Then we have, again using (8.6), that

$$\begin{aligned}
\alpha\tau(x, v) + g(y_\tau^{x,v}) &\le J(x, v) \le V(x) + \varepsilon \le J(x, w) + \varepsilon \\
&\le M(\tau(x, w)) + g(y_\tau^{x,w}) + \varepsilon \\
&\le M(T(x) + \eta) + g(y_\tau^{x,w}) + \varepsilon.
\end{aligned}$$

Therefore

$$\begin{aligned}
\alpha\tau(x, v) - M(T(x) + \eta) - \varepsilon &\le g(y_\tau^{x,w}) - g(y_\tau^{x,v}) \\
&\le G|y_\tau^{x,w} - y_\tau^{x,v}| \\
&= G\left| \int_0^{\tau(x,w)} f(y^{x,w}(s), w(s))ds - \int_0^{\tau(x,v)} f(y^{x,v}(s), v(s))ds \right| \\
&\le GN[T(x) + \eta + \tau(x, v)].
\end{aligned}$$

Since $\eta > 0$ is arbitrary, we conclude that

$$\alpha \tau(x, v) - MT(x) - \varepsilon \leq GN[T(x) + \tau(x, v)].$$

Let us set $\beta = (M + GN + 1)(\alpha - GN)^{-1}$. Then $\beta > 0$ by (8.6) and, if $\varepsilon < T(x)$, we have

$$\tau(x, v) \leq \beta T(x). \qquad \blacksquare$$

We can now give a result about the Lipschitz continuity of V for a general exit time problem.

Theorem 8.2.5 *Let our system satisfy properties* (H0), (H1), (L1), (8.6) *and* (8.13). *Then the value function V of problem* (ETP), *given by* (8.3), *is locally Lipschitz continuous in $\mathcal{R}$.*

Proof — We assume for simplicity that, for any $x \in \mathcal{R}$, there exist optimal trajectories both for the exit time problem (ETP) and for the minimum time problem. If this is not the case, we can use a standard approximation argument as in the proof of Lemma 8.2.4.

For fixed $x^* \in \mathcal{R} \setminus \mathcal{K}$, we will prove a Lipschitz estimate in $B_\rho(x^*)$, with $\rho > 0$ suitably small. First of all, we choose ρ such that $B_\rho(x^*) \subset \mathcal{R}$; then, by Theorem 8.2.1, T is bounded on $B_\rho(x^*)$ and, by Lemma 8.2.4, the exit time associated with the optimal trajectories starting from $B_\rho(x^*)$ is also bounded by some constant T^*. Let us fix $R > 0$ such that all admissible trajectories starting from points in $B_\rho(x^*)$ remain inside B_R for $t \in [0, T^* + 1]$ and let us denote by M_R the supremum of $L(x, u)$ on $B_R \times U$.

Let us now take $x_0, x_1 \in B_\rho(x^*)$. Suppose for instance $V(x_1) > V(x_0)$ and let v_0 be an optimal control for the point x_0. By estimate (7.4) there exists a constant c, depending only on T^* (which in turns depends only on x_0, ρ), such that

$$|y^{x_0, v_0}(t) - y^{x_1, v_0}(t)| \leq c|x_0 - x_1|, \qquad \forall t \in [0, T^*]. \qquad (8.28)$$

By Theorem 8.2.3 there exist δ, k such that (8.11) holds for the R chosen above. We then assume that ρ is chosen small enough to have

$$2c\rho \leq \delta. \qquad (8.29)$$

It is now convenient to distinguish two cases.

Case 1: $\tau(x_1, v_0) \geq \tau(x_0, v_0)$.

By the dynamic programming principle (8.4) we have

$$V(x_0) = \int_0^{\tau(x_0, v_0)} L(y^{x_0, v_0}(s), v_0(s))ds + g(y^{x_0, v_0}(\tau(x_0, v_0)));$$

$$V(x_1) \leq \int_0^{\tau(x_0, v_0)} L(y^{x_1, v_0}(s), v_0(s))ds + V(y^{x_1, v_0}(\tau(x_0, v_0))).$$

Set $y_1 = y^{x_1, v_0}(\tau(x_0, v_0))$ and $y_0 = y^{x_0, v_0}(\tau(x_0, v_0))$. By (8.28),

$$|y_1 - y_0| \le c|x_1 - x_0| \le 2c\rho.$$

By construction $y_0 \in \partial\mathcal{K}$ and $y_0, y_1 \in B_R$. Then, by (8.29) and (8.11),

$$T(y_1) \le kd_{\mathcal{K}}(y_1) \le kc|x_1 - x_0|. \tag{8.30}$$

Let v be a time-optimal control for y_1. Since $T(y_1) \le kc|x_1 - x_0| \le 2kc\rho$ and $|f| \le N$, the trajectory $y^{y_1, v}(t)$ remains inside B_R for $t \in [0, T(y_1)]$ if ρ is chosen small enough. Then we can estimate

$$V(y_1) \le \int_0^{T(y_1)} L(y^{y_1, v}(s), v(s))ds + g(y_\tau^{y_1, v}) \le M_R T(y_1) + g(y_\tau^{y_1, v}).$$

Therefore we obtain, using (L1), (8.28), (8.30) and (8.6),

$$V(x_1) - V(x_0)$$
$$\le \int_0^{\tau(x_0, v_0)} [L(y^{x_1, v_0}(s), v_0(s)) - L(y^{x_0, v_0}(s), v_0(s))]ds + V(y_1) - g(y_0)$$
$$\le \int_0^{\tau(x_0, v_0)} \gamma_R c|x_1 - x_0| \, ds + M_R T(y_1) + g(y_\tau^{y_1, v}) - g(y_0)$$
$$\le T^* \gamma_R c|x_1 - x_0| + M_R kc|x_1 - x_0| + G|y_\tau^{y_1, v} - y_0|.$$

Since

$$|y_\tau^{y_1, v} - y_0| \le |y_\tau^{y_1, v} - y_1| + |y_1 - y_0|$$
$$= \left| \int_0^{T(y_1)} f(y^{y_1, v}(s), v(s))ds \right| + |y_1 - y_0|$$
$$\le NT(y_1) + c|x_1 - x_0| \le (Nk + 1)c|x_1 - x_0|,$$

we conclude that

$$V(x_1) - V(x_0) \le C|x_1 - x_0|$$

for a suitable $C > 0$ depending only on x^*, ρ.

Case 2: $\tau(x_1, v_0) < \tau(x_0, v_0)$.

We have, by the dynamic programming principle, (L1), (8.28), and (8.6), that

$$V(x_1) - V(x_0) \le \int_0^{\tau(x_1, v_0)} [L(y^{x_1, v_0}(s), v_0(s)) - L(y^{x_0, v_0}(s), v_0(s))]ds$$
$$- \int_{\tau(x_1, v_0)}^{\tau(x_0, v_0)} L(y^{x_0, v_0}(s), v_0(s))ds + g(y_\tau^{x_1, v_0}) - g(y_\tau^{x_0, v_0})$$
$$\le T^* \gamma_R c|x_1 - x_0| - \alpha[\tau(x_0, v_0) - \tau(x_1, v_0)]$$
$$+ G|y_\tau^{x_1, v_0} - y_\tau^{x_0, v_0}|.$$

Let us set $y^* := y^{x_0, v_0}(\tau(x_1, v_0))$. We find, by (8.28) and (8.6), that

$$|y_\tau^{x_1,v_0} - y_\tau^{x_0,v_0}| \leq |y_\tau^{x_1,v_0} - y^*| + |y^* - y_\tau^{x_0,v_0}|$$

$$= |y_\tau^{x_1,v_0} - y^*| + \left| \int_{\tau(x_1,v_0)}^{\tau(x_0,v_0)} f(y^{x_0,v_0}(s), v_0(s)) \, ds \right|$$

$$\leq c|x_1 - x_0| + N[\tau(x_0, v_0) - \tau(x_1, v_0)]$$

and therefore

$$V(x_1) - V(x_0) \leq (GN - \alpha)[\tau(x_0, v_0) - \tau(x_1, v_0)] + (T^*\gamma_R + G)c|x_1 - x_0|.$$

Recalling (8.6), we find also in this case that, for a suitable $C > 0$,

$$V(x_1) - V(x_0) \leq C|x_1 - x_0|. \qquad \blacksquare$$

Proposition 8.2.6 *Under the hypotheses of the previous theorem we have, for any $\bar{x} \in \partial\mathcal{R}$, that $V(x) \to +\infty$ as $x \to \bar{x}$.*

Proof — Let $\{x_n\} \subset \mathcal{R}$ be any sequence converging to a point $\bar{x} \in \partial\mathcal{R}$. For any n, let u_n be a control such that $J(x_n, u_n) \leq V(x_n) + 1$. Let us set for simplicity

$$y_n(\cdot) = y^{x_n,u_n}(\cdot), \qquad \tau_n = \tau(x_n, u_n), \qquad z_n = y_n(\tau_n).$$

Then (8.6) implies that

$$|z_n - x_n| \leq N\tau_n, \qquad J(x_n, u_n) \geq \alpha\tau_n + g(z_n).$$

Therefore, if we fix any point $x_0 \in \partial\mathcal{K}$ we have

$$V(x_n) \geq J(x_n, u_n) - 1 \geq \alpha\tau_n + g(z_n) - 1$$
$$\geq \alpha\tau_n + g(x_0) - G(|z_n - x_n| + |x_n - x_0|) - 1$$
$$\geq (\alpha - GN)\tau_n + g(x_0) - G|x_n - x_0| - 1.$$

Now $\tau_n \geq T(x_n) \to +\infty$ by Theorem 8.2.1(ii). Since $\alpha - GN > 0$ by (8.6) and $|x_n - x_0| \to |\bar{x} - x_0|$, we conclude that $V(x_n) \to \infty$. $\qquad \blacksquare$

We now turn to semiconcavity results. For this purpose, we need some suitable additional hypotheses on our system and on the data of the problem. As in the semiconcavity results for finite horizon problems (Theorems 7.2.8 and 7.4.11), we assume that f satisfies (H2) and that L satisfies (L3). We also need a kind of semiconcavity requirement on g. Since the values of g that are relevant for our problem are the ones on $\partial\mathcal{K}$, which is not a linear space, we put the semiconcavity requirement in the following form: for any $R > 0$ there exists $\Gamma_R > 0$ such that

$$g(x_1) + g(x_2) - 2g(\bar{x}) \leq \Gamma_R|x_1 - x_2|^2 + G|x_1 + x_2 - 2\bar{x}|,$$
$$\forall x_1, x_2, \bar{x} \in \partial\mathcal{K} \cap B_R,$$
(8.31)

where G is the same constant as in (8.6). It is easy to see that if $g \in SCL(\mathcal{N})$ with $\mathcal{N}$ neighborhood of $\partial\mathcal{K}$, and it has Lipschitz constant G, then it satisfies the above estimate.

A suitable hypothesis on the target will be also required, namely the following locally uniform *interior sphere condition*: for any $R > 0$ there exists $r > 0$ such that

$$\forall x \in \mathcal{K} \cap B_R, \ \exists y \in \mathcal{K} \ : \ x \in \overline{B}_r(y) \subset \mathcal{K}. \tag{8.32}$$

For some comments about the sets satisfying such an assumption, see Remark 2.2.3. We recall (see Proposition 2.2.2-(iii)) that property (8.32) implies the local semiconcavity of the distance function $d_\mathcal{K}$ in $\mathbb{R}^n \setminus \mathcal{K}$. Actually, the semiconcavity of $d_\mathcal{K}$ is the property needed in the proof and thus it could be taken as an assumption instead of (8.32).

Theorem 8.2.7 *Under hypotheses* (H0), (H1), (H2), (L1), (L3), (8.6), (8.13), (8.31) *and* (8.32) *the value function V of problem* (ETP), *given by* (8.3), *belongs to* $\mathrm{SCL}_{loc}(\mathcal{R} \setminus \mathcal{K})$.

Proof — Given $x_0 \in \mathcal{R} \setminus \mathcal{K}$, we will prove the semiconcavity of V in $B_\rho(x_0)$, for a suitably small ρ. We assume for simplicity the existence of optimal trajectories for all points in the controllable set. We consider arbitrary x, h such that $x \pm h \in B_\rho(x_0)$. In what follows we construct suitable trajectories steering the points $x - h, x, x + h$ to the target; as in the proof of Theorem 8.2.5, it is easy to check that if ρ is small enough, all the trajectories that we consider are contained in some ball B_R with R only depending on x_0, ρ. Let us again set

$$M_R = \max\{L(x, u) \ : \ x \in B_R, u \in U\}.$$

By Lemma 8.2.4, we know that the exit times associated with the optimal trajectories starting from points in $B_\rho(x_0)$ are uniformly bounded. For simplicity, we shall use the letter C to denote positive constants depending only on x_0, R which may vary from one formula to another.

Let us take a control v which is optimal for x and consider the trajectories $y^{x-h,v}, y^{x,v}, y^{x+h,v}$ starting from the points $x - h, x, x + h$ associated with the same control v. We treat separately two different cases depending on which of these trajectories reaches the target first.

Case 1: $\tau(x, v) \leq \min\{\tau(x + h, v), \tau(x - h, v)\}$.

Since v is optimal for x, we have by the dynamic programming principle (8.4) that

$$V(x - h) + V(x + h) - 2V(x)$$
$$\leq V(y^{x-h,v}(\tau(x, v))) + V(y^{x+h,v}(\tau(x, v))) - 2g(y_\tau^{x,v}) \tag{8.33}$$
$$+ \int_0^{\tau(x,v)} [L(y^{x-h,v}(s), v(s)) + L(y^{x+h,v}(s), v(s)) - 2L(y^{x,v}(s), v(s))]ds.$$

Let us observe that by assumptions (8.6), (L1) and (L3),

$$L(y^{x-h,v}(s), v(s)) + L(y^{x+h,v}(s), v(s)) - 2L(y^{x,v}(s), v(s))$$

$$= 2L\left(\frac{y^{x-h,v}(s) + y^{x+h,v}(s)}{2}, v(s)\right) - 2L(y^{x,v}(s), v(s))$$

$$+ L(y^{x-h,v}(s), v(s)) + L(y^{x+h,v}(s), v(s))$$

$$- 2L\left(\frac{y^{x-h,v}(s) + y^{x+h,v}(s)}{2}, v(s)\right)$$

$$\leq \gamma_R |y^{x-h,v}(s) + y^{x+h,v}(s) - 2y^{x,v}(s)|$$

$$+ \lambda_R |y^{x-h,v}(s) - y^{x+h,v}(s)|^2.$$

Using Lemma 7.1.2 we conclude that the integral term in (8.33) can be estimated as

$$\int_0^{\tau(x,v)} [L(y^{x-h,v}(s), v(s)) + L(y^{x+h,v}(s), v(s)) - 2L(y^{x,v}(s), v(s))]ds \leq C|h|^2$$

$$(8.34)$$

for a suitable C. We now analyze the other terms. Let us set for simplicity

$$x_+ = y^{x+h,v}(\tau(x, v)), \qquad x_- = y^{x-h,v}(\tau(x, v)).$$

By Lemma 7.1.2 these points satisfy

$$|x_\pm - y_\tau^{x,v}| \leq C|h|, \qquad |x_+ + x_- - 2y_\tau^{x,v}| \leq C|h|^2. \qquad (8.35)$$

By Proposition 2.2.2-(iii) $d_{\mathcal{K}}$ is semiconcave in the closure of $\mathcal{K}^c$. Since $y_\tau^{x,v} \in \partial\mathcal{K}$ we obtain, using (8.35), that

$$d_{\mathcal{K}}(x_+) + d_{\mathcal{K}}(x_-) = d_{\mathcal{K}}(x_+) + d_{\mathcal{K}}(x_-) - 2d_{\mathcal{K}}(y_\tau^{x,v})$$

$$= d_{\mathcal{K}}(x_+) + d_{\mathcal{K}}(x_-) - 2d_{\mathcal{K}}\left(\frac{x_+ + x_-}{2}\right)$$

$$+ 2d_{\mathcal{K}}\left(\frac{x_+ + x_-}{2}\right) - 2d_{\mathcal{K}}(y_\tau^{x,v})$$

$$\leq C|x_+ - x_-|^2 + 2\left|\frac{x_+ + x_-}{2} - y_\tau^{x,v}\right|$$

$$\leq C|h|^2. \qquad (8.36)$$

Let us now take v_+, v_- controls which are time-optimal for x_+, x_-, respectively, and let us set for simplicity $y_\pm = y_\tau^{x_\pm, v_\pm}$ to denote the terminal points of the corresponding trajectories. We have, by (8.6), (8.11) and (8.36), that

$$|y_+ - x_+| \leq NT(x_+) \leq Nkd_{\mathcal{K}}(x_+) \leq C|h|^2.$$

An analogous estimate holds for $|y_- - x_-|$. Using (8.35) we also deduce that

$$|y_+ - y_-| \leq |y_+ - x_+| + |x_+ - y_\tau^{x,v}| + |x_- - y_\tau^{x,v}| + |y_- - x_-| \leq C|h|,$$

$$|y_+ + y_- - 2y_\tau^{x,v}| \leq |y_+ - x_+| + |x_+ + x_- - 2y_\tau^{x,v}| + |y_- - x_-| \leq C|h|^2.$$

$$(8.37)$$

In addition

$$V(x_+) \leq \int_0^{T(x_+)} L(y^{x_+,v_+}(s), v_+(s))ds + g(y_+)$$
$$\leq T(x_+)M_R + g(y_+) \leq C|h|^2 + g(y_+).$$

An analogous estimate holds replacing x_+ with x_-. Therefore, using (8.31) and (8.37),

$$V(x_+) + V(x_-) - 2g(y_\tau^{x,v})$$
$$\leq C|h|^2 + g(y_+) + g(y_-) - 2g(y_\tau^{x,v})$$
$$\leq C|h|^2 + \Gamma_R|y_+ - y_-|^2 + G|y_+ + y_- - 2y_\tau^{x,v}| \leq C|h|^2,$$

which yields the desired semiconcavity estimate in view of (8.33) and (8.34).

Case 2: $\tau(x - h, v) \leq \min\{\tau(x, v), \tau(x + h, v)\}.$

The analysis of this case will conclude the proof since the remaining case $\tau(x + h, v) \leq \min\{\tau(x, v), \tau(x - h, v)\}$ is entirely symmetric. By the dynamic programming principle, we have

$$V(x - h) + V(x + h) - 2V(x)$$
$$\leq \int_0^{\tau(x-h,v)} [L(y^{x-h,v}(s), v(s)) + L(y^{x+h,v}(s), v(s)) - 2L(y^{x,v}(s), v(s))]ds$$
$$+V(y^{x+h,v}(\tau(x - h, v))) - 2V(y^{x,v}(\tau(x - h, v)))$$
$$g(y^{x-h,v}(\tau(x - h, v))).$$

As in the previous case, we find

$$\int_0^{\tau(x-h,v)} [L(y^{x-h,v}(s), v(s)) + L(y^{x+h,v}(s), v(s))$$
$$-2L(y^{x,v}(s), v(s))]ds \leq C|h|^2.$$

The estimation of the remaining terms is more complex. Let us first set

$$\tau_0 = \tau(x - h, v), \quad x_0 = y^{x-h,v}(\tau_0), \quad x_1 = y^{x,v}(\tau_0), \quad x_2 = y^{x+h,v}(\tau_0).$$

Then $x_0 \in \partial \mathcal{K}$ and we have, by Lemma 7.1.2, that

$$|x_0 - x_1| \leq C|h|, \quad |x_2 - x_1| \leq C|h|, \quad |x_0 + x_2 - 2x_1| \leq C|h|^2. \quad (8.38)$$

From the previous computations we see that the estimate needed to complete the proof is

$$g(x_0) + V(x_2) - 2V(x_1) \leq C|h|^2. \quad (8.39)$$

Let us define

$$\bar{v}(t) := v(t + \tau_0).$$

Then $\bar{v}$ is an optimal control for x_1 and $\tau(x_1, \bar{v}) = \tau(x, v) - \tau_0$. By Lemma 8.2.4 and estimates (8.11), (8.38) we have that

$$\tau(x_1, \bar{v}) \leq \beta k d_{\mathcal{K}}(x_1) \leq \beta k |x_1 - x_0| \leq C|h|. \qquad (8.40)$$

We will use the control $v^*(t) := \bar{v}(t/2)$ for the point x_2. For our future purposes, let us observe that, given any $s \in [0, \tau(x_1, \bar{v})]$,

$$|y^{x_1, \bar{v}}(s) - y^{x_2, v^*}(2s)|$$
$$= \left| x_1 + \int_0^s f(y^{x_1, \bar{v}}(\sigma), \bar{v}(\sigma)) \, d\sigma - x_2 - \int_0^{2s} f(y^{x_2, v^*}(\sigma), v^*(\sigma)) \, d\sigma \right|$$
$$\leq 3N\tau(x_1, \bar{v}) + |x_2 - x_1| \leq C|h|. \qquad (8.41)$$

We have again two cases which require a separate analysis.

(i): $\tau(x_2, v^*) \geq 2\tau(x_1, \bar{v})$.

Let $z = y^{x_2, v^*}(2\tau(x_1, \bar{v}))$. Then, by the dynamic programming principle, (8.40), (8.41) and (L1),

$$g(x_0) + V(x_2) - 2V(x_1)$$
$$\leq g(x_0) + V(z) + \int_0^{2\tau(x_1, \bar{v})} L(y^{x_2, v^*}(s), v^*(s)) ds$$
$$- 2g(y_\tau^{x_1, \bar{v}}) - 2\int_0^{\tau(x_1, \bar{v})} L(y^{x_1, \bar{v}}(s), \bar{v}(s)) ds$$
$$= g(x_0) + V(z) - 2g(y_\tau^{x_1, \bar{v}})$$
$$+ 2\int_0^{\tau(x_1, \bar{v})} [L(y^{x_2, v^*}(2s), \bar{v}(s)) - L(y^{x_1, \bar{v}}(s), \bar{v}(s))] ds$$
$$\leq g(x_0) + V(z) - 2g(y_\tau^{x_1, \bar{v}}) + C|h|^2. \qquad (8.42)$$

Similarly we have, using (8.40), (8.41) and (H1),

$$|x_0 + z - 2y_\tau^{x_1, \bar{v}}|$$
$$\leq |x_0 + x_2 - 2x_1| + 2\int_0^{\tau(x_1, \bar{v})} |f(y^{x_2, v^*}(2s), \bar{v}(s)) - f(y^{x_1, \bar{v}}(s), \bar{v}(s))| ds$$
$$\leq C|h|^2. \qquad (8.43)$$

Using the semiconcavity of $d_{\mathcal{K}}$ and estimates (8.38), (8.40), and recalling that x_0 and $y_\tau^{x_1, \bar{v}}$ both belong to $\partial \mathcal{K}$ we obtain

$$d_{\mathcal{K}}(2y_\tau^{x_1, \bar{v}} - x_0) = d_{\mathcal{K}}(2y_\tau^{x_1, \bar{v}} - x_0) + d_{\mathcal{K}}(x_0) - 2d_{\mathcal{K}}(y^{x_1, \bar{v}})$$
$$\leq C|2y_\tau^{x_1, \bar{v}} - 2x_0|^2$$
$$\leq C(2|y_\tau^{x_1, \bar{v}} - x_1| + 2|x_1 - x_0|)^2$$
$$\leq 4C(N^2\tau(x_1, \bar{v})^2 + |x_1 - x_0|^2) \leq C|h|^2.$$

Therefore, by (8.43), we find that $d_K(z) \leq C|h|^2$. Let now $\tilde{v}$ be an optimal control for z. From Lemma 8.2.4 and estimate (8.11) we deduce that

$$\tau(z, \tilde{v}) \leq \beta T(z) \leq \beta k d_K(z) \leq C|h|^2, \tag{8.44}$$

and so

$$|z - y_\tau^{z,\tilde{v}}| \leq N\tau(z, \tilde{v}) \leq C|h|^2. \tag{8.45}$$

Therefore

$$V(z) = \int_0^{\tau(z,\tilde{v})} L(y^{z,\tilde{v}}(s), \tilde{v}(s))\, ds + g(y_\tau^{z,\tilde{v}})$$
$$\leq M_R \tau(z, \tilde{v}) + g(y_\tau^{z,\tilde{v}}) \leq C|h|^2 + g(y_\tau^{z,\tilde{v}}).$$

Finally, we observe that by (8.38), (8.40) and (8.44),

$$|x_0 - y_\tau^{z,\tilde{v}}| \leq |x_0 - x_2| + |x_2 - z| + |z - y_\tau^{z,\tilde{v}}|$$
$$\leq C|h| + N[2\tau(x_1, \tilde{v}) + \tau(z, \tilde{v})] \leq C|h|,$$

while, by (8.43) and (8.45)

$$|x_0 + y_\tau^{z,\tilde{v}} - 2y_\tau^{x_1,\tilde{v}}| \leq |x_0 + z - 2y_\tau^{x_1,\tilde{v}}| + |z - y_\tau^{z,\tilde{v}}| \leq C|h|^2.$$

From (8.42), using the last three estimates and assumption (8.31), we deduce that

$$g(x_0) + V(x_2) - 2V(x_1) \leq g(x_0) + V(z) - 2g(y_\tau^{x_1,\tilde{v}}) + C|h|^2$$
$$\leq g(x_0) + g(y_\tau^{z,\tilde{v}}) - 2g(y_\tau^{x_1,\tilde{v}}) + C|h|^2$$
$$\leq \Gamma_R |x_0 - y_\tau^{z,\tilde{v}}|^2 + G|x_0 + y_\tau^{z,\tilde{v}} - 2y_\tau^{x_1,\tilde{v}}| + C|h|^2 \leq C|h|^2,$$

which proves (8.39) in this case.

(ii): $\tau(x_2, v^*) < 2\tau(x_1, \tilde{v})$.

Let us set $w = y^{x_1,\tilde{v}}(\tau(x_2, v^*)/2)$. Then, similarly to (8.43), we obtain that

$$|x_0 + y_\tau^{x_2,v^*} - 2w| \leq C|h|^2. \tag{8.46}$$

In addition, by (8.38) and (8.40),

$$|y_\tau^{x_2,v^*} - x_0| \leq |x_2 - x_0| + N\tau(x_2, v^*) \leq C|h|. \tag{8.47}$$

Using (8.6), we observe that

$$|w - y_\tau^{x_1,\tilde{v}}| = \left| \int_{\tau(x_2,v^*)/2}^{\tau(x_1,\tilde{v})} f(y^{x_1,\tilde{v}}(s), \tilde{v}(s)) ds \right|$$
$$\leq \frac{1}{G} \int_{\tau(x_2,v^*)/2}^{\tau(x_1,\tilde{v})} L(y^{x_1,\tilde{v}}(s), \tilde{v}(s)) ds.$$

By the dynamic programming principle,

$$g(x_0) + V(x_2) - 2V(x_1)$$

$$\leq g(x_0) + g(y_\tau^{x_2,v^*}) + \int_0^{\tau(x_2,v^*)} L(y^{x_2,v^*}(s), v^*(s))ds$$

$$-2g(y_\tau^{x_1,\bar{v}}) - 2\int_0^{\tau(x_1,\bar{v})} L(y^{x_1,\bar{v}}(s), \bar{v}(s))ds$$

$$\leq g(x_0) + g(y_\tau^{x_2,v^*}) - 2g(y_\tau^{x_1,\bar{v}})$$

$$+2\int_0^{\tau(x_2,v^*)/2} [L(y^{x_2,v^*}(2s), \bar{v}(s)) - L(y^{x_1,\bar{v}}(s), \bar{v}(s))]ds$$

$$-2\int_{\tau(x_2,v^*)/2}^{\tau(x_1,\bar{v})} L(y^{x_1,\bar{v}}(s), \bar{v}(s))ds$$

$$\leq g(x_0) + g(y_\tau^{x_2,v^*}) - 2g(y_\tau^{x_1,\bar{v}}) - 2G|w - y_\tau^{x_1,\bar{v}}|$$

$$+2\int_0^{\tau(x_2,v^*)/2} [L(y^{x_2,v^*}(2s), \bar{v}(s)) - L(y^{x_1,\bar{v}}(s), \bar{v}(s))]ds.$$

Now we observe that by (8.31), (8.46) and (8.47),

$$g(x_0) + g(y_\tau^{x_2,v^*}) - 2g(y_\tau^{x_1,\bar{v}})$$

$$\leq \Gamma_R|x_0 - y_\tau^{x_2,v^*}|^2 + G|x_0 + y_\tau^{x_2,v^*} - 2y_\tau^{x_1,\bar{v}}|$$

$$\leq (\Gamma_R + G)C|h|^2 + 2G|w - y_\tau^{x_1,\bar{v}}|.$$

Moreover, from (8.40), (8.41) and (L1),

$$\int_0^{\tau(x_2,v^*)/2} [L(y^{x_2,v^*}(2s), \bar{v}(s)) - L(y^{x_1,\bar{v}}(s), \bar{v}(s))]ds \leq C|h|^2.$$

Gathering the above estimates, we obtain (8.39). This completes the proof. ∎

In the case of the minimum time function, we have the following semiconcavity result.

Theorem 8.2.8 *Let the control system* (f, U) *with target* $\mathcal{K}$ *satisfy assumptions* (H0), (H1), (H2), (8.13) *and* (8.32). *Then the minimum time function belongs to* $SCL_{loc}(\mathcal{R} \setminus \mathcal{K})$.

Proof — The result is not an immediate consequence of the general theorem since we are not requiring here the global boundedness of f. However, one can check that this property is not needed when the final cost g is identically zero as in this case and that the same method of proof can be applied, with some simplification. For the details see [42]. ∎

Example 8.2.9 If the interior sphere condition (8.32) is not satisfied, then T may fail to be semiconcave. Consider for instance the system

$$y' = u,$$

with $y \in \mathbb{R}^2$ and $u(t) \in U = [-1, 1] \times [-1, 1]$. We take $\mathcal{K} = \{(0, 0)\}$ as our target, and so condition (8.32) is violated.

Let us analyze the time optimal trajectories starting from a given point $(x_1, x_2) \in \mathbb{R}^2$, $(x_1, x_2) \neq (0, 0)$. If $|x_1| = |x_2|$, then there exists a unique optimal control, given by $u = (-\operatorname{sgn} x_1, -\operatorname{sgn} x_2)$. If $|x_1| \neq |x_2|$ there are infinitely many optimal controls. Suppose for instance $x_1 > x_2 \geq 0$; then we can take a control whose first component is identically -1, while the other one can be any function $u_2 : [0, x_1] \to [-1, 1]$ such that $\int_0^{x_1} u_2(t) \, dt = -x_2$. In both cases, we see that

$$T(x_1, x_2) = \max\{|x_1|, |x_2|\}.$$

Observe that T is convex and is not differentiable at the points (x_1, x_2) with $|x_1| = |x_2|$. Hence T is not semiconcave. The example also shows that in general the minimum time function can exhibit a different behavior from the distance function. We know in fact that the distance function $d_{\mathcal{K}}$ is semiconcave away from $\mathcal{K}$ for any set $\mathcal{K}$ (see Proposition 2.2.2); the interior sphere condition is only necessary to have semiconcavity up to the boundary of $\mathcal{K}$. The minimum time function of this example, instead, is not even semiconcave in sets of the form $\mathbb{R}^2 \setminus A$, with A bounded neighborhood of the target. Observe also that the points where T is not differentiable are exactly the ones where the time optimal trajectory is unique. Such a behavior is in strong contrast with the one exhibited by the problems in the previous chapters (see e.g., Theorem 6.4.9). ∎

8.3 Semiconvexity results in the linear case

In this section we restrict our attention to the minimum time problem for a linear control system of the form

$$\begin{cases} y'(t) = Ay(t) + u(t) & u(t) \in U \subset \mathbb{R}^n, \\ y(0) = x \in \mathbb{R}^n \end{cases} \tag{8.48}$$

where $A \in \mathbb{R}^{n \times n}$ and U is compact. Such a system satisfies (H0), (H1) and (H2). In our following results we assume that U is convex and that a target set $\mathcal{K}$ is given such that the Petrov condition (8.13) is satisfied. Let us mention that for a system of the above form the minimum time function for this system does not change if we substitute U with its convex hull (see e.g., [111]). Thus, our assumption that U is convex is not restrictive for the analysis of the properties of $T(x)$.

The next result shows that if the target is a convex set, then the minimum time function is semiconvex.

Theorem 8.3.1 *Let the linear system (8.48) satisfy condition (8.13), and let $\mathcal{K}$ and U be convex. Then the minimum time function T associated with the system is locally semiconvex in $\mathcal{R} \setminus \mathcal{K}$ with a linear modulus.*

Proof. We first prove the following:

$$x_1, x_2 \in \mathcal{R}, \ T(x_1) = T(x_2) \implies T\left(\frac{x_1 + x_2}{2}\right) \leq T(x_1). \tag{8.49}$$

In fact, taken u_1, u_2 optimal controls for x_1, x_2, we can define

$$\bar{u}(t) = \frac{u_1(t) + u_2(t)}{2},$$

which is an admissible control, by the convexity of U. Let us set for simplicity $x_{1/2} := \frac{x_1 + x_2}{2}$. Since the state equation (8.48) is linear, we find

$$y^{x_{1/2}, \bar{u}}(t) = \frac{1}{2}(y^{x_1, u_1}(t) + y^{x_2, u_2}(t)).$$

Therefore, since $\mathcal{K}$ is convex,

$$y^{x_{1/2}, \bar{u}}(T(x_1)) \in \mathcal{K},$$

and therefore

$$T\left(\frac{x_1 + x_2}{2}\right) \leq T(x_1).$$

Let us now take x_1, x_2 in a fixed convex set $Q \subset\subset \mathcal{R} \setminus \mathcal{K}$, and suppose for instance $T(x_1) \leq T(x_2)$. Let u_2 be an optimal control for x_2. We define

$$\bar{u}(t) = u_2(2t),$$

$$\tilde{y}(\cdot) = y^{x_2, u_2}(\cdot) \qquad \bar{y}(\cdot) = y^{x_{1/2}, \bar{u}}(\cdot),$$

$$\tilde{x} = \tilde{y}(T(x_2) - T(x_1)) \qquad \bar{x} = \bar{y}\left(\frac{T(x_2) - T(x_1)}{2}\right).$$

Then, by the dynamic programming principle,

$$T(\tilde{x}) = T(x_1) \qquad T\left(\frac{x_1 + x_2}{2}\right) \leq T(\bar{x}) + \frac{T(x_2) - T(x_1)}{2}. \tag{8.50}$$

Since T is locally Lipschitz, $|T(x_2) - T(x_1)| \leq C|x_2 - x_1|$ (with C we denote a positive constant which only depends on Q, and which may be different from one formula to another). Moreover, the time-optimal trajectories starting from Q are uniformly bounded and hence their speed is also uniformly bounded. Therefore, for every $t \in [0, T(x_2) - T(x_1)]$,

$$|\bar{y}(t/2) - \tilde{y}(t)| \leq |\bar{y}(0) - \tilde{y}(0)| + 3Ct/2 \leq C|x_2 - x_1|.$$

Furthermore,

$$
|2\bar{x} - x_1 - \tilde{x}| = \left| 2 \int_0^{\frac{T(x_2)-T(x_1)}{2}} \bar{y}'(t)\, dt - \int_0^{T(x_2)-T(x_1)} \tilde{y}'(t)\, dt \right|
$$

$$
= \left| \int_0^{T(x_2)-T(x_1)} A(\bar{y}(t/2) - \tilde{y}(t))\, dt \right|
$$

$$
\leq (T(x_2) - T(x_1)) \|A\| \max_{t \in [0, T(x_2)-T(x_1)]} |\bar{y}(t/2) - \tilde{y}(t)|
$$

$$
\leq C|x_2 - x_1|^2.
$$

Using again the Lipschitz continuity of T we obtain

$$
T(\bar{x}) \leq T\left(\frac{x_1 + \tilde{x}}{2}\right) + C|x_2 - x_1|^2.
$$

On the other hand, by (8.49) and (8.50),

$$
T\left(\frac{x_1 + \tilde{x}}{2}\right) \leq T(x_1).
$$

Thus, by (8.50), we conclude that

$$
T\left(\frac{x_1 + x_2}{2}\right) \leq T(\bar{x}) + \frac{T(x_2) - T(x_1)}{2} \leq \frac{T(x_2) + T(x_1)}{2} + C|x_2 - x_1|^2. \quad \blacksquare
$$

Remark 8.3.2 It is easy to see that even under the assumptions of Theorem 8.3.1, the minimum time function is not convex in general. Let us consider for instance the one-dimensional system with state equation

$$
y' = -y + u, \qquad u \in U = [-1, 1]
$$

and target $\mathcal{K} = \{0\}$. Clearly, the time-optimal control for a point $x < 0$ (resp. $x > 0$) is given by $u \equiv 1$ (resp. $u \equiv -1$). The corresponding optimal trajectory is

$$
x(t) = \mathrm{sgn}(x)[(1 + |x|)e^{-t} - 1],
$$

which reaches the target at a time $T(x) = \ln(1 + |x|)$. Thus T is semiconvex (it is even smooth outside the target) but is not convex. $\blacksquare$

The semiconvexity result of Theorem 8.3.1 cannot be extended to nonlinear systems, as shown by the example below.

Example 8.3.3 For any $(x_1, x_2) \in \mathbb{R}^2$ consider the minimum time problem with target $\mathcal{K} = \{0\}$ for the control system

$$\begin{cases} y_1'(t) = -[y_2(t)]^2 + u_1(t), & y_1(0) = x_1 \\ y_2'(t) = u_2(t), & y_2(0) = x_2. \end{cases} \tag{8.51}$$

We take $U = [-1, 1]^2$ as a control set. For this problem it is easy to see that $\mathcal{R} = \mathbb{R}^2$. Moreover, by Theorem 8.1.2, there exists a time-optimal control for any initial point.

Let $x_1 > 0$, $x_2 \geq 0$. We define, for any $\tau \geq 0$, a control $u_\tau = (u_{\tau,1}, u_{\tau,2})$ as follows:

$$u_{\tau,1}(t) = -1 \qquad t \geq 0,$$

$$u_{\tau,2}(t) = \begin{cases} 1 & 0 \leq t \leq \tau \\ -1 & t > \tau. \end{cases}$$

The trajectory of (8.51) associated with u_τ is $y_\tau = (y_{\tau,1}, y_{\tau,2})$, where

$$y_{\tau,2}(t) = \begin{cases} x_2 + t, & 0 \leq t \leq \tau \\ x_2 + 2\tau - t, & t \geq \tau, \end{cases}$$

$$y_{\tau,1}(t) = x_1 - t - \int_0^t [y_{\tau,2}(s)]^2 ds.$$

Now we keep x_1 fixed and show that if x_2 is smaller than some constant $\varepsilon(x_1)$, the curve y_τ hits the origin at time $x_2 + 2\tau$, for a suitable τ. In fact

$$y_{\tau,2}(x_2 + 2\tau) = 0,$$

while

$$y_{\tau,1}(x_2 + 2\tau) = x_1 - x_2 - 2\tau - \frac{2}{3}(x_2 + \tau)^3 + \frac{x_2^3}{3}.$$

The right-hand side is a decreasing function of τ and tends to $-\infty$ as $\tau \to +\infty$. It is positive for $\tau = 0$, provided x_2 is small enough. Hence there exists $\bar{\tau}$ such that $y_{\bar{\tau},1}(x_2 + 2\bar{\tau}) = 0$.

We claim that $y_{\bar{\tau}}$ is optimal. Indeed, let y^* be an optimal trajectory at $x = (x_1, x_2)$. Since

$$\left| \frac{dy_2^*}{dt} \right| \leq 1, \; y_2^*(0) = x_2, \; y_2^*(T(x)) = 0,$$

it follows that

$$|y_2^*(t)| \leq y_{\bar{\tau},2}(t), \; \forall t \in [0, T(x)].$$

Therefore,

$$0 \leq y_{\bar{\tau},1}(t) \leq y_1^*(t), \; \forall t \in [0, T(x)],$$

which yields $y_{\bar{\tau}}(T(x)) = 0$. Hence $y_{\bar{\tau}}$ is optimal and $\bar{\tau} = (T(x) - x_2)/2$.

From the above formula for $y_{\bar{\tau},1}$, it follows that $T(x)$ is a solution of

$$0 = x_1 + \frac{x_2^3}{3} - T - \frac{2}{3}\left(\frac{T + x_2}{2}\right)^3.$$

for $x_1 > 0$ and $0 < x_2 < \varepsilon(x_1)$. Then, by the implicit function theorem, T is of class C^1 at such points and

$$\frac{\partial T}{\partial x_2}(x) = -\frac{\left(\frac{T(x)+x_2}{2}\right)^2 - x_2^2}{1 + \left(\frac{T(x)+x_2}{2}\right)^2}.$$

Hence,

$$\lim_{x_2 \downarrow 0} \frac{\partial T}{\partial x_2}(x_1, x_2) = -\frac{T^2(x_1, 0)}{4 + T^2(x_1, 0)} < 0. \tag{8.52}$$

Since the problem is symmetric with respect to the x_2 variable, we have that $T(x_1, -x_2) = T(x_1, x_2)$. Thus,

$$\lim_{x_2 \uparrow 0} \frac{\partial T}{\partial x_2}(x_1, x_2) = -\lim_{x_2 \downarrow 0} \frac{\partial T}{\partial x_2}(x_1, x_2) > 0. \tag{8.53}$$

In particular,

$$\lim_{h \downarrow 0} \frac{T(x_1, h) + T(x_1, -h) - 2T(x_1, 0)}{h} < 0,$$

showing that T fails to be semiconvex in any neighborhood of $(x_1, 0)$.   ∎

We now show that the results of the previous sections, together with some hypotheses on the target or the control set, can be used to obtain regularity results for the minimum time function. We begin with an immediate consequence of Corollary 8.2.8 and Theorem 8.3.1.

Theorem 8.3.4 *Let the linear system* (8.48) *satisfy assumption* (8.13), *let U be compact and convex, let $\mathcal{K}$ be convex, and let $\partial \mathcal{K}$ be of class $C^{1,1}$. Then the minimum time function T belongs to $C^{1,1}_{loc}(\mathcal{R} \setminus \mathcal{K})$.*

Proof. By Corollary 8.2.8 and Theorem 8.3.1, T is both semiconcave and semiconvex with a linear modulus. Thus, by Corollary 3.3.8, it is locally of class $C^{1,1}$.   ∎

If $\mathcal{K}$ has not a smooth boundary, it is still possible to obtain a bound on the dimension of $D^- T$ depending on the regularity of the control set U, as shown by the next theorem.

Theorem 8.3.5 *Let $\mathcal{K}$ and U be convex sets, let U be compact, and let the linear system* (8.48) *satisfy assumption* (8.13). *Define*

$$\nu = \max_{\bar{u} \in U} \dim N_U(\bar{u}).$$

Then, for all $x \in \mathcal{R} \setminus \mathcal{K}$, the dimension of the subdifferential $D^- T(x)$ is strictly less than ν.

Proof — Since T is a viscosity solution of equation (8.10), we have, recalling Proposition 5.3.4, that

$$\sigma_U(-p) - Ax \cdot p = 1, \qquad \forall x \in \mathcal{R} \setminus \mathcal{K}, \ \forall \, p \in D^- T(x). \qquad (8.54)$$

Such an equality implies that, for all $x \in \mathcal{R} \setminus \mathcal{K}$, the set $M := D^- T(x)$ satisfies property (ii) of Lemma A. 1.24 (with $V = U$), and therefore is contained in some normal cone to U. Again by (8.54), M satisfies condition (A. 8). Hence, by Lemma A. 1.25, its dimension is less than or equal to $\nu - 1$. The conclusion follows. ∎

Corollary 8.3.6 *Under the hypotheses of Theorem 8.3.5, the subdifferential of T has dimension strictly less than n at every point.*

Corollary 8.3.7 *Under the hypotheses of Theorem 8.3.5, if ∂U is of class C^1, then $T \in C^1(\mathcal{R} \setminus \mathcal{K})$.*

Proof — Since T is semiconvex and $D^- T$ is a singleton for all $x \in \mathcal{R} \setminus \mathcal{K}$, we deduce from Proposition 3.3.4(e) that $T \in C^1(\mathcal{R} \setminus \mathcal{K})$. ∎

8.4 Optimality conditions

In this section we analyze some optimality conditions for control problems of the form (ETP). Our aim is to obtain results analogous to those for the finite horizon problems given in the previous chapters (see Section 6.4 or 7.3); the statements and the methods of proof require suitable adaptations due to the presence of the exit time.

Unless otherwise stated, we assume throughout the section properties (H0), (H1), (H2), the compatibility condition (8.6), the Petrov condition (8.13), and the following additional regularity requirements:

$$L_x(x, u) \text{ exists everywhere and is continuous w.r.t. } x; \qquad (8.55)$$

$$\partial \mathcal{K} \text{ is an } (n - 1)\text{–dimensional manifold of class } C^{1,1}; \qquad (8.56)$$

$$g \in \mathrm{SCL}_{loc}(\mathcal{N}), \text{ where } \mathcal{N} \text{ is a neighborhood of } \mathcal{K}. \qquad (8.57)$$

Let us observe that property (8.56) easily implies the interior sphere condition (8.32). In addition, the set of proximal normals to a point $z \in \partial \mathcal{K}$ which appears in the Petrov condition (8.13) reduces to the usual outer normal to $\mathcal{K}$ at z and its positive multiples. Observe also that (8.57), together with (8.6), imply that g satisfies (8.31).

We begin by proving a version of Pontryagin's maximum principle. Let us mention that it is possible to give more general statements of this principle, which include cases where the target is a point or a nonsmooth set (see e.g., [87, 111, 54]). We start with two preliminary results.

Lemma 8.4.1 *Let $z \in \partial K$ and let v be the outer normal to K at z. Let $u^* \in U$ be such that $f(z, u^*) \cdot v < 0$ and consider the exit time $\tau(x, u^*)$ under the constant control u^*. Then there exists a neighborhood of z, which we denote by A, such that*

$$\tau(x, u^*) = -\frac{v \cdot (x - z)}{v \cdot f(z, u^*)} + o(|x - z|), \qquad \forall x \in A \setminus K. \qquad (8.58)$$

Proof – Let us consider the signed distance function from K, defined as

$$\bar{d}_K(x) := d_K(x) - d_{K^c}(x).$$

Our smoothness assumption (8.56) on ∂K implies that $\bar{d}_K$ is differentiable near z and $D\bar{d}_K(z) = v$. Let us define

$$F(x, t) = \bar{d}_K(y^{x, u^*}(t)).$$

Then F is differentiable for (x, t) near $(z, 0)$ and satisfies

$$F(z, 0) = 0, \qquad F_t(z, 0) = v \cdot f(z, u^*) \neq 0.$$

By the implicit function theorem we can find a neighborhood A of z, a number $\delta > 0$ and a function $s : A \to [-\delta, \delta]$ such that for all $x \in A$ and $t \in [-\delta, \delta]$,

$$\bar{d}_K(y^{x, u^*}(t)) = 0 \iff t = s(x). \qquad (8.59)$$

We also take A suitably small in such a way that $A \setminus \partial K$ consists of two connected components A_+ and A_-, with $A_+ = A \setminus K$ and $A_- \subset K$.

From (8.59) we see that $s(x)$ coincides with $\tau(x, u^*)$ for all $x \in A_+$ such that $s(x) > 0$. In addition $s(x) = 0$ if and only if $x \in \partial K$, and therefore $s(x)$ has constant sign on A_+. Moreover we have

$$Ds(z) = -\frac{F_x(z, 0)}{F_t(z, 0)} = -\frac{v}{v \cdot f(z, u^*)},$$

which implies

$$s(x) = Ds(z) \cdot (x - z) + o(|x - z|) = -\frac{v \cdot (x - z)}{v \cdot f(z, u^*)} + o(|x - z|).$$

In particular, since we assume $f(z, u^*) \cdot v < 0$, we have that $s(x) > 0$ if $x = z + \varepsilon v$ with $\varepsilon > 0$ small enough. By the previous remarks, we deduce that $s(x)$ is positive on A_+. Thus, it coincides with $\tau(x, u^*)$ for $x \in A_+ = A \setminus K$. This proves (8.58). ∎

Lemma 8.4.2 *Given $z \in \partial K$, let v be the outer normal to K at z. Then, for any $q \in D^+g(z)$ there exists a unique $\mu > 0$ such that $H(z, q + \mu v) = 0$.*

Proof — Since g is Lipschitz continuous with constant G, any vector $q \in D^+g(z)$ satisfies $|q| \leq G$. Therefore, by (8.6),

$$H(z, q) = \max_{u \in U}[-f(z, u) \cdot q - L(z, u)] \leq NG - \alpha < 0.$$

On the other hand, by (8.13) there exists $v \in U$ such that $-f(z, v) \cdot v > 0$. We have, for any $\mu \geq 0$,

$$H(z, q + \mu v) \geq -f(z, v) \cdot q - L(z, v) - \mu f(z, v) \cdot v.$$

Therefore,

$$H(z, q + \mu v) \to +\infty \quad \text{as} \quad \mu \to +\infty.$$

This implies that there exists $\mu > 0$ such that $H(z, q + \mu v) = 0$.

Now we prove the uniqueness of μ. Arguing by contradiction we suppose that there exist μ_1, μ_2, with $0 < \mu_1 < \mu_2$ such that $H(z, q + \mu_1 v) = H(z, q + \mu_2 v) = 0$. Let us denote by u^* an element of U such that

$$H(z, q + \mu_1 v) = -f(x, u^*) \cdot (q + \mu_1 v) - L(z, u^*).$$

Then

$$
\begin{aligned}
0 &= -f(z, u^*) \cdot q - L(z, u^*) - \mu_1 f(z, u^*) \cdot v \\
&\leq NG - \alpha - \mu_1 f(z, u^*) \cdot v < -\mu_1 f(z, u^*) \cdot v.
\end{aligned}
$$

Therefore $f(z, u^*) \cdot v < 0$, and this implies

$$
\begin{aligned}
H(z, q + \mu_2 v) &\geq -f(z, u^*) \cdot (q + \mu_2 v) - L(z, u^*) \\
&= -f(z, u^*) \cdot (q + \mu_1 v) - L(z, u^*) - (\mu_2 - \mu_1) f(z, u^*) \cdot v \\
&> 0,
\end{aligned}
$$

which contradicts our assumption that $H(z, q + \mu_2 v) = 0$. ∎

We are now ready to state the maximum principle.

Theorem 8.4.3 *Let properties* (H0)–(H2), (L1), (8.6), (8.13), (8.55)–(8.57) *hold, let* $x \in \mathcal{R} \setminus \mathcal{K}$ *and let* $\bar{u}$ *be an optimal control for* x. *Set for simplicity*

$$y(t) := y^{x, \bar{u}}(t), \qquad \tau := \tau(x, \bar{u}), \qquad z := y_\tau^{x, \bar{u}},$$

and denote by v *the outer normal to* $\mathcal{K}$ *at* z. *Given* $q \in D^+g(z)$, *let* $\mu > 0$ *be such that* $H(z, q + \mu v) = 0$ (μ *is uniquely determined by the previous lemma). Let* $p : [0, \tau] \to \mathbb{R}^n$ *be the solution to the system*

$$
\begin{cases}
p'(t) = -f_x^T(y(t), \bar{u}(t))p(t) - L_x(y(t), \bar{u}(t)) \\
p(\tau) = q + \mu v,
\end{cases}
\tag{8.60}
$$

where f_x^T *denotes the transpose of the matrix* f_x. *Then* p *satisfies, for a.e.* $t \in [0, \tau]$,

$$
\begin{aligned}
&-p(t) \cdot f(y(t), \bar{u}(t)) - L(y(t), \bar{u}(t)) \\
&= \max_{u \in U}[-p(t) \cdot f(y(t), u) - L(y(t), u)].
\end{aligned}
\tag{8.61}
$$

Proof — Let $t \in \,]0, \tau[$ be a Lebesgue point for the functions $f(y(\cdot), \bar{u}(\cdot))$ and $L(y(\cdot), \bar{u}(\cdot))$, i.e., a value such that

$$\lim_{h \to 0} \frac{1}{h} \int_{t-h}^{t+h} |f(y(s), \bar{u}(s)) - f(y(t), \bar{u}(t))| \, ds = 0, \qquad (8.62)$$

$$\lim_{h \to 0} \frac{1}{h} \int_{t-h}^{t+h} |L(y(s), \bar{u}(s)) - L(y(t), \bar{u}(t))| \, ds = 0. \qquad (8.63)$$

Let us fix $u \in U$. For $\varepsilon > 0$ small we define

$$u_\varepsilon(s) = \begin{cases} u & s \in [t - \varepsilon, t] \\ \bar{u}(s) \ s \in [0, \tau]/[t - \varepsilon, t] \\ u^* & s > \tau, \end{cases} \qquad (8.64)$$

where $u^* \in U$ is such that

$$-f(z, u^*) \cdot (q + \mu v) - L(z, u^*) = H(z, q + \mu v) = 0. \qquad (8.65)$$

We set $y_\varepsilon(\cdot) := y^{x, u_\varepsilon}(\cdot)$ and $\tau_\varepsilon := \tau(x, u_\varepsilon)$. From Lemma 8.4.1 we deduce that τ_ε is finite if ε is small enough.

Since $\bar{u}$ is an optimal control for x, we have

$$0 \le J(x, u_\varepsilon) - J(x, \bar{u})$$
$$= \int_0^{\tau_\varepsilon} L(y_\varepsilon(s), u_\varepsilon(s)) ds - \int_0^{\tau} L(y(s), \bar{u}(s)) ds + g(y_\varepsilon(\tau_\varepsilon)) - g(z)$$
$$= \int_{t-\varepsilon}^{\tau} [L(y_\varepsilon(s), \bar{u}(s)) - L(y(s), \bar{u}(s))] ds + \int_{\tau}^{\tau_\varepsilon} L(y_\varepsilon(s), u_\varepsilon(s)) ds$$
$$- \int_{t-\varepsilon}^{t} [L(y_\varepsilon(s), \bar{u}(s)) - L(y_\varepsilon(s), u)] ds + g(y_\varepsilon(\tau_\varepsilon)) - g(z). \qquad (8.66)$$

Taking into account (8.62), we find

$$y_\varepsilon(t) - y(t) = \int_{t-\varepsilon}^{t} [f(y_\varepsilon(s), u) - f(y(s), \bar{u}(s))] ds$$
$$= \varepsilon(f(y(t), u) - f(y(t), \bar{u}(t))) + o(\varepsilon).$$

Therefore, by well-known results about ordinary differential equations (see Theorem A. 4.4),

$$y_\varepsilon(s) = y(s) + \varepsilon v(s) + o(\varepsilon), \quad s \in [t, \tau], \qquad (8.67)$$

where $v(\cdot)$ is the solution to the linearized system

$$\begin{cases} v'(s) = f_x(y(s), \bar{u}(s)) v(s) \\ v(t) = f(y(t), u) - f(y(t), \bar{u}(t)). \end{cases} \qquad (8.68)$$

Let us observe that

$$\frac{d}{ds}\, p(s) \cdot v(s) = -L_x(y(s), \bar{u}(s)) \cdot v(s),$$

and therefore

$$\int_t^\tau L_x(y(s), \bar{u}(s)) \cdot v(s)\, ds = p(t) \cdot v(t) - p(\tau) \cdot v(\tau)$$
$$= p(t)[f(y(t), u) - f(y(t), \bar{u}(t)] - (q + \mu v)v(\tau). \qquad (8.69)$$

It is now convenient to treat separately the cases $\tau_\varepsilon \geq \tau$ and $\tau_\varepsilon < \tau$.

(i) $\tau_\varepsilon < \tau$

We first observe that, by (8.67), the point $y(\tau_\varepsilon)$ has distance of order $O(\varepsilon)$ from $y_\varepsilon(\tau_\varepsilon) \in \partial K$, and so $d_K(y(\tau_\varepsilon)) = O(\varepsilon)$. Since the optimal control $\bar{u}$ steers $y(\tau_\varepsilon)$ to the target in a time $\tau - \tau_\varepsilon$, we obtain from Lemma 8.2.4 and estimate (8.11) that

$$\tau - \tau_\varepsilon = O(\varepsilon). \qquad (8.70)$$

Since $q \in D^+ g(z)$, we have

$$g(y_\varepsilon(\tau_\varepsilon)) - g(z) \leq q \cdot (y_\varepsilon(\tau_\varepsilon) - z) + o(|y_\varepsilon(\tau_\varepsilon) - z|).$$

By (8.6), (8.67) and (8.70) we have, for any $s \in [\tau_\varepsilon, \tau]$,

$$|y_\varepsilon(s) - z| \leq \int_s^\tau |f(y_\varepsilon(s), u_\varepsilon(s))|ds + |y_\varepsilon(\tau) - z|$$
$$\leq N(\tau - s) + |\varepsilon v(\tau) + o(\varepsilon)| = O(\varepsilon).$$

Therefore,

$$g(y_\varepsilon(\tau_\varepsilon)) - g(z) \leq q \cdot (y_\varepsilon(\tau_\varepsilon) - z) + o(\varepsilon)$$
$$= q \cdot \left(-\int_{\tau_\varepsilon}^\tau f(y_\varepsilon, u_\varepsilon(s))ds + \varepsilon v(\tau) \right) + o(\varepsilon)$$
$$= q \cdot \left(-\int_{\tau_\varepsilon}^\tau f(z, \bar{u}(s))ds + \varepsilon v(\tau) \right) + o(\varepsilon). \qquad (8.71)$$

By the smoothness of ∂K we have, since $y_\varepsilon(\tau_\varepsilon) \in \partial K$, that

$$v \cdot (z - y_\varepsilon(\tau_\varepsilon)) = O(|z - y_\varepsilon(\tau_\varepsilon)|^2) = o(\varepsilon).$$

Recalling also that

$$0 = H(z, q + \mu v) = \max_{u \in U}[-f(z, u) \cdot (q + \mu v) - L(z, u)],$$

we obtain

$$- \int_{\tau_\varepsilon}^{\tau} L(z, \bar{u}(s)) ds - q \cdot \int_{\tau_\varepsilon}^{\tau} f(z, \bar{u}(s)) ds$$

$$\leq \mu v \cdot \int_{\tau_\varepsilon}^{\tau} f(z, \bar{u}(s)) ds$$

$$= \mu v \cdot (z - y(\tau_\varepsilon)) + \mu v \cdot \int_{\tau_\varepsilon}^{\tau} [f(z, \bar{u}(s)) - f(y(s), \bar{u}(s))] ds$$

$$= \mu v \cdot (z - y(\tau_\varepsilon)) + o(\varepsilon)$$

$$= \mu v \cdot (z - y_\varepsilon(\tau_\varepsilon)) + \mu v \cdot (y_\varepsilon(\tau_\varepsilon) - y(\tau_\varepsilon)) + o(\varepsilon)$$

$$= \varepsilon \mu v \cdot v(\tau_\varepsilon) + o(\varepsilon) = \varepsilon \mu v \cdot v(\tau) + o(\varepsilon).$$

Thus (8.66) implies, using also (8.71)

$$0 \leq \int_{t-\varepsilon}^{\tau} [L(y_\varepsilon(s), \bar{u}(s)) - L(y(s), \bar{u}(s))] ds + \varepsilon (q + \mu v) \cdot v(\tau)$$

$$- \int_{t-\varepsilon}^{t} [L(y_\varepsilon(s), \bar{u}(s)) - L(y_\varepsilon(s), u)] ds + o(\varepsilon).$$

Dividing by ε and letting $\varepsilon \to 0^+$ we obtain, by (8.63), (8.67) and (8.69),

$$0 \leq \int_t^{\tau} L_x(y(s), \bar{u}(s)) \cdot v(s) ds + (q + \mu v) \cdot v(\tau)$$

$$+ L(y(t), u) - L(y(t), \bar{u}(t))$$

$$= p(t) \cdot [f(x(t), u) - f(x(t), \bar{u}(t))] + L(x(t), u) - L(x(t), \bar{u}(t)).$$

Since $u \in U$ is arbitrary, this proves (8.61).

(ii) $\tau_\varepsilon \geq \tau$

By (8.67), $y_\varepsilon(\tau) - z = \varepsilon v(\tau) + o(\varepsilon)$; thus, for ε small enough we can apply Lemma 8.4.1 to obtain

$$\tau_\varepsilon - \tau = \frac{(y_\varepsilon(\tau) - z) \cdot v}{-f(z, u^*) \cdot v} + o(\varepsilon) = \frac{\varepsilon v(\tau) \cdot v}{-f(z, u^*) \cdot v} + o(\varepsilon). \qquad (8.72)$$

This implies in particular that

$$\tau_\varepsilon - \tau = O(\varepsilon),$$

$$|y_\varepsilon(s) - z| \leq |y_\varepsilon(s) - y_\varepsilon(\tau)| + |y_\varepsilon(\tau) - z| = O(\varepsilon) \qquad \forall t \in [\tau, \tau_\varepsilon].$$

Thus we have

$$\int_\tau^{\tau_\varepsilon} L(y_\varepsilon(s), u_\varepsilon(s)) ds = (\tau_\varepsilon - \tau) L(z, u^*) + \int_\tau^{\tau_\varepsilon} [L(y_\varepsilon(s), u^*) - L(z, u^*)] ds$$

$$= (\tau_\varepsilon - \tau) L(z, u^*) + o(\varepsilon),$$

$$y_\varepsilon(\tau_\varepsilon) - z = \int_\tau^{\tau_\varepsilon} f(y_\varepsilon(s), u^*) ds + y_\varepsilon(\tau) - z$$

$$= (\tau_\varepsilon - \tau) f(z, u^*) + \varepsilon v(\tau) + o(\varepsilon).$$

Taking also into account that $q \in D^+g(z)$, we obtain from (8.66),

$$
0 \leq \int_{t-\varepsilon}^{\tau} [L(y_\varepsilon(s), \bar{u}(s)) - L(y(s), \bar{u}(s))]ds + (\tau_\varepsilon - \tau)L(z, u^*)
$$
$$
- \int_{t-\varepsilon}^{t} [L(y_\varepsilon(s), \bar{u}(s)) - L(y(s), \bar{u}(s))] ds
$$
$$
+ (\tau_\varepsilon - \tau)q \cdot f(z, u^*) + \varepsilon q \cdot v(\tau) + o(\varepsilon). \tag{8.73}
$$

By definition of u^* we have

$$
L(z, u^*) + q \cdot f(z, u^*) = -\mu v \cdot f(z, u^*).
$$

Recalling (8.72) we obtain

$$
(\tau_\varepsilon - \tau)(L(z, u^*) + q \cdot f(z, u^*)) = -\mu(\tau_\varepsilon - \tau)f(z, u^*) \cdot v = \varepsilon\mu v(\tau) \cdot v + o(\varepsilon).
$$

Thus (8.73) implies that

$$
0 \leq \int_{t-\varepsilon}^{\tau} [L(y_\varepsilon(s), \bar{u}(s)) - L(y(s), \bar{u}(s))]ds
$$
$$
- \int_{t-\varepsilon}^{t} [L(y_\varepsilon(s), \bar{u}(s)) - L(y(s), \bar{u}(s))] ds
$$
$$
+ \varepsilon(\mu v + q) \cdot v(\tau) + o(\varepsilon).
$$

Now we can divide by ε and let $\varepsilon \to 0$ to obtain the conclusion as in the first step. ∎

Given an optimal trajectory y, we will say that p is a *dual arc* associated with y if it satisfies the properties of Theorem 8.4.3, that is, if it solves problem (8.60) for some $q \in D^+g(z)$ and $\mu > 0$ such that $H(z, q+\mu v) = 0$. We have seen an existence result for dual arcs; we will see later that under a suitable additional condition the dual arc associated to an optimal trajectory is unique. We now prove that the dual arc p of the previous theorem is included in the superdifferential of the value function.

Theorem 8.4.4 *Under the assumptions of Theorem 8.4.3, the arc p solution of system (8.60) satisfies*

$$
p(t) \in D^+V(y(t)), \quad \forall t \in [0, \tau).
$$

Proof — For simplicity of notation we restrict ourselves to the case $t = 0$, the proof in the general case being entirely analogous. The statement for $t = 0$ becomes $p(0) \in D^+V(x)$; to prove this we will show that for all $h \in \mathbb{R}^n$ with $|h| = 1$,

$$
V(x + \varepsilon h) \leq V(x) + \varepsilon p(0) \cdot h + o(\varepsilon) \tag{8.74}
$$

with $o(\varepsilon)$ independent of h. Let us define a control strategy by setting

$$
\hat{u}(s) = \begin{cases} \bar{u}(s) & s \in [0, \tau] \\ u^* & s > \tau, \end{cases} \tag{8.75}
$$

where u^* is such that

$$-f(z, u^*) \cdot (q + \mu v) - L(z, u^*) = H(z, q + \mu v) = 0. \qquad (8.76)$$

Let us set for simplicity

$$y_\varepsilon(\cdot) := y^{x + \varepsilon h, \hat{u}}(\cdot), \qquad \tau_\varepsilon := \tau(x + \varepsilon h, \hat{u}).$$

Then we have, for all $s \in [0, \tau]$,

$$y_\varepsilon(s) = y(s) + \varepsilon v(s) + o(\varepsilon), \qquad (8.77)$$

where v is the solution of the linearized problem

$$\begin{cases} v'(s) = f_x(y(s), \bar{u}(s))v(s) & s \in [0, t] \\ v(0) = h \end{cases}$$

and $o(\varepsilon)$ does not depend on h. Now we observe that

$$\begin{aligned} V(x + \varepsilon h) - V(x) &\leq J(x + \varepsilon h, \hat{u}) - V(x) \\ &= \int_0^\tau (L(y_\varepsilon(s), \bar{u}(s)) - L(y(s), \bar{u}(s))) \, ds \\ &\quad + \int_\tau^{\tau_\varepsilon} L(y_\varepsilon(s), \hat{u}(s)) ds + g(y_\varepsilon(\tau_\varepsilon)) - g(z). \end{aligned}$$

The terms on the right-hand side can be estimated as the corresponding terms in formula (8.66). Following the computations of the proof of Theorem 8.4.3 we obtain

$$\begin{aligned} V(x + \varepsilon h) - V(x) &\leq \varepsilon \int_0^\tau L_x(y(s), \bar{u}(s)) \cdot v(s) \, ds \\ &\quad + \varepsilon(q + \mu v)v(\tau) + o(\varepsilon) \\ &= \varepsilon p(0) \cdot h + o(\varepsilon), \end{aligned}$$

which is the desired inequality (8.74). ■

In the case of the minimum time problem, the two previous results take the following form.

Theorem 8.4.5 *Consider the minimum time problem for system (f, U) with target K. Assume properties (H0)–(H2), (8.13) and (8.55). Let $x \in \mathcal{R}\backslash K$ and let $\bar{u}$ be a time-optimal control for x. Let $y(\cdot) = y^{x, \bar{u}}(\cdot)$ be the corresponding optimal trajectory and let $z = y(T(x))$. Let v be the outer normal to ∂K at z and set*

$$\mu = \frac{1}{\max_{u \in U}[-v \cdot f(z, u)]} \qquad (8.78)$$

(by (8.13) μ is well defined and positive). Then the solution of the system

$$p'(t) = -f_x^T(y(t), u(t)) \, p(t) \tag{8.79}$$

with the final condition $p(T(x)) = \mu\nu$ is such that

$$-p(t) \cdot f(y(t), u(t)) = \max_{u \in U} -p(t) \cdot f(y(t), u) \tag{8.80}$$

for a.e. $t \in [0, T(x)]$. In addition,

$$p(t) \in D^+ T(y(t)) \qquad \forall t \in [0, T(x)[. \tag{8.81}$$

Proof — The result is a special case of the previous Theorems 8.4.3 and 8.4.4. We are no longer requiring the global boundedness of f as in (8.6), but this makes no substantial difference since all arguments are of a local nature. The definition of μ in (8.78) is equivalent to the requirement that $H(z, \mu\nu) = 0$ which appears in Theorem 8.4.3. For a direct proof of these results (under more general hypotheses) see [35]. ∎

Theorem 8.4.6 *Let the hypotheses of Theorem 8.2.7 be satisfied. Suppose in addition that the hamiltonian H is such that*

$$\text{for all } x \in \mathbb{R}^n, \text{ if } H(x, p) = 0 \text{ for all } p \text{ in a convex set } C, \\ \text{then } C \text{ is a singleton.} \tag{8.82}$$

Let $y^{x,u}(\cdot)$ be an optimal trajectory for a point $x \in \mathcal{R}$. Then V is differentiable at $y^{x,u}(s)$ for all $s \in \,]0, \tau(x, u)[$.

Proof — By Proposition 8.1.9 and assumption (8.82), the superdifferential $D^+V(y(s))$ is a singleton for any $s \in \,]0, \tau(x, u)[$. Since V is semiconcave, this implies that V is differentiable at $(y(s))$ for these values of s. ∎

Remark 8.4.7 Assumption (8.82) is satisfied when H is strictly convex, but is actually a less restrictive hypothesis. Consider for instance the case when L does not depend on u (e.g., the minimum time problem). Then H is not strictly convex, but (8.82) is satisfied provided $f(x, U)$ is a convex set of class C^1 for all $x \in \mathbb{R}^n$. In fact, we have in this case

$$H(x, p) = \sigma_{f(x,U)}(-p) - L(x), \tag{8.83}$$

where $\sigma(\cdot)$ denotes the support function. Thus, H is affine along radial directions; on the other hand, if $f(x, U)$ is smooth, we can apply Corollary A.1.26 to conclude that property (8.82) is satisfied.

Proposition 8.4.8 *Let the hypotheses of Theorems 8.2.7 and 8.4.3 hold, and let property (8.82) also be satisfied. Let $y = y^{x,u}$ be an optimal trajectory for a point x. Then there exists a unique dual arc p associated to y as in Theorem 8.4.3. Moreover,*

$$p(t) = DV(y(t)) \qquad \forall t \in \,]0, \tau(x, u)[$$

$$p(0) \in D^*V(x)$$

$$p(t) \neq 0 \qquad \forall t \in [0, \tau(x, u)].$$

Proof — By Theorem 8.4.6, V is differentiable at all points of the form $y(t)$ with $t \in]0, \tau(x, u)[$. Therefore at these points the superdifferential of V contains only the gradient $DV(y(t))$, and so $p(t) = DV(y(t))$ by Theorem 8.4.4. This also implies that $p(0) \in D^*V(x)$ by the definition of D^*.

These properties show that there cannot exist two distinct dual arcs since they should both coincide with $DV(y(t))$. Finally, since V is a viscosity solution of (8.9), it satisfies

$$H(y(t), p(t)) = H(y(t), DV(y(t))) = 0$$

for all $t \in]0, \tau(x, u)[$. By continuity, this also holds for $t = 0, t = \tau(x, u)$. Since by (8.6)

$$H(y(t), 0) = \max_{u \in U} -L(y(t), u) \le -\alpha < 0,$$

we deduce that $p(t) \ne 0$ for all $t \in [0, \tau(x, u)]$. ∎

When hypothesis (8.82) is satisfied, we have also the following result about the propagation of singularities of V.

Theorem 8.4.9 *Let the hypotheses of Theorem 8.2.7 be satisfied, and let in addition the hamiltonian H satisfy (8.82). Let $x \in \mathcal{R} \setminus \mathcal{K}$ be a point where V is not differentiable and such that $\dim D^+V(x) < n$. Then there exists a Lipschitz singular arc for V starting from x as in Theorem 4.2.2.*

Proof — We recall that the Hamilton–Jacobi equation (8.9) holds as an equality at all points of differentiability of V. By definition of D^*V, this implies

$$H(x, p) = 0, \qquad \forall\, p \in D^*V(x).$$

In addition, since V is semiconcave, we have that $D^*V(x) \subset D^+V(x)$. Suppose $D^*V(x) = D^+V(x)$. Then, since $D^+V(x)$ is a convex set, property (8.82) implies that $D^+V(x)$ is a singleton. But then V is differentiable at x, in contradiction with our assumptions.

Therefore $D^+V(x) \setminus D^*V(x) \ne \emptyset$. Since $\dim D^+V(x) < n$, we also have that $D^+V(x) = \partial D^+V(x)$ and so Theorem 4.2.2 can be applied. ∎

Remark 8.4.10 The hypothesis that $D^+V(x)$ has dimension less than n in the previous result is essential. Consider in fact the distance function from the set $\mathcal{K} = \mathbb{R}^n \setminus B_1$, which is a special case of the minimum time function. We have that $d_{\mathcal{K}}(x) = 1 - |x|$ for all $x \in B_1$, and so $x = 0$ is an isolated singularity with superdifferential of dimension n. It is easy to check that all other hypotheses of Theorem 8.4.9 are satisfied in this example.

In the remainder of the section we assume that

$$H \in C^{1,1}_{loc}(\mathbb{R}^n \times (\mathbb{R}^n \setminus \{0\})). \tag{8.84}$$

Remark 8.4.11 As in Theorem 7.3.6 we find that if assumption (8.84) holds, then the derivatives of H are given by

$$H_p(x, p) = -f(x, u^*(x, p)), \tag{8.85}$$

$$H_x(x, p) = -f_x^T(x, u^*(x, p))p - L_x(x, u^*(x, p)), \tag{8.86}$$

where $u^*(x, p)$ is any element of U such that

$$-f(x, u^*) \cdot p - L(x, u^*) = \max_{u \in U} -f(x, u) \cdot p - L(x, u).$$

Remark 8.4.12 In the cases where L does not depend on u (e.g., the minimum time problem) the hamiltonian takes the form (8.83) and so property (8.84) is satisfied if, for instance, $f(x, U)$ is a family of uniformly convex sets of class C^2 (see Definition A. 1.23). Observe also that in this case H is not differentiable for $p = 0$, (except for the trivial case where U is a singleton); this is the reason why we exclude that point in condition (8.84). The lack of differentiability of H for $p = 0$ is not a real complication for our purposes, since we are going to evaluate H along dual arcs, which are nonzero by Proposition 8.4.8.

In view of equalities (8.85), (8.86) and of the property $p(t) \neq 0$ stated in Proposition 8.4.8 we can restate the maximum principle as follows.

Corollary 8.4.13 *Assume that properties* (H0)–(H2), (L0)–(L2), (8.6), (8.13), (8.55)–(8.57), (8.82) *and* (8.84) *are satisfied. Let* y *be an optimal trajectory and let* p *be the associated dual arc. Then the pair* (y, p) *solves the system*

$$\begin{cases} y'(t) = -H_p(y(t), p(t)) \\ p'(t) = H_x(y(t), p(t)). \end{cases} \tag{8.87}$$

Consequently y *and* p *are of class* C^1.

Now we can obtain a correspondence between the optimal trajectories starting at a point $x \in \mathcal{R} \setminus \mathcal{K}$ and the elements of $D^*V(x)$, similarly to the case of the calculus of variations (see Theorem 6.4.9).

Theorem 8.4.14 *Assume that properties* (H0)–(H2), (L0)–(L2), (8.6), (8.13), (8.31), (8.55), (8.56), (8.82) *and* (8.84) *hold. Let* $x \in \mathcal{R} \setminus \mathcal{K}$ *and* $q \in D^*V(x)$. *Consider the solution* (y, p) *of* (8.87) *with initial conditions*

$$\begin{cases} y(0) = x \\ p(0) = q. \end{cases} \tag{8.88}$$

Then y *is an optimal trajectory for* x *and* p *is the dual arc associated with* y. *The correspondence between vectors in* $D^*V(x)$ *and optimal trajectories for* x *defined in this way is one-to-one.*

Proof — First we consider the case when V is differentiable at x. By Theorem 8.1.4, there exists a trajectory $y(\cdot)$ which is optimal for x and, by Theorem 8.4.3, there exists a dual arc p associated with y. By Proposition 8.4.8 and Corollary 8.4.13, the pair (y, p) is solution of (8.87) with initial conditions $y(0) = x$, $p(0) = DV(x)$. This proves the assertion in this case.

Next we consider the general case. Let $q \in D^*V(x)$. Then there exists a sequence $\{x_k\}_{k\in\mathbb{N}} \subset \mathcal{R} \setminus \mathcal{K}$ such that V is differentiable at x_k and such that

$$x_k \to x, \quad DV(x_k) \to q \quad \text{as } k \to \infty.$$

We denote by (y_k, p_k) the solution of (8.87) with initial conditions

$$\begin{cases} y(0) = x_k \\ p(0) = DV(x_k). \end{cases}$$

By the first part of the proof, y_k is an optimal trajectory for x_k and p_k is the associated dual arc. In addition, (y_k, p_k) converges to (y, p) locally uniformly, where (y, p) is the solution to system (8.87) with initial conditions (8.88). By Theorem 8.1.7, y is an optimal trajectory for x. Moreover

$$p(t) = \lim_{k\to\infty} p_k(t) = \lim_{k\to\infty} DV(y_k(t)), \quad t \in [0, \tau[,$$

where τ is the exit time of y. Since

$$\lim_{k\to\infty} DV(y_k(t)) \in D^*V(y(t)),$$

we have, by Theorem 8.4.6, that $p(t) = DV(y(t)), t \in]0, \tau[$. Therefore p is the dual arc associated to y. This proves that any solution to system (8.87) with initial conditions (8.88) coincides with an optimal trajectory and its associated dual arc.

Conversely, let y be an optimal trajectory for x. Then, by Proposition 8.4.8 and Corollary 8.4.13, there exists a dual arc p such that the pair (y, p) solves system (8.87) with initial conditions (8.88) for some $q \in D^*V(y(0))$. Observe also that q is uniquely determined, because $q = p(0)$, where p is the unique dual arc associated with y. Thus, the correspondence defined in the theorem is one-to-one. ∎

Corollary 8.4.15 *Under the hypotheses of the previous theorem, V is differentiable at a point $x \in \mathcal{R} \setminus \mathcal{K}$ if and only if the optimal trajectory for x is unique.*

Remark 8.4.16 In Example 8.2.9 we have seen a system where the time-optimal control problem does not satisfy the conclusion of the above corollary. This is not surprising, because some of our assumptions do not hold for that system. In fact the target is a single point and does not satisfy the smoothness assumption (8.56). In addition, the hamiltonian of that system is given by

$$H(p_1, p_2) = |p_1| + |p_2|, \quad (p_1, p_2) \in \mathbb{R}^2,$$

and so assumptions (8.82) and (8.84) are both violated.

Bibliographical notes

Most of the general references on optimal control problems quoted in the previous chapter include the case of problems with free terminal time. We recall also the monographs [87, 62] dealing specifically with the minimum time problem.

It is a well-known feature of exit time problems (see [110, 20]) that some suitable compatibility conditions on the data are needed in order to avoid irregular behaviors of the value function along the boundary. Our formulation of the compatibility condition follows [41]; it is more restrictive but simpler to state than the general ones in [110, 20].

The uniqueness result of Theorem 8.1.10 was first obtained in [19] in a weaker form and then generalized in [73] and [131] (see also [141] and the references therein). The property called "Petrov condition" in Theorem 8.2.3, which is crucial for the Lipschitz continuity of the minimum time function, can be regarded as a generalization of the "positive basis condition" introduced in [115] in the case of a target consisting of a single point. The case of a target with a smooth boundary was considered in [21]. Our presentation here follows [42, 41]. More results concerning the Lipschitz continuity of the value function of problems related to the ones considered here can be found in [20, 136, 141] and in the references therein.

The semiconcavity results of Section 8.2 have been obtained in [42] (see also [128]) in the case of the minimum time function and in [41] in the general case. The semiconvexity results of Section 8.3 are also taken from [42]. In the case $\mathcal{K} = \{0\}$, Corollary 8.3.7 was first conjectured in [84] and then proved in [27] (by a different approach from the one used here). The optimality conditions of Section 8.4 have been obtained in [35, 43] in the case of the minimum time function and in [41] for general exit time problems. A different kind of semiconcavity result for the minimum time function has been proved in [50]; the hypotheses on the system are more general than ours, but semiconcavity holds only on a dense subset of the controllable set.

Other semiconcavity results for exit time problems have been recently obtained in [36, 130]. In contrast to Theorem 8.2.7, the target set in these results can be a general closed set; on the other hand, the set $f(x, U)$ is required to be convex and to satisfy an interior sphere condition. It is also shown that the semiconcavity of the minimum time function is related to the interior sphere condition for the attainable sets of the control system at a given time.

Let us mention that, for exit time problems, an analysis of the singular set similar to that of Section 6.6 for problems in the calculus of variations has been done in [116]. It is shown there that the closure of the singular set of the value function is Hausdorff $(n-1)$-rectifiable, like the singular set itself. A key step in the analysis is the study of conjugate points. Such a result holds under suitable smoothness assumptions on the target set and on the data of the problem.

As we have remarked, the distance function from a smooth submanifold in $\mathbb{R}^n$ (or in a riemannian manifold) can be regarded as the value function of a minimum time problem. In this case the closure of the singular set coincides with a classical object in riemannian geometry, the so-called *cut-locus*. Some fine regularity results can be found e.g. in [97, 112]

The minimum time problem has been widely studied also for infinite dimensional systems, see e.g., the monograph [109]. Some generalizations to the infinite dimensional case of the results of this chapter have been obtained in [44, 3, 4, 7, 8].

Appendix

A. 1 Convex sets and convex functions

In this section we collect some basic results about convex sets and convex functions. These topics have a great importance in many fields of mathematics, and their study has become a branch of its own called *convex analysis*. We do not aim at a comprehensive introduction to the subject, but we will only present the definitions and theorems that are important for the purposes of this book. For a more detailed study the reader can consult some of the monographs on convex analysis, e.g., [122, 124, 126, 91].

For the sake of completeness, we start by recalling the definition of convexity.

Definition A. 1.1

(i) A set $C \subset \mathbb{R}^n$ is called convex *if, for any $x_0, x_1 \in C$, the segment $[x_0, x_1]$ is contained in C. We say that C is* strictly convex *if, for any $x_0, x_1 \in \overline{C}$, all points of the segment $[x_0, x_1]$, except possibly the endpoints, are contained in the interior of C.*

(ii) A function $L : C \to \mathbb{R}$, with $C \subset \mathbb{R}^n$ convex, is called convex *if*

$$\lambda L(x) + (1 - \lambda)L(y) \geq L(\lambda x + (1 - \lambda)y), \quad \forall x, y \in C, \ \lambda \in [0, 1]. \quad \text{(A. 1)}$$

We say that $L : C \to \mathbb{R}$ is strictly convex *if the above inequality is strict except when $\lambda = 0, 1$. A function $L : C \to \mathbb{R}$ is called* concave *if $-L$ is convex.*

It is easy to see that $L : C \to \mathbb{R}$, with $C \subset \mathbb{R}^n$ convex, is a convex function if and only if the epigraph of L is a convex subset of $\mathbb{R}^n \times \mathbb{R}$; in addition, the epigraph of L is strictly convex if and only if L is strictly convex and C is a strictly convex set. Also, it is easily checked that the supremum of any family of convex functions is a convex function (in the set where it is finite).

We observe that convex (concave) functions can be regarded as a particular case of semiconvex (semiconcave) functions, corresponding to a modulus $\omega \equiv 0$. Thus, the results of this book apply in particular to convex functions; of course, many of them admit a more direct proof or a stronger statement in the context of convex functions.

It is sometimes useful to observe that in order to prove the convexity of a function, it suffices to consider the midpoint of the segments.

Proposition A. 1.2 *A function $L : A \to \mathbb{R}$, with $A \subset \mathbb{R}^n$ open convex, is convex if and only if it is continuous and satisfies*

$$L(x + h) + L(x - h) - 2L(x) \geq 0 \qquad (A. 2)$$

for all x, h such that $x \pm h \in A$.

Proof — The above equivalence is well known, but can also be deduced as a special case of some properties of semiconcave functions which are proved in the text. Observe that inequality (A. 2) is a special case of (A. 1); on the other hand, in Definition A. 1.1 the continuity of L is not explicitly required. However, the continuity of L is a consequence of the convexity (see Theorem 2.1.7); on the other hand, a continuous function which satisfies (A. 2) for all x, h also satisfies (A. 1), as one can deduce from Theorem 2.1.10 in the case $\tilde{\omega} \equiv 0$. ∎

Definition A. 1.3 *A function $L : C \to \mathbb{R}$, with $C \subset \mathbb{R}^n$ convex is called* uniformly convex *if there exists $k > 0$ such that*

$$\lambda L(x) + (1 - \lambda)L(y) \geq L(\lambda x + (1 - \lambda)y) + k\lambda(1 - \lambda)|x - y|^2$$

for all $x, y \in C$, $\lambda \in [0, 1]$.

Remark A. 1.4 Uniform convexity implies strict convexity and is a more restrictive property; for instance, the function $L(x) = x^4$ is strictly convex but not uniformly convex. Arguing as in Proposition 1.1.3 one finds that a function L is uniformly convex if and only if $L(x) - k|x|^2$ is convex. It is well known that a function $L \in C^2$ is convex if and only if its hessian D^2L is positive definite everywhere. Thus, $L \in C^2$ is uniformly convex if and only if $D^2L \geq 2kI > 0$ everywhere. Therefore, if $L \in C^2$ satisfies $D^2L > 0$ everywhere, it is uniformly convex on all compact convex subsets of its domain.

A *convex combination* of k points $x_1, x_2, \ldots, x_k \in \mathbb{R}^n$ is a point $\bar{x}$ of the form

$$\bar{x} = \lambda_1 x_1 + \lambda_2 x_2 + \cdots + \lambda_k x_k$$

where $\lambda_1, \lambda_2, \ldots, \lambda_k$ are nonnegative numbers such that $\sum \lambda_i = 1$.

Definition A. 1.5 *Given $S \subset \mathbb{R}^n$, the* convex hull *of S is the set of all convex combinations of points of S.*

It is easy to see that co S is convex and is the smallest convex set containing S. A remarkable result, known as *Carathéodory's Theorem*, says that it suffices to take combinations of $n + 1$ points.

Theorem A. 1.6 *If $S \subset \mathbb{R}^n$, then any point in co S can be written as the convex combination of at most $n + 1$ points in S.*

Proof — See [122, Theor. 17.1]. ∎

Using Carathéodory's theorem one easily obtains the following property.

Corollary A. 1.7 *The convex hull of a compact set $K \subset \mathbb{R}^n$ is compact.*

There is a natural notion of dimension and topology associated to a convex set, as explained in the next definition.

Definition A. 1.8 *Let $C \subset \mathbb{R}^n$ be a convex set and let Π be the minimal affine subspace containing C. We define the* dimension *of C to be equal to the dimension of Π. The* relative topology *on C is the topology induced by the topology on Π (in contrast with the standard topology induced by the one on $\mathbb{R}^n$).*

Example A. 1.9 *Let $C \subset \mathbb{R}^3$ be the disk*

$$C = \{(x, y, z) : x^2 + y^2 \leq 1, z = 1\}.$$

Then the minimal affine subspace containing C is the plane $z = 1$ and the dimension of C is 2. The relative boundary and the relative interior of C are respectively

$$\{(x, y, z) : x^2 + y^2 = 1, z = 1\}, \qquad \{(x, y, z) : x^2 + y^2 < 1, z = 1\}.$$

Observe that in the standard topology C has empty interior and $\partial C = C$.

The picture is similar when we consider any convex set C whose dimension is strictly less than n: the interior is empty but the relative interior is nonempty, as shown by the next result.

Proposition A. 1.10 *If C is a nonempty convex set, then its relative interior is nonempty.*

Proof — See e.g., [126, Theor.1.1.12]. ∎

Let us recall a fundamental property of convex sets called the *separation theorem*.

Theorem A. 1.11 *Let $C_1, C_2 \subset \mathbb{R}^n$ be two convex sets.*

(i) *If the relative interiors of C_1 and C_2 are disjoint, then there exists $p \in \mathbb{R}^n$ such that*
$$p \cdot x \leq p \cdot y \qquad \forall x \in C_1, y \in C_2.$$

(ii) *If C_1 and C_2 are disjoint, if C_1 is closed and C_2 is compact, then there exists $p \in \mathbb{R}^n$ and $\varepsilon > 0$ such that*

$$p \cdot x + \varepsilon \leq p \cdot y \qquad \forall x \in C_1, y \in C_2.$$

Proof — See e.g., [126, Theor. 1.3.7 and 1.3.8]. ∎

Definition A. 1.12 *Let $V \subset \mathbb{R}^n$ be a convex set. If $\bar{v} \in V$, the* normal cone *to V at $\bar{v}$ (in the sense of convex analysis) is the set*

$$N_V(\bar{v}) = \{p \in \mathbb{R}^n \ : \ p \cdot (\bar{v} - v) \geq 0, \ \forall v \in V\}.$$

The separation theorem implies that the normal cone to any $\bar{v} \in \partial V$ contains some nonzero vector p. In fact, it suffices to take $C_1 = V$, $C_2 = \{\bar{v}\}$ and apply Theorem A. 1.11(i). It is also easy to see that if the convex set V possesses a normal vector v at $\bar{v}$ in the classical (smooth) sense, then $N_V(\bar{v}) = \{\lambda v \ : \ \lambda \geq 0\}$. Thus, if V has a C^1 boundary, then all normal cones at the boundary points of V have dimension one.

Let us now turn to some properties of convex functions. Given $L : C \to \mathbb{R}$ convex and $x \in \mathbb{R}^n$, we recall that the *subdifferential* (in the sense of convex analysis) of L at x is the set of all $p \in \mathbb{R}^n$ such that

$$L(y) \geq L(x) + p \cdot (y - x) \qquad \forall y \in C. \tag{A.3}$$

Equivalently, p belongs to the superdifferential of L at x if $(p, -1)$ is a normal vector to the epigraph of L at the point $(x, L(x))$.

The subdifferential of L at x is denoted by $D^- L(x)$, the same symbol which is used throughout the book to denote the Fréchet subdifferential. However, as observed in Remark 3.3.2, for a convex function the two notions are equivalent, so there is no ambiguity in using the same symbol. Let us recall some well-known properties of $D^- L$.

Theorem A. 1.13 *Let $L : \Omega \to \mathbb{R}$ be a convex function, with Ω open convex. Then $D^- L(x)$ is nonempty for every $x \in \Omega$. In addition, L is differentiable at x if and only if $D^- L(x)$ is a singleton. If L is differentiable at all $x \in \Omega$, then its differential $DL(x)$ is continuous in Ω.*

Proof — The above properties can be found for instance in [122, 126], or can be obtained as a special case of Proposition 3.3.4 on semiconcave functions. Observe that the statement about $D^- L(x)$ being nonempty for every x is a direct consequence of the separation theorem, like the analogous property of the normal cone to a convex set. ∎

Corollary A. 1.14 *Let $L : \Omega \to \mathbb{R}$, with Ω open convex. Then L is a convex function if and only if it can be written as $L(x) = \sup_{i \in \mathcal{I}} \gamma_i(x)$, where $\{\gamma_i\}_{i \in \mathcal{I}}$ is a family of linear functions.*

Proof — The property that the infimum of linear functions is a convex function follows easily from the definition. Suppose instead that $L : \Omega \to \mathbb{R}$ is convex. By the previous theorem, for any $y \in \Omega$ we can find p^y such that

$$L(x) \geq L(y) + p^y \cdot (x - y), \qquad x \in \Omega.$$

Let us set, for any $x, y \in \Omega$, $\gamma_y(x) = L(y) + p^y \cdot (x - y)$. Then, for any $y \in \Omega$, $\gamma_y(x) \leq L(x)$ with equality if $x = y$. Thus $L(x) = \sup_{y \in \Omega} \gamma_y(x)$, and so L is the supremum of linear functions. ∎

Remark A. 1.15 A more elegant representation of a convex function as the supremum of linear functions is provided by the Legendre transform (see the next section).

Definition A. 1.16 *The* support function *to a convex set V is given by*

$$\sigma_V(p) = \sup_{v \in V} v \cdot p, \qquad p \in \mathbb{R}^n. \tag{A.4}$$

The exposed face *of a convex set V in direction p, where $p \in \mathbb{R}^n \setminus \{0\}$, is defined by $\{v \in V : \sigma_V(p) = v \cdot p\}$.*

The definition easily implies that σ_V is convex and homogeneous of degree one, and that the exposed faces of V are convex and contained in ∂V. In addition

$$p \in N_V(\bar{v}) \iff \sigma_V(p) = p \cdot \bar{v}. \tag{A.5}$$

For the sake of simplicity we restrict our analysis to the case where V is compact; then the supremum in (A. 4) is a maximum and $\sigma_V(p)$ is finite for every p.

Proposition A. 1.17 *Let $S \subset \mathbb{R}^n$ be a compact set and let $T = \mathrm{co}\, S$. Then*

$$\max_{x \in S} x \cdot p = \sigma_T(p), \qquad \forall p \in \mathbb{R}^n.$$

Proof — Since $S \subset T$, we have, for any $p \in \mathbb{R}^n$,

$$\max_{x \in S} x \cdot p \leq \max_{x \in T} x \cdot p = \sigma_T(p).$$

On the other hand, any $y \in T$ has the form $y = \sum \lambda_i x_i$ with $x_i \in S$ and $\sum \lambda_i = 1$. Therefore

$$\langle p, y \rangle = \sum \lambda_i \langle p, x_i \rangle \leq \max_{x \in S} \langle x, p \rangle \sum \lambda_i = \max_{x \in S} \langle x, p \rangle,$$

and so $\sigma_T(p) \leq \max_{x \in S} x \cdot p$. ∎

Proposition A. 1.18 *Two compact convex sets with the same support function coincide.*

Proof — We will show that if $V \subset \mathbb{R}^n$ is a convex compact set, then

$$V = \{w \in \mathbb{R}^n : w \cdot p \leq \sigma_V(p) \,\forall\, p \in \mathbb{R}^n\}, \tag{A.6}$$

which implies that V is uniquely determined by σ_V.

Let us denote by W the set on the right-hand side of (A. 6). The definition of σ_V immediately implies that $V \subset W$. On the other hand, let w be any point not in V. By the separation theorem A. 1.11(ii), there exists $p \in \mathbb{R}^n$ and $\varepsilon > 0$ such that

$$p \cdot v + \varepsilon \leq p \cdot w \qquad \forall v \in V.$$

Thus, $\sigma_V(p) \leq p \cdot w - \varepsilon$, and so $w \notin W$.   ∎

We have seen that in many cases the hamiltonian function associated with an optimal control problem is a support function. In control theory it is important to know if the hamiltonian is smooth and if it is strictly convex, so we will investigate here these properties for a support function σ_V. Since σ_V is homogeneous of degree one, it is not differentiable at the origin (except for the trivial case where V is a singleton and so σ_V is linear) and it is linear in the radial directions. The best that one can hope for, therefore, is to have differentiability for $p \neq 0$ and strict convexity in nonradial directions. Before giving some results let us consider an example.

Example A. 1.19 Let $V \subset \mathbb{R}^2$ be the set

$$V = \{(x, y) \; : \; x \in [-1, 1], |y| \leq 1 + \sqrt{1 - x^2}\}.$$

Such a set is a square with two half-circles on the top and on the bottom. For a given (p_1, p_2) it is easily checked that

$$
\max_{(x,y) \in V} (x, y) \cdot (p_1, p_2) = \left(\frac{p_1}{\sqrt{p_1^2 + p_2^2}}, \; \text{sgn}(p_2) + \frac{p_2}{\sqrt{p_1^2 + p_2^2}} \right) \cdot (p_1, p_2)
$$
$$
= \sqrt{p_1^2 + p_2^2} + |p_2|.
$$

Observe that V has a $C^{1,1}$ boundary, while σ_V is nondifferentiable on the whole line where $p_2 = 0$. It turns out, as the next results will show, that the smoothness of σ_V is related to the strict convexity of V rather than to its smoothness. In fact, in this example V is not strictly convex since its boundary has two flat parts.   ∎

Theorem A. 1.20 *Let $V \subset \mathbb{R}^n$ be a convex compact set. For any $p \in \mathbb{R}^n \setminus \{0\}$, let us consider the exposed face*

$$V^*(p) = \{v \in V \; : \; p \cdot v = \sigma_V(p)\}.$$

Then the following properties are equivalent.

(i) *V is strictly convex.*
(ii) *$V^*(p)$ is a singleton for all $p \neq 0$.*
(iii) *$\sigma_V \in C^1(\mathbb{R}^n \setminus \{0\})$.*

If the above properties hold, then for all $p \neq 0$ we have $D\sigma_V(p) = v^*(p)$, where $v^*(p)$ is the unique element of $V^*(p)$.

Proof — We begin by proving that (i) implies (ii). We argue by contradiction and suppose that there exists $p \neq 0$ such that $V^*(p)$ contains two distinct points v_1, v_2. Then, for any $\lambda \in [0, 1]$,

$$p \cdot (\lambda v_1 + (1 - \lambda)v_2) = \lambda p \cdot v_1 + (1 - \lambda)p \cdot v_2$$
$$= \lambda \sigma_V(p) + (1 - \lambda)\sigma_V(p) = \sigma_V(p).$$

This shows that the whole segment $[v_1, v_2]$ is contained in $V^*(p)$. On the other hand, it is easy to see that any exposed face $V^*(p)$ is contained in ∂V if $p \neq 0$. Therefore, V is not strictly convex. This proves that (i) implies (ii).

To show the converse, suppose that (i) does not hold. Then there exist two distinct $v_1, v_2 \in V$ and $\lambda \in \,]0, 1[$ such that $\bar{v} := \lambda v_1 + (1 - \lambda)v_2 \in \partial V$. Then $N_V(\bar{v})$ contains some nonzero vector p, and by (A. 5) this is equivalent to $\sigma_V(p) = p \cdot \bar{v}$. But then

$$\sigma_V(p) = p \cdot \bar{v} = \lambda p \cdot v_1 + (1 - \lambda)p \cdot v_2$$
$$\leq \lambda \sigma_V(p) + (1 - \lambda)\sigma_V(p) = \sigma_V(p).$$

This is possible only if $p \cdot v_1 = p \cdot v_2 = \sigma_V(p)$. Thus, $v_1, v_2 \in V^*(p)$ and (ii) is violated.

Let us now turn to property (iii). We observe that $-\sigma_V$ is a marginal function in the sense of Section 3.4. Applying Theorem 3.4.4 we find that

$$D^-\sigma_V(p) = \text{co} V^*(p).$$

In addition, since σ_V is a convex function, Theorem A. 1.13 implies that it is differentiable exactly at those points where the subdifferential is a singleton. Thus, σ_V is differentiable at p if and only if $V^*(p)$ is a singleton, and in this case $D\sigma_V(p)$ coincides with the unique element of $V^*(p)$. This completes the proof. ∎

Let us now study the existence of the higher order derivatives of σ_V.

Theorem A. 1.21 *Let V be a strictly convex compact set. If V coincides locally with the epigraph of a function of class C^k for some $k \geq 2$ with strictly positive definite hessian, then $\sigma_V \in C^k(\mathbb{R}^n \setminus \{0\})$.*

Proof — By the previous theorem we only need to show that the map $v^*(p)$ is of class C^{k-1}. We argue locally in a neighborhood of some $\bar{p} = (\bar{p}_1, \ldots, \bar{p}_n) \neq 0$. Assume for definiteness $\bar{p}_n < 0$. Let us set $\bar{v} = v^*(\bar{p})$. Then, by (A. 5), $\bar{p} \in N_V(\bar{v})$. Using our hypotheses we see that in a neighborhood of $\bar{v}$ we can write V as the epigraph of a function $v_n = \phi(v_1, \ldots, v_{n-1})$, with ϕ of class C^k. In addition $D^2\phi > 0$, which implies that $D\phi$ is a C^{k-1}-diffeomorphism.

Let us set for simplicity $v' = (v_1, \ldots, v_{n-1})$. We have that $(D\phi(v'), -1)$ generates the normal cone to V at the point $(v', \phi(v'))$. Thus, if p is any vector homothetic to $(D\phi(v'), -1)$ by a positive factor, then $v^*(p) = (v', \phi(v'))$, and vice-versa.

In particular, $\bar{p}$ is homothetic to $(D\phi(\bar{v}'), -1)$. Let us consider $p = (p_1, \ldots, p_n)$ close to $\bar{p}$. Then $(-p_1/p_n, \ldots, -p_{n-1}/p_n)$ is close to $(-\bar{p}_1/\bar{p}_n, \ldots, -\bar{p}_{n-1}/\bar{p}_n) = D\phi(\bar{v}')$ and, since $D\phi$ is a local diffeomorphism, we deduce that $v^*(p) = (v'(p), \phi(v'(p)))$, where

$$v'(p) = (D\phi)^{-1}\left(-\frac{p_1}{p_n}, \ldots, -\frac{p_{n-1}}{p_n}\right).$$

It follows that v^* is of class C^{k-1} as desired.   ∎

By similar arguments one can prove

Theorem A. 1.22 *Let U_y be a family of strictly convex compact sets depending on the parameter $y \in \Omega \subset \mathbb{R}^m$. Suppose that the boundary of the sets U_y can be locally described by functions of the form $\phi \in C^k(\Omega \times A)$, with $A \subset \mathbb{R}^{n-1}$ and $k \geq 2$, in the sense that, for all $y \in \Omega$, the graph of $\phi(y, \cdot)$ locally coincides, up to an isometry, with ∂U_y. Suppose that the functions ϕ have strictly positive definite hessian with respect to the second group of variables. Then, setting*

$$H(y, p) = \sigma_{U_y}(p),$$

we have that $H \in C^k(A \times (\mathbb{R}^n \setminus \{0\}))$.

Definition A. 1.23 *If U satisfies the hypothesis of Theorem A. 1.21 we say that U is a uniformly convex set of class C^k. If the family U_y satisfies the hypothesis of Theorem A. 1.22 we say that U_y is a family of uniformly convex sets of class C^k.*

The next result gives information about the strict convexity of σ_V, showing that σ_V is linear on some set M if and only if M is contained in some normal cone to V.

Theorem A. 1.24 *Let $V, M \subset \mathbb{R}^n$ be convex sets, with V compact. The following two properties are equivalent:*

(i) $\exists \bar{v} \in V : M \subset N_V(\bar{v})$;
(ii) $\sigma_V(tp_0 + (1-t)p_1) = t\sigma_V(p_0) + (1-t)\sigma_V(p_1), \ \forall p_0, p_1 \in M, \ \forall t \in [0,1]$.

Proof — The implication from (i) to (ii) is an easy consequence of (A. 5). Conversely, let us suppose that (ii) holds. If M is a singleton, say $M = \{\bar{p}\}$, then (i) holds taking any v achieving the maximum in the definition of $\sigma_V(\bar{p})$. If M is not a singleton, we proceed as follows. We take any $\bar{p}$ in the relative interior of M and we choose again $\bar{v} \in V$ such that $\sigma_V(\bar{p}) = \bar{v} \cdot \bar{p}$. By definition

$$\sigma_V(p) \geq \bar{v} \cdot p \qquad \forall p \in M. \tag{A. 7}$$

Let us suppose that there exists $p_0 \in M$ such that $\sigma_V(p_0) > \bar{v} \cdot p_0$. Since $\bar{p}$ is in the relative interior of M, there exist $p_1 \in M, t \in {]}0, 1{[}$ such that $\bar{p} = tp_0 + (1-t)p_1$. Then

$$\sigma_V(p_1) = (1-t)^{-1}(\sigma_V(\bar{p}) - t\sigma_V(p_0)) < (1-t)^{-1}(\bar{p} - tp_0) \cdot \bar{v} = p_1 \cdot \bar{v},$$

in contradiction with (A. 7). It follows that (A. 7) holds as an equality for every $p \in M$, and therefore $M \subset N_V(\bar{v})$. ∎

Lemma A. 1.25 *Let the convex set M satisfy*

$$p \in M \implies tp \notin M, \ \forall t \neq 1. \tag{A. 8}$$

Define

$$K_M = \{tp : p \in M, t > 0\}.$$

Then K_M is convex and

$$\dim K_M = \dim M + 1.$$

Proof — Observe first that (A. 8) implies in particular that $0 \notin M$. If M is a singleton then K_M is a half-line and the result is straightforward, so we assume that $\dim M \geq 1$. Let us first show that K_M is convex. Given arbitrary $p_1, p_2 \in M$, $t_1, t_2 > 0$ and $\lambda \in]0, 1[$, we have to show that $\lambda t_1 p_1 + (1 - \lambda)t_2 p_2 \in K_M$. Let us set

$$\tau = \lambda t_1 + (1 - \lambda)t_2, \qquad \mu = \frac{\lambda t_1}{\tau}.$$

Then $\tau > 0$, $0 < \mu < 1$ and we have

$$\lambda t_1 p_1 + (1 - \lambda)t_2 p_2 = \tau \left(\frac{\lambda t_1}{\tau} p_1 + \frac{(1 - \lambda)t_2}{\tau} p_2 \right)$$
$$= \tau(\mu p_1 + (1 - \mu)p_2),$$

which belongs to K_M since $\mu p_1 + (1 - \mu)p_2 \in M$.

Let us now prove the second assertion. For a given p_0 in the relative interior of M, we set

$$V = \text{span } (M - p_0), \qquad W = \text{span } K_M.$$

Then $\dim M = \dim V$, $\dim K_M = \dim W$. It is easily seen that

$$W = \text{span } (M) = \text{span } (V \cup p_0).$$

We argue by contradiction and suppose that $p_0 \in V$. Since 0 is in the relative interior of $M - p_0$, and $V = \text{span } (M - p_0)$, we have $\lambda p_0 \in M - p_0$ for some $\lambda \in \mathbb{R} \setminus \{0\}$. But then $(1 + \lambda)p_0 \in M$, contradicting our assumptions. Hence $p_0 \notin V$ and $\dim W = \dim V + 1$, which proves our conclusion. ∎

Corollary A. 1.26 *Let $V \subset \mathbb{R}^n$ be a convex set whose boundary is a smooth $(n-1)$-dimensional surface. If $M \subset \mathbb{R}^n$ is convex and is such that $\sigma_V(p) = 1$ for all $p \in M$, then M is a singleton.*

Proof — The set M satisfies property (ii) of Theorem A. 1.24, and so is contained in $N_V(\bar{v})$ for some $\bar{v} \in V$. Defining K_M as in (A. 8) we have that K_M is also contained in $N_V(\bar{v})$ since $N_V(\bar{v})$ is a cone. The smoothness assumption on the boundary of V implies that any normal cone to V has dimension at most one, and therefore, by Lemma A. 1.25,

$$\dim M = \dim K_M - 1 \le \dim N_V(\bar{v}) - 1 = 0. \qquad \blacksquare$$

We conclude this section with a property of the integral of a convex-valued function.

Theorem A. 1.27 *Let $v \in L^1([0, T], \mathbb{R}^n)$ be such that $v(t) \in C$ a.e., where $C \subset \mathbb{R}^n$ is a closed and convex set. Then*

$$\frac{1}{T} \int_0^T v(t)\, dt \in C.$$

Proof — Let us set $\bar{v} = T^{-1} \int_0^T v(t)$. Suppose that $\bar{v} \notin C$. Setting $C_1 = C$ and $C_2 = \{\bar{v}\}$ in the separation theorem A. 1.11(ii) we obtain that there exist $p \in \mathbb{R}^n$ and $\varepsilon > 0$ such that

$$p \cdot x + \varepsilon \le p \cdot \bar{v}, \qquad \forall x \in C.$$

It follows that

$$p \cdot \bar{v} = \frac{1}{T} \int_0^T p \cdot v(t)\, dt \le \frac{1}{T} \int_0^T (p \cdot \bar{v} - \varepsilon)\, dt = p \cdot \bar{v} - \varepsilon,$$

which is a contradiction. $\qquad \blacksquare$

A. 2 The Legendre transform

The *Legendre transform* (also called *Fenchel transform* or *convex conjugate*) is a classical topic of convex analysis. In this section we give a short and self-contained exposition of the properties of this transform which are needed in our book; a more comprehensive treatment can be found in the references quoted in the previous section. We consider the Legendre transform of convex functions which are defined in the whole space and are superlinear, a class which is general enough for our purposes.

Definition A. 2.1 *Let $L : \mathbb{R}^n \to \mathbb{R}$ be a convex function which satisfies*

$$\lim_{|q| \to \infty} \frac{L(q)}{|q|} = +\infty. \tag{A. 9}$$

The Legendre transform of L is the function

$$L^*(p) = \sup_{q \in \mathbb{R}^n} [q \cdot p - L(q)], \qquad p \in \mathbb{R}^n. \tag{A. 10}$$

Example A. 2.2 Let $L(q) = \frac{1}{2}|q|^2$. Then, for any given $p \in \mathbb{R}^n$, the map $q \mapsto p \cdot q - L(q)$ attains its maximum when $q = p$, hence $L^*(p) = p \cdot p - L(p) = \frac{1}{2}|p|^2$. In general, if A is any symmetric, positive definite matrix, and if L is defined by $L(q) = \frac{1}{2}Aq \cdot q$, one finds that $q \mapsto p \cdot q - L(q)$ achieves its maximum at the point $q = A^{-1}p$ and that $L^*(p) = \frac{1}{2}A^{-1}p \cdot p$.

Theorem A. 2.3 *Let $L : \mathbb{R}^n \to \mathbb{R}$ be a convex function satisfying* (A. 9).

(a) *For every p there exists at least one point q_p where the supremum in* (A. 10) *is attained. In addition, for every bounded set $C \subset \mathbb{R}^n$ there exists $R > 0$ such that $|q_p| < R$ for all $p \in C$.*

(b) *The function L^* is convex and satisfies* $\lim\limits_{|p| \to \infty} \dfrac{L^*(p)}{|p|} = +\infty.$

(c) $L^{**} = L.$

(d) *Given $\bar{q}, \bar{p} \in \mathbb{R}^n$ we have*

$$\bar{p} \in D^- L(\bar{q}) \iff \bar{q} \in D^- L^*(\bar{p}) \iff L(\bar{q}) + L^*(\bar{p}) = \bar{q} \cdot \bar{p}.$$

Proof — (a) The claimed properties are a straightforward consequence of the convexity of L and of assumption (A. 9).
(b) The conjugate function L^* is convex since it is the supremum of linear functions. In addition, for all $M > 0$ and $p \in \mathbb{R}^n$ we have

$$L^*(p) \geq M\frac{p}{|p|} \cdot p - L\left(M\frac{p}{|p|}\right) \geq M|p| - \max_{|q|=M} L(q),$$

and so, for all $M > 0$,

$$\liminf_{|p| \to \infty} \frac{L^*(p)}{|p|} \geq M.$$

(c) We have by definition $L(q) \geq q \cdot p - L^*(p)$ for all q, p, which implies that $L \geq L^{**}$. To prove the converse inequality, fix any $\bar{q} \in \mathbb{R}^n$ and take $\bar{p} \in D^- L(\bar{q})$. Let $\hat{q}$ be such that $L^*(\bar{p}) = \hat{q} \cdot \bar{p} - L(\hat{q})$. Since $\bar{p} \in D^- L(\bar{q})$, we have $L(\hat{q}) \geq L(\bar{q}) + \bar{p} \cdot (\hat{q} - \bar{q})$ and therefore

$$L(\bar{q}) \leq L(\hat{q}) - \bar{p} \cdot (\hat{q} - \bar{q}) = -L^*(\bar{p}) + \bar{p} \cdot \bar{q},$$

which implies that $L(\bar{q}) \leq L^{**}(\bar{q})$.
(d) By the definition of subdifferential (A. 3),

$$\bar{p} \in D^- L(\bar{q}) \iff \bar{p} \cdot \bar{q} - L(\bar{q}) = \max_{q \in \mathbb{R}^n}[\bar{p} \cdot q - L(q)].$$

Therefore $\bar{p} \in D^- L(\bar{q})$ if and only if $L^*(\bar{p}) = \bar{q} \cdot \bar{p} - L(\bar{q})$. Analogously, $\bar{q} \in D^- L^*(\bar{p})$ if and only if $L^{**}(\bar{q}) = \bar{q} \cdot \bar{p} - L^*(\bar{p})$. Since $L^{**} = L$, we obtain the desired equivalence. ∎

Theorem A. 2.4 *Let $L : \mathbb{R}^n \to \mathbb{R}$ be convex and superlinear. Then the following properties are equivalent.*

(i) L is strictly convex.
(ii) $D^-L(q_1) \cap D^-L(q_2) = \emptyset$ *for all* $q_1, q_2 \in \mathbb{R}^n$, $q_1 \neq q_2$.
(iii) L* is of class C^1 *on* $\mathbb{R}^n$.

Proof — Let us first show that (i) implies (ii). We argue by contradiction and assume that there exist $q_1 \neq q_2$ such that $D^-L(q_1) \cap D^-L(q_2)$ contains some vector p. Then we have

$$L(q) \geq L(q_1) + p \cdot (q - q_1), \qquad \forall q \in \mathbb{R}^n, \tag{A. 11}$$

$$L(q) \geq L(q_2) + p \cdot (q - q_2), \qquad \forall q \in \mathbb{R}^n. \tag{A. 12}$$

Taking $q = q_2$ in (A. 11) and $q = q_1$ in (A. 12), we obtain that $L(q_1) = L(q_2) + p \cdot (q_1 - q_2)$, which gives

$$tL(q_1) + (1 - t)L(q_2) = L(q_2) + tp \cdot (q_1 - q_2), \qquad \forall t \in [0, 1]. \tag{A. 13}$$

Moreover, taking $q = tq_1 + (1 - t)q_2$ in (A. 12),

$$L(tq_1 + (1 - t)q_2) \geq L(q_2) + tp \cdot (q_1 - q_2). \tag{A. 14}$$

On the other hand, by the convexity of L, we know that

$$L(tq_1 + (1 - t)q_2) \leq tL(q_1) + (1 - t)L(q_2). \tag{A. 15}$$

Comparing (A. 13), (A. 14) and (A. 15) we conclude that

$$L(tq_1 + (1 - t)q_2) = L(q_2) + tp \cdot (q_1 - q_2) = tL(q_1) + (1 - t)L(q_2),$$

a contradiction to the strict convexity of L.

To prove the converse, suppose now that L is not strictly convex, i.e., there exists $q_1 \neq q_2 \in \mathbb{R}^n$ such that $L(q_t) = tL(q_1) + (1 - t)L(q_2)$ for all $t \in [0, 1]$, where $q_t = tq_1 + (1 - t)q_2$. If we choose any $p \in D^-L(q_{1/2})$, we then have

$$L(q) \geq L(q_{1/2}) + p \cdot (q - q_{1/2})$$
$$= \frac{1}{2}(L(q_1) + L(q_2)) + p \cdot \left(q - \frac{q_1 + q_2}{2}\right) \tag{A. 16}$$

for all $q \in \mathbb{R}^n$; in particular, for $q = q_1$ we obtain

$$\frac{L(q_1) - L(q_2)}{2} \geq p \cdot \frac{q_1 - q_2}{2}. \tag{A. 17}$$

Adding inequalities A. 16 and A. 17 we conclude that $L(q) - L(q_2) \geq p \cdot (q - q_2)$ for all $q \in \mathbb{R}^n$, which implies $p \in D^-(q_2) \cap D^-(q_{1/2})$. This completes the equivalence of (i) and (ii).

By part (d) of the previous theorem, property (ii) holds if and only if $D^-L^*(p)$ is a singleton for all $p \in \mathbb{R}^n$. By Theorem A. 1.13, this is equivalent to the continuous differentiability of L^* on $\mathbb{R}^n$. ∎

Theorem A. 2.5 *Suppose that $L \in C^k(\mathbb{R}^n)$, with $k \geq 2$, satisfies assumption (A. 9) and*

$$D^2 L(q) \text{ is positive definite for all } q \in \mathbb{R}^n. \tag{A. 18}$$

Then L^* belongs to $C^k(\mathbb{R}^n)$, DL is a C^{k-1} diffeomorphism from $\mathbb{R}^n$ to $\mathbb{R}^n$ and

$$DL^*(p) = (DL)^{-1}(p), \tag{A. 19}$$

$$D^2 L^*(p) = \left[D^2 L \left(DL^*(p) \right) \right]^{-1}, \tag{A. 20}$$

$$L^*(p) = p \cdot DL^*(p) - L(DL^*(p)). \tag{A. 21}$$

Proof — By the previous theorem and by the property $L^{**} = L$, if L is differentiable and strictly convex, then so is L^*; in this case, by statement (d) in Theorem A. 2.3, the differentials DL and DL^* are reciprocal inverse. Therefore DL is a diffeomorphism provided $D^2 L$ is everywhere nondegenerate, which is ensured by (A. 18). This proves (A. 19) and (A. 20). To obtain (A. 21) we observe that, under our assumptions, the maximum in (A. 10) is achieved at a unique point q_p which satisfies $DL(q_p) = p$. ∎

Let now $\Omega \subset \mathbb{R}^m$ be an open set and let $L \in C(\Omega \times \mathbb{R}^n)$. Suppose that L is convex in the second argument and satisfies

$$\lim_{|q| \to \infty} \inf_{x \in \Omega} \frac{L(x, u)}{|u|} = +\infty. \tag{A. 22}$$

Then we can define the Legendre transform of L with respect to the second group of variables as follows:

$$H(x, p) = \sup_{q \in \mathbb{R}^n} \left[p \cdot q - L(x, q) \right], \qquad (x, q) \in \Omega \times \mathbb{R}^n. \tag{A. 23}$$

Theorem A. 2.6

(a) *For every $(x, p) \in \Omega \times \mathbb{R}^n$ there exists at least one point $q(x, p)$ where the supremum in (A. 23) is attained. In addition, for every bounded set $C \subset \mathbb{R}^n$ there exists $R > 0$ such that every q associated to $(x, p) \in \Omega \times C$ satisfies $|q| < R$.*

(b) *The transformed function H is continuous, convex in the second argument and satisfies*

$$\lim_{|p| \to \infty} \inf_{x \in \Omega} \frac{H(x, p)}{|p|} = +\infty.$$

(c) *If L is strictly convex in q, then $\dfrac{\partial H}{\partial p}(x, p)$ exists and is continuous for $(x, p) \in \Omega \times \mathbb{R}^n$.*

(d) Suppose that for every $R > 0$ there exists $\alpha_R > 0$ such that

$$|L(x, q) - L(y, q)| \leq \alpha_R |x - y|, \qquad x, y \in \Omega, q \in B_R.$$

Then for every $\rho > 0$ there exists $\beta_\rho > 0$ such that

$$|H(x, p) - H(y, p)| \leq \beta_\rho |x - y|, \qquad x, y \in \Omega, u \in B_\rho.$$

Proof — **(a)** It is an easy consequence of the properties of L.

(b) Convexity and superlinearity are proved like in the case where L does not depend on x. To prove continuity, let us consider a sequence $(x_n, p_n) \subset \Omega \times \mathbb{R}^n$ converging to some $(\bar{x}, \bar{p})$. We have

$$H(x_n, p_n) = p_n \cdot q_n - L(x_n, q_n), \qquad H(\bar{x}, \bar{p}) = \bar{p} \cdot \bar{q} - L(\bar{x}, \bar{q})$$

for suitable $q_n, \bar{q}$. Then

$$H(x_n, p_n) \geq p_n \cdot \bar{q} - L(x_n, \bar{q}).$$

Letting $n \to \infty$ we obtain that $\liminf_{n\to\infty} H(x_n, p_n) \geq H(\bar{x}, \bar{p})$. To prove the converse inequality, let us choose a subsequence (x_{n_k}, p_{n_k}) such that $\lim_{k\to\infty} H(x_{n_k}, p_{n_k}) = \limsup_{n\to\infty} H(x_n, p_n)$. Since, by part (a), $\{q_n\}$ is bounded, we can assume that the subsequence q_{n_k} converges to some value q^*. Then we have

$$H(\bar{x}, \bar{p}) \geq \bar{p} \cdot q^* - L(\bar{x}, q^*) = \lim_{k\to\infty} p_{n_k} \cdot q_{n_k} - L(x_{n_k}, q_{n_k})$$

$$= \limsup_{n\to\infty} H(x_n, p_n).$$

This proves the continuity of H.

(c) If L is strictly convex, then the supremum in (A. 23) is attained for a unique value of q, which we will denote by $q(x, p)$. From Theorem A. 2.4 and Theorem A. 2.3(d) it follows that H is differentiable with respect to p and that $H_p(x, p) = q(x, p)$. Let us prove the continuity of $q(x, p)$. Recalling (a), it suffices to show that, if $(x_n, p_n) \to (\bar{x}, \bar{p})$ and $q(x_n, p_n) \to \bar{q}$, then $\bar{q} = q(\bar{x}, \bar{p})$. This follows from the continuity of H, since we can pass to the limit in the equality $H(x_n, p_n) = p_n \cdot q_n - L(x_n, q_n)$ to obtain $H(\bar{x}, \bar{p}) = \bar{p} \cdot \bar{q} - L(\bar{x}, \bar{q})$, which implies that $\bar{q} = q(\bar{x}, \bar{p})$ by the uniqueness of the maximizer in (A. 23).

(d) Let us fix $\rho > 0$. By part (a) there exists $R > 0$ with the following property: for every $(x, p) \in \Omega \times B_\rho$, any q which maximizes the expression in (A. 23) belongs to B_R. Let us now consider $(x, p), (y, p) \in \Omega \times B_\rho$ and suppose, for instance, that $H(x, p) \geq H(y, p)$. Let $\bar{q}$ be such that $H(y, p) = p \cdot \bar{q} - L(y, \bar{q})$. Then $\bar{q} \in B_R$ and we can use our hypothesis on L to obtain

$$|H(x, p) - H(y, p)| = H(x, p) - H(y, p)$$
$$\leq p \cdot \bar{q} - L(x, \bar{q}) - p \cdot \bar{q} + L(y, \bar{q}) \leq \alpha_R |x - y|.$$

Choosing $\beta_\rho = \alpha_R$ the proof is complete. ∎

To conclude we consider the case when L is uniformly convex and smooth.

Corollary A. 2.7 *Let $L \in C^k(\Omega \times \mathbb{R}^n)$ with $k \geq 2$. Assume that L satisfies* (A. 22) *and*

$$\frac{\partial^2 L}{\partial q^2}(x, q) \text{ is positive definite for all } (x, q) \in \Omega \times \mathbb{R}^n. \qquad (A. 24)$$

Then H belongs to $C^k(\Omega \times \mathbb{R}^n)$. Moreover, if we denote by $q(x, p)$ the unique value of q at which the infimum in (A. 23) *is attained, we have*

$$\frac{\partial H}{\partial p}(x, p) = q(x, p), \qquad (A. 25)$$

$$\frac{\partial H}{\partial x}(x, p) = -\frac{\partial L}{\partial x}(x, q(x, p)) \qquad (A. 26)$$

$$\frac{\partial^2 H}{\partial p^2}(x, p) = \left[\frac{\partial^2 L}{\partial q^2}(x, q(x, p)) \right]^{-1}. \qquad (A. 27)$$

In addition,

$$q = \frac{\partial H}{\partial p}(x, p) \text{ if and only if } p = \frac{\partial L}{\partial q}(x, q). \qquad (A. 28)$$

Proof — For any $x \in \Omega$, Theorem A. 2.5 implies that $H(x, \cdot) \in C^k(\mathbb{R}^n)$ and (A. 28) holds. Since $q(x, p)$ satisfies

$$p - \frac{\partial L}{\partial q}(x, q(x, p)) = 0 \qquad (A. 29)$$

we see that (A. 25) follows from (A. 28) and (A. 27) is a consequence of (A. 20).

From the implicit function theorem and (A. 24) we obtain that $q(x, p) \in C^{k-1}(\Omega \times \mathbb{R}^n)$. Hence, recalling that $H(x, p) = p \cdot q(x, p) - L(x, q(x, p))$, we see that H is at least of class $C^{k-1}(\Omega \times \mathbb{R}^n)$, with $k - 1 \geq 1$. Consider now $x \in \Omega$, $y \in \mathbb{R}^m$ and $\lambda > 0$ small enough. We have

$$\frac{H(x + \lambda y, p) - H(x, p)}{\lambda}$$
$$\geq \frac{p \cdot q(x, p) - L(x + \lambda y, q(x, p)) - p \cdot q(x, p) + L(x, q(x, p))}{\lambda}$$
$$= \frac{L(x, q(x, p)) - L(x + \lambda y, q(x, p))}{\lambda}.$$

Letting $\lambda \to 0$ we obtain

$$\frac{\partial H}{\partial x}(x, p) \cdot y \geq -\frac{\partial L}{\partial x}(x, q(x, p)) \cdot y$$

for all $y \in \mathbb{R}^m$, which implies (A. 26). The C^k regularity of H is then a consequence of (A. 25), (A. 26). ∎

A. 3 Hausdorff measure and rectifiable sets

We recall the definition of the ρ-dimensional Hausdorff measure of a set $S \subset \mathbb{R}^n$ for a given $\rho \in [0, \infty[$. First we define, for $\delta > 0$,

$$\mathcal{H}_\delta^\rho(S) = \frac{\alpha(\rho)}{2^\rho} \inf \left\{ \sum_{i=1}^{+\infty} (\text{diam } C_i)^\rho \; : \; C_i \subset \mathbb{R}^n, \; S \subset \bigcup_{i=1}^{+\infty} C_i, \; \text{diam } C_i < \delta \right\},$$

where

$$\alpha(\rho) = \frac{\pi^{\rho/2}}{\Gamma\left(\frac{\rho}{2} + 1\right)},$$

and $\Gamma(t) = \int_0^\infty e^{-x} x^{t-1} \, dx$ is Euler's Gamma function. We recall that if ρ is an integer, then $\alpha(\rho)$ is the measure of a ρ-dimensional unit sphere.

Definition A. 3.1 *Given $S \subset \mathbb{R}^n$ and $\rho \geq 0$, the* (outer) ρ-dimensional Hausdorff measure *of S is defined as*

$$\mathcal{H}^\rho(S) = \lim_{\delta > 0} \mathcal{H}_\delta^\rho(S) = \sup_{\delta > 0} \mathcal{H}_\delta^\rho(S).$$

It can be proved (see e.g., [72, 14]) that the measure defined above satisfies the following properties.

Theorem A. 3.2

(i) *The measure $\mathcal{H}^\rho$ is Borel regular for all $\rho > 0$.*

(ii) *$\mathcal{H}^n$ coincides with the Lebesgue measure.*

(iii) *$\mathcal{H}^0$ is the counting measure.*

(iv) *If $A \subset \mathbb{R}^n$ is a k-dimensional smooth manifold and $\mu(A)$ is the k-dimensional area of A, then $\mathcal{H}^k(A) = \mu(A)$.*

(v) *If $\mathcal{H}^\rho(S) < +\infty$ for some $\rho < n$, then $\mathcal{H}^\sigma(S) = 0$ for any $\sigma > \rho$.*

(vi) *If $\mathcal{H}^\rho(S) > 0$ for some $\rho > 0$, then $\mathcal{H}^\sigma(S) = +\infty$ for any $\sigma \in [0, \rho[$.*

(vii) *If $f \in \text{Lip}(\mathbb{R}^n; \mathbb{R}^m)$, then*

$$\mathcal{H}^\rho(f(A)) \leq (\text{Lip}(f))^\rho \mathcal{H}^\rho(A) \qquad \forall A \subset \mathbb{R}^n. \tag{A. 30}$$

The previous properties justify the following definition.

Definition A. 3.3 *Given $S \subset \mathbb{R}^n$, we define its* Hausdorff dimension *to be*

$$\mathcal{H}\text{--}\dim(S) = \inf\{\rho \geq 0 \; : \; \mathcal{H}^\rho(S) = 0\}.$$

Observe that $\mathcal{H}\text{--}\dim(S) \leq n$ by parts (ii) and (v) of the previous theorem. We also have, by parts (v) and (vi),

$$\mathcal{H}\text{--}\dim(S) = d \quad \Longrightarrow \quad \mathcal{H}^\rho(S) = 0, \, \forall \rho > s, \quad \mathcal{H}^\rho(S) = +\infty, \, \forall \rho < s.$$

Therefore

$$\mathcal{H}\text{-}\dim(S) = \inf\{\rho \geq 0 \ : \ \mathcal{H}^\rho(S) < +\infty\}$$
$$= \sup\{\rho \geq 0 \ : \ \mathcal{H}^\rho(S) > 0\}.$$

Let us conclude the section by recalling the definition of k-rectifiable sets, which can be regarded as a generalization of k-dimensional manifolds.

Definition A. 3.4 *Let $k \in \{0, 1, \ldots, n\}$ and let $C \subset \mathbb{R}^n$.*

(i) *C is called a k-rectifiable set if there exists a Lipschitz continuous function f : $\mathbb{R}^k \rightarrow \mathbb{R}^n$ such that $C \subset f(\mathbb{R}^k)$.*
(ii) *C is called a countably k-rectifiable set if it is the union of a countable family of k-rectifiable sets.*
(iii) *C is called a countably $\mathcal{H}^k$-rectifiable set if there exists a countably k-rectifiable set $E \subset \mathbb{R}^n$ such that $\mathcal{H}^k(S \setminus E) = 0$.*

A. 4 Ordinary differential equations

Given an open set $\Omega \subset \mathbb{R} \times \mathbb{R}^n$ and a function $F : \Omega \rightarrow \mathbb{R}^n$, we consider the ordinary differential equation

$$x'(t) = F(t, x(t)). \tag{A. 31}$$

For the purposes of control theory it is important to allow a discontinuous dependence on t on the right-hand side. Thus our assumptions on F will be as follows.

(C) The function $x \rightarrow F(t, x)$ is continuous for any fixed t and the function $t \rightarrow F(t, x)$ is measurable for any fixed x.

We say that $x : I \rightarrow \mathbb{R}^n$, where $I \subset \mathbb{R}$ is an interval, is a solution of (A. 31) (in the Carathéodory sense) if it is absolutely continuous and if it satisfies

$$(t, x(t)) \in \Omega, \qquad \forall t \in I, \qquad x'(t) = F(t, x(t)) \quad t \in I \text{ a.e.}.$$

All the results of this section are well known in the case of F continuous and can be found in any textbook on ordinary differential equations (e.g., [58, 86]); in the framework considered here, a proof can be found for instance in [103, 28]. We start by recalling the result about the local existence and uniqueness of solutions for the Cauchy problem.

Theorem A. 4.1 *Let $F : \Omega \rightarrow \mathbb{R}^n$ satisfy (C) and suppose in addition that for every compact set $K \subset \Omega$ there exists $M_K, L_K > 0$ such that*

$$|F(t, x)| \leq M_K \qquad\qquad \forall (t, x) \in K,$$
$$|F(t, x_1) - F(t, x_2)| \leq L_K |x_1 - x_2| \qquad \forall (t, x_1), (t, x_2) \in K.$$

Then for any given $(t_0, x_0) \in \Omega$ there exists $\delta > 0$ such that the Cauchy problem

$$x' = F(t, x), \qquad x(t_0) = x_0 \tag{A. 32}$$

has a unique solution in the interval $t \in \,]t_0 - \delta, t_0 + \delta[$.

We denote by $x(t; t_0, x_0)$ the solution to (A. 32). As in the classical theory, such a solution can be extended up to a maximal interval of existence and one can characterize the possible behaviors of $x(t; t_0, x_0)$ as t tends to the endpoints of the maximal interval. We recall a well-known criterion giving global existence when F is globally Lipschitz continuous w.r.t. x.

Theorem A. 4.2 *Let $I \subset \mathbb{R}$ be an interval and let $F : I \times \mathbb{R}^n \to \mathbb{R}^n$ satisfy* **(C)**. *Assume in addition that F is locally bounded and that for any compact subinterval $K \subset I$ there exists $L_K > 0$ such that*

$$|F(t, x) - F(t, y)| \le L_K |x - y|, \qquad \forall t \in K, \ x, y \in \mathbb{R}^n.$$

Then, for any $(t_0, x_0) \in I \times \mathbb{R}^n$, the Cauchy problem (A. 32) possesses a solution $x(t; t_0, x_0)$ defined for all $t \in I$.

The next result is known as *Gronwall's Lemma* or *Gronwall's inequality*. It is often used to derive a priori estimates for solutions to differential equations or differential inequalities.

Theorem A. 4.3 *Let $z : [t_0, t_1] \to \mathbb{R}$ be an absolutely continuous nonnegative function satisfying*

$$z(t) \le k(t) + \int_{t_0}^t z(s) v(s)\, ds, \qquad t \in [t_0, t_1],$$

where $k \in C^1([t_0, t_1])$, $v \in C([t_0, t_1])$, $k, v \ge 0$. Then

$$z(t) \le k(t_0) \exp\left(\int_{t_0}^t v(s)\, ds \right) + \int_{t_0}^t k'(s) \exp\left(\int_s^t v(r)\, dr \right) ds$$

for all $t \in [t_0, t_1]$. In particular, if $k(t) \equiv k$ and $v(t) \equiv L$ we have

$$z(t) \le k e^{L(t - t_0)}, \qquad \forall t \in [t_0, t_1].$$

The dependence of the solution on the initial value is described by the next result.

Theorem A. 4.4 *Let $F : \Omega \to \mathbb{R}^n$ satisfy the conditions of Theorem A. 4.1 and assume in addition that the jacobian F_x with respect to the x variables exists and is continuous w.r.t. x. Let $(t_0, x_0) \in \Omega$ be given and set $\hat{x}(\cdot) = x(\cdot; t_0, x_0)$. Let I be a compact interval where $\hat{x}(t)$ is defined. For a given $\bar{v} \in \mathbb{R}^n$, call $v(t)$ the solution to the linear Cauchy problem*

$$v'(t) = F_x(t, \hat{x}(t)) v(t) \tag{A. 33}$$

with initial condition $v(t_0) = \bar{v}$. Then, for any $t \in I$ we have

$$\lim_{\varepsilon \to 0} \left| \frac{x(t; t_0, x_0 + \varepsilon \bar{v}) - \hat{x}(t)}{\varepsilon} - v(t) \right| = 0$$

the limit being uniform for $t \in I$, $|\bar{v}| \le 1$.

The differential system (A. 33) is called the *linearization* of system (A. 31) around the solution $\hat{x}$.

Finally, let us recall some notions from the theory of linear systems. Consider a linear differential system of the form

$$x'(t) = A(t)x(t), \qquad t \in I, \tag{A. 34}$$

where I is an interval and $A(t)$ is a $n \times n$ matrix whose entries are measurable and locally bounded for $t \in I$. The solutions to such a system are defined globally in I by Theorem A. 4.2.

The *adjoint system* associated to (A. 34) is given by

$$w'(t) = -A^T(t)w(t), \qquad t \in I \tag{A. 35}$$

where A^T denotes the transpose matrix. It is immediate to check that the solutions of the two systems satisfy the relation

$$\frac{d}{dt}[x(t) \cdot w(t)] = 0.$$

The *fundamental matrix solution* to (A. 34) is the $n \times n$ matrix $M(t, s)$ depending on $(t, s) \in I \times I$, which solves the problem

$$\frac{\partial M}{\partial t}(t, s) = A(t)M(t, s), \qquad M(s, s) = I$$

for any $s \in I$. In other words, $M(t, s)$ is the matrix whose i-th column is the value at time t of the solution to (A. 34) with initial condition $x(s) = e_i$, where e_i is the i-th vector in the standard basis of $\mathbb{R}^n$. It is easy to see that the fundamental matrix solution enjoys the following properties.

Theorem A. 4.5

(i) $M(t, s)$ satisfies the semigroup property

$$M(t, s)M(s, r) = M(t, r), \qquad \forall t, s, r \in I.$$

(ii) The solution to (A. 34) with initial conditions $x(t_0) = x_0$ is given by

$$x(t) = M(t, t_0)x_0.$$

(iii) If we set $\hat{M}(t, s) = M(s, t)$, then $\hat{M}$ is the fundamental matrix solution to the adjoint system (A. 35).

(iv) If $h \in L^1(I)$, then the solution to the nonhomogeneous problem

$$x'(t) = A(t)x(t) + h(t), \qquad x(t_0) = x_0$$

is given by

$$x(t) = M(t, t_0)x_0 + \int_{t_0}^{t} M(t, s)h(s)\, ds.$$

A. 5 Set-valued analysis

A *multifunction*, or *set-valued function* Γ from $\mathbb{R}^m$ to $\mathbb{R}^n$ is a function which associates to every $x \in \mathbb{R}^m$ a set $\Gamma(x) \subset \mathbb{R}^n$ (possibly empty). The set

$$D(\Gamma) = \{x \in \mathbb{R}^m : \Gamma(x) \neq \emptyset\}$$

is called the *domain* of Γ.

Definition A. 5.1 *Let Γ be a multifunction from $\mathbb{R}^m$ to $\mathbb{R}^n$.*

 (i) *Γ is called closed (resp. convex, compact) if $\Gamma(x)$ is closed (resp. convex, compact) for every $x \in \mathbb{R}^m$.*
(iii) *Γ is called measurable if, for every open set $A \subset \mathbb{R}^n$, its counterimage*

$$\Gamma^{-1}(A) := \{x \in \mathbb{R}^m : \Gamma(x) \cap A \neq \emptyset\}$$

 is measurable.
(iii) *Γ is called upper semicontinuous if, for every $x \in \mathbb{R}^m$ and $\varepsilon > 0$, there exists $\delta > 0$ such that*

$$\Gamma(x') \subset \Gamma(x) + B_\varepsilon, \qquad \forall x' \in B_\delta(x).$$

A *selection* of a given multifunction Γ is a (single-valued) function $\gamma : D(\Gamma) \to \mathbb{R}^m$ such that $\gamma(x) \in \Gamma(x)$ for all $x \in D(\Gamma)$. The existence of a selection is obvious by definition; however, one is interested in having a selection with some regularity properties. Existence theorems of selections are a central topic in set-valued analysis; here we recall a result about measurable selections which we use for the existence of optimal controls (see Chapters 7 and 8).

Theorem A. 5.2 *Let Γ be a measurable and closed multifunction from $\mathbb{R}^m$ to $\mathbb{R}^n$. Then Γ admits a measurable selection.*

Proof — See e.g. [55, Ch.3, Th 5.3].   ∎

Let us also recall the following simple criterion for the measurability of multifunctions.

Proposition A. 5.3 *Let Γ be a multifunction from $\mathbb{R}^m$ to $\mathbb{R}^n$ and suppose that the graph of Γ, defined by*

$$\{(x, y) : x \in \mathbb{R}^m, y \in \Gamma(x)\},$$

is closed. Then Γ is measurable and closed.

Proof — The closedness of Γ is an immediate consequence of the closedness of the graph, so it suffices to prove the measurability. Let $K \subset \mathbb{R}^n$ be any compact set. We claim that $\Gamma^{-1}(K)$ is closed. Indeed, let $\{x_k\} \subset \Gamma^{-1}(K)$ be a sequence converging to some $\bar{x} \in \mathbb{R}^m$. Then for every k there exists $y_k \in \Gamma(x_k) \cap K$; by compactness, a subsequence y_{k_h} converges to some $\bar{y} \in K$. Since the graph of Γ is closed, the pair $(\bar{x}, \bar{y}) = \lim(x_{k_h}, y_{k_h})$ belongs to the graph of Γ. Thus, $\bar{y} \in \Gamma(\bar{x})$ and so $\bar{x} \in \Gamma^{-1}(K)$. This proves that $\Gamma^{-1}(K)$ is closed.

Now any open set $A \subset \mathbb{R}^n$ can be written as $A = \cup_{h=1}^{\infty} K_h$, where $\{K_h\}$ are compact sets. Then $\Gamma^{-1}(A) = \cup_{h=1}^{\infty} \Gamma^{-1}(K_h)$ is the countable union of closed sets and so it is measurable. ∎

For a comprehensive treatment of set-valued analysis, the reader may consult for instance [17, 18].

A. 6 BV functions

We recall here the definition and some basic properties of the class BV of functions of bounded variation. Given an open set $\Omega \subset \mathbb{R}^N$ and a vector-valued function $w \in L^1(\Omega, \mathbb{R}^M)$, we denote by $D^{wk}w$ the distributional derivative of w, while $Dw(y)$ denotes, as usual, the gradient of w at a point y at which w is differentiable.

Definition A. 6.1 *We say that a function $w \in L^1(\Omega, \mathbb{R}^M)$ belongs to the class* BV $(\Omega, \mathbb{R}^M)$ *of functions of bounded variation in Ω if the first partial derivatives of w in the sense of distributions are measures with finite total variation in Ω.*

The derivative of a function of bounded variation is usually decomposed into three parts. To introduce this decomposition we first need to recall some definitions from measure theory.

Definition A. 6.2 *Given $w \in L^1(\Omega, \mathbb{R}^M)$ and $x_0 \in \Omega$, we say that $\bar{w} \in \mathbb{R}^M$ is the* approximate limit *of w at y_0 if*

$$\lim_{\rho \to 0^+} \frac{\operatorname{meas}(\{y \in B_\rho(y_0) \cap \Omega \ : \ |w(y) - \bar{w}| > \varepsilon\})}{\rho^N} = 0$$

for any $\varepsilon > 0$. In this case we write $\bar{w} = \operatorname{ap} \lim_{y \to y_0} w(y)$. More generally, given $v \in \mathbb{R}^N$, $|v| = 1$, we denote by $w_v(y_0)$ the value $\bar{w} \in \mathbb{R}^M$ (if it exists) such that

$$\lim_{\rho \to 0^+} \frac{\operatorname{meas}(\{y \in B_\rho(y_0) \cap \Omega_v(y_0) \ : \ |w(y) - \bar{w}| > \varepsilon\})}{\rho^N} = 0$$

for any $\varepsilon > 0$, where

$$\Omega_v(y_0) = \{y \in \Omega \mid \langle y - y_0, v \rangle \geq 0\}.$$

Clearly, if $\operatorname{ap} \lim_{y \to y_0} w(y)$ exists, then $w_v(y_0) = \operatorname{ap} \lim_{y \to y_0} w(y)$ for all v. An interesting case is when the approximate limit does not exist, but we can find v such that the two limits $w_v(y_0)$ and $w_{-v}(y_0)$ exist and are different. Such a behavior means, roughly speaking, that w has an approximate jump discontinuity at y_0 along the direction v.

Definition A. 6.3 *If $w \in BV(\Omega, \mathbb{R}^M)$ we call the* jump set *of w the set*

$$S_w = \{y_0 \in \Omega \mid \not\exists \text{ ap} \lim_{y \to y_0} w(y)\}.$$

Let $D^{wk}w = D^a w + D^s w$ be the Radon–Nikodym decomposition of $D^{wk}w$ in its absolutely continuous and singular part with respect to Lebesgue measure. We denote by Jw the restriction of $D^s w$ to the jump set S_w and define $Cw = D^s w - Jw = D^{wk}w - D^a w - Jw$. The terms $D^a w$, Jw and Cw are called respectively the regular part, *the* jump part *and the* Cantor part *of $D^{wk}w$.*

Definition A. 6.4 *We say that a function $w \in BV(\Omega, \mathbb{R}^M)$ belongs to the class $SBV(\Omega, \mathbb{R}^M)$ of* special functions of bounded variation *in Ω if the Cantor part Cw of its first derivatives is zero.*

Theorem A. 6.5 *Let $w \in BV(\Omega, \mathbb{R}^M)$. Then S_w is countably $\mathcal{H}^{N-1}$-rectifiable. Furthermore:*

(i) *if $Dw(y)$ exists for a.e. $y \in \Omega$, then it coincides with the density of $D^a w$;*
(ii) *there exists $\hat{\nu} : S_w \to \partial B_1$ such that $w_{\hat{\nu}(y)}(y)$, $w_{-\hat{\nu}(y)}(y)$ exist for $\mathcal{H}^{N-1}$–a.e. $y \in S_w$ and*

$$Jw = (w_{\hat{\nu}} - w_{-\hat{\nu}}) \otimes \hat{\nu} \, d\mathcal{H}^{N-1}|_{S_w}.$$

The unit vector field $\hat{\nu}$ is uniquely defined up to a set of zero $\mathcal{H}^{N-1}$ measure, and may be regarded as an approximate normal to S_w in the sense that if G is a C^1 hypersurface with normal vector ν, then $\nu = \hat{\nu}$ on $S_w \cap G$ ($\mathcal{H}^{N-1}$ a.e.);
(iii) *if $F \subset \Omega$ is a countably $\mathcal{H}^{N-1}$-rectifiable set, then the restriction of Cw to F is zero.*

Proof — See [72, 14]. ∎

Proposition A. 6.6 *Let $w \in BV(\Omega, \mathbb{R}^M)$ be given. Suppose that there exists a closed $\mathcal{H}^{N-1}$-rectifiable set $F \subset \Omega$ such that $w \in C^1(\Omega \setminus F)$. Then $w \in SBV(\Omega, \mathbb{R}^M)$.*

Proof — Since F is closed and $w \in C^1(\Omega \setminus F)$, we have $Cw = Jw = 0$ in $\Omega \setminus F$. On the other hand, $Cw = 0$ in F by Theorem A. 6.5(iii). ∎

References

1. ALBANO P., Some properties of semiconcave functions with general modulus, *J. Math. Anal. Appl.* **271** (2002), 217–231.
2. ALBANO P., On the singular set for solutions to a class of Hamilton–Jacobi equations, *NoDEA Nonlinear Differential Equations Appl.* **9** (2002), 473–497.
3. ALBANO P., CANNARSA P., Singularities of the minimum time function for semilinear parabolic systems, in *Control and partial differential equations (Marseille-Luminy, 1997)*, ESAIM Proc. **4**, Soc. Math. Appl. Indust., Paris, (1998).
4. ALBANO P., CANNARSA P., Singularities of semiconcave functions in Banach spaces, in *Stochastic Analysis, Control, Optimization and Applications*, W.M. McEneaney, G.G. Yin, Q. Zhang (eds.), Birkhäuser, Boston, (1999), 171–190.
5. ALBANO P., CANNARSA P., Structural properties of singularities of semiconcave functions, *Annali Scuola Norm. Sup. Pisa Sci. Fis. Mat.* **28** (1999), 719–740.
6. ALBANO P., CANNARSA P., Propagation of singularities for concave solutions of Hamilton–Jacobi equations, in *International Conference on Differential Equations, (Berlin, 1999)*, World Sci. Publishing, River Edge, NJ, 2000.
7. ALBANO P., CANNARSA P., Propagation of singularities for solutions of nonlinear first order partial differential equations, *Arch. Ration. Mech. Anal.* **162** (2002), 1–23.
8. ALBANO P., CANNARSA P., SINESTRARI C., Regularity results for the minimum time function of a class of semilinear evolution equations of parabolic type, *SIAM J. Control Optim.*, **38** (2000), 916–946.
9. ALBERTI G., On the structure of singular sets of convex functions, *Calc. Var. Partial Differential Equations* **2** (1994), 17–27.
10. ALBERTI G., AMBROSIO L., CANNARSA P., On the singularities of convex functions, *Manuscripta Math.* **76** (1992), 421–435.
11. AMBROSETTI A., PRODI G., *A primer of nonlinear analysis*, Cambridge University Press, Cambridge, 1993.
12. AMBROSIO L., Geometric evolution problems, distance function and viscosity solutions, in *Calculus of variations and partial differential equations (Pisa, 1996)*, (G. Buttazzo, A. Marino, M.K.V. Murthy, eds.) Springer, Berlin, (2000).
13. AMBROSIO L., CANNARSA P., SONER H.M., On the propagation of singularities of semi-convex functions, *Ann. Scuola Normale Sup. Pisa Ser. IV* **20** (1993), 597–616.
14. AMBROSIO L., FUSCO N., PALLARA D., *Functions of bounded variation and free discontinuity problems*, Oxford Mathematical Monographs. Oxford University Press, New York, 2000.

15. ANZELLOTTI G., OSSANNA E., Singular sets of convex bodies and surfaces with generalized curvatures, *Manuscripta Math.* **86** (1995), 417–433.

16. ARONSON D.G., The porous medium equation, in *Some problems on nonlinear diffusion*, (Fasano A. and Primicerio M., eds.), Lect. Notes Math. **1224**, Springer, 1986, 1–46.

17. AUBIN J.P., CELLINA A., *Differential Inclusions*, Springer-Verlag, Grundlehren der Math. Wiss., 1984.

18. AUBIN J.P., FRANKOWSKA H., *Set-valued Analysis*, Birkhäuser, Boston, 1990.

19. BARDI M., A boundary value problem for the minimum time function, *SIAM J. Control Optim.* **27** (1989), 776–785.

20. BARDI M., CAPUZZO DOLCETTA I., *Optimal control and viscosity solutions of Hamilton–Jacobi equations*, Birkhäuser, Boston, 1997.

21. BARDI M., FALCONE M., An approximation scheme for the minimum time function, *SIAM J. Control Optim.* **28** (1990), 950–965.

22. BARLES G., *Solutions de viscosité des équations de Hamilton–Jacobi*, Springer Verlag, Berlin, 1994.

23. BARRON E.N., CANNARSA P., JENSEN R., SINESTRARI C., Regularity of Hamilton–Jacobi equations when forward is backward, *Indiana Univ. Math. J.* **48** (1999), 385–409.

24. BARTKE K., BERENS H., Eine Beschreibung der Nichteindeutigkeitsmenge für die beste Approximation in der euklidischen Ebene, *J. Approx. Theory* **47** (1986), 54–74.

25. BELLMAN R., *Dynamic programming*, Princeton University Press, Princeton, 1957.

26. BERKOVITZ, L.D., *Optimal Control Theory,* Applied Mathematical Sciences, **12**. Springer-Verlag, New York, Heidelberg, 1974.

27. BRESSAN A., On two conjectures by Hájek., *Funkcial. Ekvac.* **23**(1980), 221–227.

28. BRESSAN A., *Lecture notes on the mathematical theory of control*, International School for Advanced Studies, Trieste, 1993.

29. BURCH B.C., A semigroup treatment of the Hamilton–Jacobi equation in several space variables, *J. Differential Equations* **23** (1977), 107–124.

30. BUTTAZZO G., GIAQUINTA M., HILDEBRANDT, S., *One-dimensional variational problems. An introduction*, Oxford Lecture Series in Mathematics and its Applications, **15**. Oxford University Press, New York, 1998.

31. CANNARSA P., Regularity properties of solutions to Hamilton–Jacobi equations in infinite dimensions and nonlinear optimal control, *Differential Integral Equations* **2** (1989), 479–493.

32. CANNARSA P., CARDALIAGUET P., Representation of equilibrium solutions to the table problem for growing sandpiles, to appear in *J. Eur. Math. Soc. (JEMS).*

33. CANNARSA P., FRANKOWSKA H., Some characterizations of optimal trajectories in control theory, *SIAM J. Control Optim.* **29** (1991), 1322–1347.

34. CANNARSA P., FRANKOWSKA H., Value function and optimality conditions for semilinear control problems. *Appl. Math. Optim.* **26** (1992), 139–169.

35. CANNARSA P., FRANKOWSKA H., Value function and optimality condition for semilinear control problems. II. Parabolic case. *Appl. Math. Optim.* **33** (1996), 1–33.

36. CANNARSA P., FRANKOWSKA H., SINESTRARI C., Optimality conditions and synthesis for the minimum time problem, *Set-Valued Anal.* **8** (2000), 127–148.

37. CANNARSA P., FRANKOWSKA H., Interior sphere property of attainable sets and time optimal control problems, preprint (2004).

38. CANNARSA P., MENNUCCI A., SINESTRARI C., Regularity results for solutions of a class of Hamilton–Jacobi equations, *Arch. Rat. Mech. Analysis* **140** (1997), 197–223.

39. CANNARSA P., PEIRONE R., Unbounded components of the singular set of the distance function in $\mathbb{R}^n$, *Trans. Amer. Math. Soc.* **353** (2001), 4567–4581.

40. CANNARSA P., PIGNOTTI C., Semiconcavity of the value function for an exit time problem with degenerate cost, *Matematiche (Catania)* **55** (2000), suppl. 2, 71–108.

41. CANNARSA P., PIGNOTTI C., SINESTRARI C., Semiconcavity for optimal control problems with exit time, *Discrete Contin. Dynam. Systems* **6** (2000), 975–997.

42. CANNARSA P., SINESTRARI C., Convexity properties of the minimum time function, *Calc. Var. Partial Differential Equations* **3** (1995), 273–298.

43. CANNARSA P., SINESTRARI C., On a class of nonlinear time optimal control problems, *Discrete Cont. Dynam. Systems* **1** (1995), 285–300.

44. CANNARSA P., SINESTRARI C., An infinite-dimensional time optimal control problem, in *Optimization methods in partial differential equations (South Hadley, MA, 1996)*, 29–41, Contemp. Math., 209, Amer. Math. Soc., Providence, RI, 1997.

45. CANNARSA P., SONER H.M., On the singularities of the viscosity solutions to Hamilton–Jacobi–Bellman equations, *Indiana Univ. Math. J.* **36** (1987), 501–524.

46. CANNARSA P., SONER H.M., Generalized one-sided estimates for solutions of Hamilton–Jacobi equations and applications, *Nonlinear Anal.* **13** (1989), 305–323.

47. CANNARSA P., TESSITORE M.E., On the behaviour of the value function of a Mayer optimal control problem along optimal trajectories, in *Control and estimation of distributed parameter systems (Vorau, 1996)* 81–88, Internat. Ser. Numer. Math., **126**, Birkhäuser, Basel, (1998).

48. CAPUZZO DOLCETTA I., ISHII I., Approximate solutions of the Bellman equation of deterministic control theory, *Appl. Math. Optimization* **11** (1984), 161–181.

49. CARATHÉODORY C., *Calculus of variations and partial differential equations of the first order*, Teubner, Berlin, 1935.

50. CARDALIAGUET P., On the regularity of semipermeable surfaces in control theory with application to the optimal exit-time problem, II, *SIAM J. Control Optim.* **35** (1997), 1653–1671.

51. CAROFF N., Caractéristiques de l'équation d'Hamilton–Jacobi et conditions d'optimalité en contrôle optimal non linéaire, Ph.D.Thesis, Université Paris IX Dauphine, 1994.

52. CASELLES V., Scalar conservation laws and Hamilton–Jacobi equations in one-space variable, *Nonlinear Anal.* **18** (1992), 461–469.

53. CESARI L., *Optimization—theory and applications. Problems with ordinary differential equations*, Applications of Mathematics **17**, Springer-Verlag, New York, 1983.

54. CLARKE F.H., *Optimization and Nonsmooth Analysis*, Wiley, New York, 1983.

55. CLARKE F.H., LEDYAEV Y.S., STERN R.J., WOLENSKI P.R., *Nonsmooth Analysis and Control Theory*, Graduate Texts in Mathematics, Springer-Verlag, New York, 1998.

56. CLARKE F.H., STERN R.J., WOLENSKI P.R., Proximal smoothness and the lower-C^2 property, *J. Convex Anal.* **2** (1995), 117–144.

57. CLARKE F.H., VINTER R.B., The relationship between the maximum principle and dynamic programming, *SIAM J. Control Optim.* **25** (1987), 1291–1311.

58. CODDINGTON E.A., LEVINSON N., *Theory of ordinary differential equations*, McGraw-Hill, New York, Toronto, London, 1955.

59. COLE J., On a quasilinear parabolic equation occurring in aerodynamics, *Q. Appl. Math.* **9** (1951), 225–236.

60. COLESANTI A., HUG D., Steiner type formulae and weighted measures of singularities for semi-convex functions, *Trans. Amer. Math. Soc.* **352** (2000), 3239–3263.

61. COLESANTI A., PUCCI C., Qualitative and quantitative results for sets of singular points of convex bodies, *Forum Math.* **9** (1997), 103–125.

62. CONTI R., *Processi di controllo lineari in* $\mathbb{R}^n$, Quad. Unione Mat. Italiana **30**, Pitagora, Bologna (1985).

63. CONWAY E.D., HOPF E., Hamilton's theory and generalized solutions of the Hamilton–Jacobi equation, *J. Math. Mech.* **13** (1964), 939–986.

64. CRANDALL M.G., Viscosity solutions: A primer, in *Viscosity solutions and applications*, (I. Capuzzo, Dolcetta, P.L. Lions, eds.), Springer-Verlag, Berlin, 1997.

65. CRANDALL M.G., EVANS L.C., LIONS P.L., Some properties of viscosity solutions of Hamilton–Jacobi equations, *Trans. Amer. Math. Soc.* **282** (1984), 487–502.

66. CRANDALL M.G., ISHII H., LIONS P.L., User's guide to viscosity solutions of second order partial differential equations, *Bull. Amer. Math. Soc.* **27** (1992), 1–67.

67. CRANDALL M.G., LIONS P.L., Viscosity solutions of Hamilton–Jacobi equations, *Trans. Amer. Math. Soc.* **277** (1983), 1–42.

68. DAFERMOS C.M., Generalized characteristics and the structure of solutions of hyperbolic conservation laws, *Indiana Univ. Math. J.* **26** (1977), 1097–1119.

69. DOUGLIS A., The continuous dependence of generalized solutions of non–linear partial differential equations upon initial data, *Comm. Pure Appl. Math.* **14** (1961), 267–284.

70. ERDÖS P., Some remarks on the measurability of certain sets, *Bull. Amer. Math. Soc.* **51** (1945), 728–731.

71. EVANS L.C., *Partial Differential Equations*, A.M.S., Providence, 1998.

72. EVANS L.C., GARIEPY R.F., *Measure Theory and Fine Properties of Functions*, Studies in Advanced Mathematics, CRC Press, Ann Arbor, 1992.

73. EVANS L.C., JAMES M.R., The Hamilton–Jacobi–Bellman equation for time optimal control, *SIAM J. Control Optim.* **27** (1989), 1477–1489.

74. FALCONER K., *Fractal Geometry*, Mathematical Foundations and Applications, Wiley, New York, 1990.

75. FEDERER, H., Curvature measures, *Trans. Amer. Math. Soc.* **93** (1959) 418–491.

76. FEDERER H., *Geometric Measure Theory*, Springer-Verlag, Berlin, 1969.

77. FILIPPOV, A.F., On certain questions in the theory of optimal control, *SIAM J. Control* **1** (1962), 76–84.

78. FLEMING W.H., The Cauchy problem for a nonlinear first order partial differential equation, *J. Diff. Eq.* **5** (1969), 515–530.

79. FLEMING W.H., MCENEANEY W.M., A max-plus based algorithm for an HJB equation of nonlinear filtering, *SIAM J. Control Optim.* **38** (2000), 683–710.

80. FLEMING W.H., RISHEL R. W., *Deterministic and stochastic optimal control*, Springer Verlag, New York, 1975.

81. FLEMING W.H., SONER H.M., *Controlled Markov processes and viscosity solutions*, Springer Verlag, Berlin, 1993.

82. FU J.H.G., Tubular neighborhoods in Euclidean spaces, *Duke Math. J.* **52** (1985), 1025–1046.

83. GOLDSTEIN J.A., SOEHARYADI Y., Regularity of perturbed Hamilton–Jacobi equations, *Nonlinear Anal.* **51** (2002), 239–248.

84. HÁJEK O., On differentiability of the minimal time function, *Funkcial. Ekvac.* **20** (1977), 97–114.

85. HAMILTON R.S., A matrix Harnack estimate for the heat equation, *Comm. Anal. Geom.* **1** (1993), 113–126.

86. HARTMAN P., *Ordinary Differential Equations*, Wiley, New York (1964).

87. HERMES H., LASALLE J.P., *Functional Analysis and Time Optimal Control*, Academic Press, New York, 1969.

88. HESTENES M.R., *Calculus of Variations and Optimal Control Theory*, Wiley, New York, 1966.

89. HOPF E., The partial differential equation $u_t + uu_x = \mu u_{xx}$, *Comm. Pure. Appl. Math.* **3** (1950), 201–230.

90. HOPF E., Generalized solutions of nonlinear equations of first order, *J. Math. Mech.* **14** (1965), 951–973.

91. HÖRMANDER L., *Notions of Convexity*, Birkhäuser, Boston, 1994.

92. HRUSTALEV M.M., Necessary and sufficient optimality conditions in the form of Bellman's equation, *Soviet Math. Dokl.* **19** (1978), 1262–1266.

93. ISHII H., Uniqueness of unbounded viscosity solutions of Hamilton–Jacobi equations, *Indiana Univ. Math. J.* **33** (1984), 721–748.

94. ISHII H., Perron's method for Hamilton–Jacobi equations, *Duke Math. J.* **55** (1987), 369–384.

95. ISHII H., LIONS P.-L., Viscosity solutions of fully nonlinear second-order elliptic partial differential equations, *J. Differential Equations* **83** (1990), 26–78.

96. ISHII H., RAMASWAMY M., Uniqueness results for a class of Hamilton–Jacobi equations with singular coefficients, *Comm. Partial Differential Equat.* **20** (1995), 2187–2213.

97. ITOH, J., TANAKA, M., A Sard theorem for the distance function, *Math. Ann.* **320** (2001), 1–10.

98. JENSEN R., The maximum principle for viscosity solutions of fully nonlinear second order partial differential equations, *Arch. Rat. Mech. Anal.* **101** (1988), 1–27.

99. KRUZHKOV S.N., The Cauchy problem in the large for certain nonlinear first order differential equations, *Soviet. Math. Dokl.* **1** (1960), 474–477.

100. KRUZHKOV S.N., The Cauchy problem in the large for nonlinear equations and for certain quasilinear systems of the first order with several variables, *Soviet. Math. Dokl.* **5** (1964), 493–496.

101. KRUZHKOV S.N., First order quasilinear equations in several independent variables, *Mat. Sb.* **81** (1970), 228–255 (English tr.: *Math. USSR-Sb.* **10** (1970), 217–243).

102. KRUZHKOV S.N., Generalized solutions of the Hamilton–Jacobi equations of the eikonal type I, *Math. USSR Sb.* **27** (1975), 406–445.

103. KURZWEIL J., *Ordinary Differential Equations*, Elsevier, Amsterdam, 1986.

104. KUZNETZOV, N.N., ŠIŠKIN, A.A., On a many dimensional problem in the theory of quasilinear equations, *Z. Vyčisl. Mat. i Mat. Fiz.* **4** (1964), 192–205.

105. LASRY J.M., LIONS P.L., A remark on regularization in Hilbert spaces, *Israel J. Math* **55** (1986), 257–266.

106. LAX P.D., Hyperbolic systems of conservation laws II, *Comm. Pure Appl. Math.* **10** (1957), 537–566.

107. LI Y. Y., NIRENBERG L., The distance function to the boundary, Finsler geometry and the singular set of viscosity solutions of some Hamilton-Jacobi equations, to appear in *Comm. Pure Appl. Math.*

108. LI P., YAU S.-T., On the parabolic kernel of the Schrödinger operator, *Acta Math.* **156** (1968), 153–201.

109. LI X.J., YONG J.M. *Optimal Control Theory for Infinite-dimensional Systems*, Systems & Control: Foundations & Applications. Birkhäuser Boston, Boston, 1995.

110. LIONS P.L., *Generalized solutions of Hamilton–Jacobi equations*, Pitman, Boston, 1982.

111. LEE E.B., MARKUS L., *Foundations of Optimal Control Ttheory*, John Wiley, New York, 1968.

112. MANTEGAZZA C., MENNUCCI A.C., Hamilton–Jacobi equations and distance functions on Riemannian manifolds, *Appl. Math. Optim.* **47** (2003), 1–25.

113. MOTZKIN T., Sur quelques propriétés caractéristiques des ensembles convexes, *Atti Accad. Naz. Lincei Rend. Cl. Sci. Fis. Mat. Natur.* **21** (1935), 562–567.

114. OLEINIK O., Discontinuous solutions of nonlinear differential equations. *Usp. Mat. Nauk* **12** (1957), 3–73. English transl. in *Amer. Math. Soc. Transl. Ser. 2* **26**, 95–172.

115. PETROV N.N., On the Bellman function for the time-optimal process problem, *J. Appl. Math. Mech.* **34** (1970), 785–791.

116. PIGNOTTI C., Rectifiability results for singular and conjugate points of optimal exit time problems, *J. Math. Anal. Appl.* **270** (2002), 681–708.

117. POLIQUIN R.A., ROCKAFELLAR R.T., THIBAULT L., Local differentiability of distance functions, *Trans. Amer. Math. Soc.* **352** (2000), 5231–5249.

118. PONTRYAGIN L.S., BOLTYANSKII V.G., GAMKRELIDZE R.V., MISHCHENKO, E.F., *The Mathematical Theory of Optimal Processes.* Interscience Publishers, John Wiley & Sons, Inc., New York, London, 1962.

119. RIFFORD L., Existence of Lipschitz and semiconcave control-Lyapunov functions, *SIAM J. Control Optim.* **39** (2000), 1043–1064.

120. RIFFORD L., Semiconcave control-Lyapunov functions and stabilizing feedbacks, *SIAM J. Control Optim.* **41** (2002), 659–681.

121. RIFFORD L., Singularities of viscosity solutions and the stabilization problem in the plane, preprint (2002).

122. ROCKAFELLAR R.T., *Convex Analysis*, Princeton University Press, Princeton, 1970.

123. ROCKAFELLAR R.T., Favorable classes of Lipschitz continuous functions in subgradient optimization, in *Progress in Nondifferential Optimization*, Nurminski E., (ed.), IIASA Collaborative Proceedings Series, Laxenburg, **125** (1982).

124. ROCKAFELLAR R.T., WETS R.J.-B.,*Variational Analysis*, Grundlehren der Math. Wiss., Springer-Verlag, Berlin, 1998.

125. SCHAEFFER D., A regularity theorem for conservation laws, *Adv. in Math.* **11** (1973), 368–386.

126. SCHNEIDER R., *Convex bodies: the Brunn–Minkowski theory*, Encyclopedia of Mathematics and its applications **44**, Cambridge University Press, Cambridge, 1993.

127. SINESTRARI C., Semiconcavity of solutions of stationary Hamilton–Jacobi equations, *Nonlinear Anal.* **24** (1995), 1321–1326.

128. SINESTRARI C., Local regularity properties of the minimum time function, in *PDE methods in control and shape analysis*, Da Prato G., Zolesio J.P., (eds.), Marcel Dekker, New York, 1996.

129. SINESTRARI C., Regularity along optimal trajectories of the value function of a Mayer problem, to appear in *ESAIM Control Optim. Calc. Var.*

130. SINESTRARI C.. Semiconcavity of the value function for exit time problems with nonsmooth target, to appear in *Commun. Pure Appl. Anal.*

131. SORAVIA P., Pursuit-evasion problems and viscosity solutions of Isaacs equations, *SIAM J. Control Optim.* **31** (1993), 604–623.

132. SPINGARN J., Submonotone subdifferentials of Lipschitz functions, *Trans. Amer. Math. Soc.* **264** (1981), 77–89.

133. STEIN E.M., *Singular Integrals and Ddifferentiability Properties of Functions*, Princeton University Press, Princeton, 1970.

134. SUBBOTIN A.I., *Generalized Solutions of First Order PDEs: the Dynamic Optimization Perspective*, Birkhäuser, Boston, 1995.

135. TIERNO G., The paratingent space and a characterization of C^1-maps defined on arbitrary sets, *J. Nonlinear Convex Anal.* **1** (2000), 129–154.

136. VELIOV V.M., Lipschitz continuity of the value function in optimal control. *J. Optim. Theory Appl.* **94** (1997), 335–363.

137. VESELÝ L., On the multiplicity points of monotone operators on separable Banach spaces, *Comment. Math. Univ. Carolin.* **27** (1986), 551–570.

138. VESELÝ L., On the multiplicity points of monotone operators on separable Banach spaces II, *Comment. Math. Univ. Carolin.* **28** (1987), 295–299.

139. VESELÝ L., A connectedness property of maximal monotone operators and its application to approximation theory, *Proc. Amer. Math. Soc.* **115** (1992), 663–667.

140. WESTPHAL U., FRERKING J., On a property of metric projections onto closed subsets of Hilbert spaces, *Proc. Amer. Math. Soc.* **105** (1989), 644–651.

141. WOLENSKI P., ZHUANG, Y., Proximal analysis and the minimal time function *SIAM J. Control Optim.* **36** (1998), 1048–1072.

142. WU Z.-Q., The ordinary differential equation with discontinuous right-hand members and the discontinuous solutions of the quasilinear partial differential equations, *Acta Math. Sinica* **13** (1963), 515–530. English translation: *Scientia Sinica* **13** (1964), 1901–1907.

143. YOUNG L.C., *Lectures on the calculus of variations and optimal control theory*, Chelsea Publishing Company, New York, 1980.

144. ZAJÍČEK L., On the points of multiplicity of monotone operators, *Comment. Math. Univ. Carolin.* **19** (1978), 179–189.

145. ZAJÍČEK L., On the differentiation of convex functions in finite and infinite dimensional spaces, *Czechoslovak Math. J.* **29** (1979), 340–348.

Index